1 MONTH OF
FREE
READING

at

www.ForgottenBooks.com

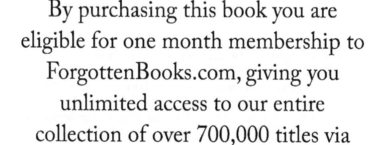

By purchasing this book you are eligible for one month membership to ForgottenBooks.com, giving you unlimited access to our entire collection of over 700,000 titles via our web site and mobile apps.

To claim your free month visit: www.forgottenbooks.com/free59468

ISBN 978-0-428-96625-6
PIBN 10059468

Communications and Exchanges intended for the Club
should be addressed to the Honorary Secretary,
Port-of-Spain, Trinidad, B.W.I.

Price 6d.   Annual Subscription, 3/.

Vol. 2.   APRIL, 1894.   No. 1.

*J'engage donc tous à éviter dans leurs écrits toute personnalité,
toute allusion dépassant les limites de la discussion la plus
sincère et la plus courtoise.*—LABOULBÈNE

Trinidad Field Naturalists' Club.

NATURA MAXIME MIRANDA IN MINIMIS

Publication Committee :

H. CARACCIOLO, F.E.S , *President.*

P. CARMODY, F.I.C., F.C.S. ; SYL. DEVENISH, M.A.

B. N. RAKE, M D,   R. R. MOLE,

F. W. URICH, F.E.S , *Hon. Secretary.*

CONTENTS :—

Report of Club Meetings—

     February   1

     March   2

GOLDNEY Prize Competition—Remaining Papers—

     Notes by the Wayside   4

     A Ramble on a Cocoa Estate   ...   17

Protection of Wild Birds   20

☞ All Communications and Exchanges intended for the Club should be addressed to the Honorary Secretary, Port-of-Spain, Trinidad, B.W.I.

JOURNAL

OF THE

Field Naturalists' Club.

VOL. 2. APRIL, 1894. NO. I.

REPORT OF CLUB MEETINGS.

2ND FEBRUARY, 1894.

PRESENT: Mr. H. Caracciolo, President; Messrs. W. E. Broadway, Charles Libert, T. I. Potter, J. Russell Murray, H. J. Baldamus, T. W. Carr, Henry Tate, R. R. Mole, and F. W. Urich, Hon. Sec. and Treasurer. The following gentlemen were elected Town Members of the Club: Hon. Dr. Lovell, C.M.G., Messrs. J. Graham Taylor, P. Taaffe O'Connor and H. A. Alcazar. Mr. Mole read a paper on "A Visit to the High-woods of Caparo," which described lappe, agouti and quenk hunting as carried on in the forests of Trinidad. The President showed specimens of a mealy bug, *Dactylopius* sp. affecting coffee, and a peculiar Mantis. He also made some remarks on a tick found on cattle in the neighbourhood of Port-of-Spain which had been determined as *Rhipicephalus americanus* (Marx). Messrs. Mole and Urich laid on the table a Preliminary List of Trinidad reptiles showing a total of 73 species viz:

Tortoises ...	7	species
Alligators	1	,,
Lizards ...	21	,,
Snakes ...	34	,,
Frogs and toads	10	,,

73 species

The compilation of this list had been the work of the last four years and there were still a great many more species to be added. Mr. Urich stated that the list was drawn up in the

same style and with the same intention as the Preliminary list of
the Mammals published by Mr. Oldfield Thomas sometime ago.
The Reptile list had been compiled from all accessible literature
and records based upon many specimens collected by Mr. Mole
and himself and ·it would be found that a' great many of them
were thus recorded for the first time from Trinidad.   The total
number of species at present was only 73 but there was every
probability that when certain specimens, recently despatched to
Europe, were identified that this figure would be considerably
increased.   On the other hand there were in the list the names
of certain snakes recorded from Trinidad at the British Museum,
and   elsewhere,   the   occurrence   here   of   which   appeared
to be exceedingly doubtful.   They had refrained from giving any
decision in this list as their experience in Reptiles was a very
short one—and new species were turning up now and again.
It was exceedingly difficult to say positively that such and
such a species did not occur in a certain country and as an
illustration of this . he   would mention that it took 50 years of
careful record to establish the fact that *Viperus berus*, the common
English viper, did not occur in a certain district in Germany.
The list was long and as it only consisted of a number of
names it was unnecessary to read it.   It would be published
in the Journal when it would be better understood.   In
answer to certain recent criticisms he would point out that it
was the duty of the Club to obtain correct lists as far as possible
of the fauna of the Island and that such lists were of much value
scientifically.—Mr. Tate presented some copies of the Annual
Report of the Belfast Naturalists' Field Club and the *Irish
Naturalist* and said during his recent absence from the Island
he had seen several members of the Belfast Field Club and
they would be glad to exchange proceedings.   The *Irish
Naturalist* would prove very useful to any members who sub-
scribed to it.—The question whether the Club should subscribe
was referred to the Business Committee.—After deciding that
members should be asked to send in proposals for excursions to
be considered at the next monthly meeting the Club adjourned at
9.30 p.m.· The additions to the Club collection since last meeting
are : A banded tree snake locally called "Lora"—*Ahœtulla
liocercus*—presented by the Rev. Father Clunes, Princes Town,
and an "American Barn owl" *Strix pratinicola* from River
Estate, Diego Martin, presented by Mr. E. West Lack.

MARCH 2ND, 1894.

PRESENT : Mr. H. Caracciolo, President ; Sir John Tankerville
    Goldney, Professor Carmody, Dr. Rake, Messrs. T. I. Potter,
T. W. Carr, A. B. Carr, W. S. Tucker, W. E. Broadway, John

Hoadley, S. A. Cumberland, Alfred Taitt, Henry Tate, R. R. Mole and F. W. Urich, Hon. Secretary. Mr. Ponsonby (London) Rev. E. D. Wright (late Jamaica), Mr. G. A. Urich and Master Rake were the visitors present. The following elections took place : Honorary member : Professor C, V. Riley, Washington ; Town members : Hon. G. L. Garcia, Drs. G. S. Seccombe and E. Prada, Rev. J. Williams, B.A., Messrs. E. C. Wilson, Leon Agostini, J. T. Rousseau, V. L. Wehekind, Carl Saurmann, Jules Anduze, J. R. Llaños.—Notes from the Jamaica Institute were read by the Secretary.—Dr. Rake read a paper upon "The Schizomycetes" and illustrated it with a number of cultures of bacteria and slides ; amongst them one of leprosy bacilli by Mr. Andrew Pringle was conspicuous ; these were shown under microscopes and lent additional interest to a highly instructive paper, which the author explained with many drawings on the blackboard. At the conclusion a unanimous vote of thanks was accorded to Dr. Rake for his paper.—Mr. Potter read a few notes on, and exhibited specimens of, a borer found destroying cacao trees at St. Ann's.—Mr. A. B. Carr also showed some borers from Caparo ; one looking very much like the *Xyleborus perforans* had killed five cacao trees. The other was found in a dead tree but it was doubtful whether it was the cause of the tree dying. Mr. Carr also placed on the table horns of two species of Trinidad deer—one being of *Cariacus nemorivagus,* (F. Cuv.) and the other undetermined ; also the skull of a quenk (peccary—*Dicotyles labiatus* Cuv.)—Dr. Rake exhibited a young tree porcupine, *Synetheres prehensilis* (Linn.) also a species of *Lycopodium* found growing at Carenage.—The President showed a specimen of lantern fly from Maracas and made some remarks on the light these insects are supposed to emit.—A species of bat *Chiroderma villosum* (Peters) was also exhibited.—Mr. Urich laid on the table, on behalf of Mr. J. Guilbert, a nest of the corn bird *Ostinops decumanus* (Pall) and a locust said to be injurious to young cacao.—The President mentioned that mole crickets were injurious to cacao nurseries by eating away the tap roots.—The Rev. E. D. Wright made some remarks on *Hypolimnas misippus* and pointed out differences in the females of this butterfly found here and in Jamaica.—Mr. S. A. Cumberland showed a young alligator caught by himself on the Caroni after a terrific struggle in which he received many and very serious wounds, and read a few humorous notes thereon.—The additions to the Club collection since last meeting are : A Sphinx Moth, presented by the Rev. Father Clunes, Princes Town ; three caterpillars and a beetle presented by the Rev. C. H. Stoker, Tunapuna ; a caterpillar presented by the Rev. E. F. Tree, Couva. —The meeting closed at 10.15 p.m.

CLUB PAPERS.

THE GOLDNEY PRIZE COMPETITION.

NOTES BY THE WAYSIDE.

THERE is perhaps no locality in the neighbourhood of Port-of-Spain which is more easily accessible or more likely to yield good results to lovers of Natural History than the vicinity of St. Ann's Valley. It was here that the writer of the following notes, and another member of the Club, elected to spend an off day in the early part of last year. A pleasant ride by one of the morning trams brought us very close to the scene of our day's work and pleasure, for Natural History means both to those who are really interested in it. Passing rapidly along over the first few hundred yards of the well kept road, bordered by grassy Savannas in which graze the ugly Indian buffaloes and sleepy but graceful young zebus, the off-spring of a stock which a paternal government has successfully introduced to improve the breed of Trinidad cattle, we could not but admire the clumps of bamboo* which adorn the banks of the Dry River and the high ground behind, still crowned by the ruins of the residence of the old Spanish Governors. Near the back of the Lunatic Asylum we crossed the fence separating the Savanna from the road and then the bed of the river, which was as dry as possible, though, probably, a week before the course of a raging torrent of storm-water, as perhaps it would be again to-morrow. The wire fence which divides the Savanna from the thick brush-wood covering the North-eastern part of the hill-side, first attracted our notice. Several large black ants† were on the wires and uprights, busily running backwards and forwards in quest of food but what that food consisted of we could not then spare the time to discover. A little further along upon one of the wooden posts of the fence we found a small centipede.‡ His great length and enormous number of legs, extraordinary even in a centipede, attracted more than usual attention. He appeared to be in trouble, for strange to say, even centipedes have their trials and do not always find the world "all beer and skittles"—perhaps it would be more correct to say "all cockroaches and rotten wood."

---

*Bambusa vulgaris SCHRAD.
†Cryptocerus atratus.
‡Geophilus sp.

Really, when one enquires into the life-history of the centipede, he is after all not nearly the desperate ruffian which mankind makes him out to be. His is a very inoffensive life and one which is devoted to doing good service to the human race, in that he makes constant war upon the cockroaches which devour our clothing and contaminate with their filthy smell our food. True the centipede boasts a pair of sharp claws which nature has endowed with poisonous properties, but he rarely uses them, except in self-defence, just as we should use our fists if we were hurt or molested when about the duties of daily life and no Policeman is at hand. Like many other things the poor centipede, because of an uncanny appearance (his numerous legs and segmented body) has only to show himself and he is greeted with cries of " Centipede ! St. Peter !! St. Paul !!!" for popular superstition believes that the invocation of these Saints has a paralyzing effect upon his poor hundred legs as they hurry him over the floor in search of the seclusion of a dark hole, and so give his arch-enemy, Man, time to get a weapon with which to despatch him. But centipedes have other foes besides mankind—the fowls, the ants and the toads, and, as we have said before, this poor fellow was in the greatest trouble. A long thin white worm nearly as long again as its victim, was inserting itself into the poor arthropod, which did not seem able to defend itself at all. We watched the struggle a few moments and then both centipede and parasite were consigned to spirits for examination at some more convenient time. Before we leave centipedes, however, it is interesting to note that the popular superstition in Trinidad with regard to centipedes and scorpions, is that the former " stings" with the two large claws in the last segment of the body, and the scorpion " bites" with the pincer claws. As it happens the facts are just the reverse, the centipede *bites* with the poison claws directly under its head, and the scorpion "stings" with the sharp sting on the final joint of the tail.

There was a bare chance of finding out what the black ants were after and where their nest was, so we ascended the hill a few yards further and finally found the rotten remnants of a tree trunk. Rotten logs are oftentimes rich harvest fields to entomological collectors, and forthwith we began to turn over, to cut up, and otherwise examine them. Our first find was one of those small short-tailed ground snakes,* which are said to prey exclusively on insects, but which always seem to have empty stomachs whenever they are examined, so that it is somewhat of a puzzle how they do get their living, though their jaws appear to be fit for nothing else but the capture of minute worms, molusks, or

---

* Geophis lineatus.

insects. One thing however is well known about them, and that is they are a favourite article in the diet of the coral snakes.* The specimen was bagged as a matter of course, for no doubt he would be welcomed, only too warmly, by the Elaps in the glass box at home. These logs appeared to be rather the reverse of rich in insect life, however, for with the exception of one or two cockroaches—has it ever occurred to a member of the Club what a number of species there are in this Island?—and one or two spirally shaped snails† there was little else. But stricter search revealed the horny elytra of a beetle, which probably came to an untimely end, also one or two of the ants locally known as tick-tacks‡ from the snapping noise they make with their jaws. My companion wished to obtain some males and females of this species and so we persevered and presently, turning over another fragment of the trunk, we came upon the whole nest. At the risk of many a bite we picked up the winged individuals— wings are one of the distinguishing features of the males and females—the workers or neuters being destitute of these appendages. Although we looked carefully we were not able to find the Queen-ant needed to complete the series in my companion's collection. While taking the ants we noticed a peculiar habit in them, which we have not hitherto seen recorded—when threatened with a stick, or the point of a cutlass they have a trick of jumping backwards with a sharp jerky movement. A little further on we found, on the under side of a piece of bamboo, a land planaria, a slug-like-looking animal, which preys on land snails and whose methods of dealing with these unfortunate molusks has been so ably described by Dr. von Kennel in his Biological and Faunistic notes on Trinidad. The extract was reprinted in a recent number of the Journal. But we had stayed long enough on the hill-side and we felt we must get on to Fonds Amandes to begin our excursion in real earnest, so we quickly regained the road and hurrying past the little hamlet and over the tiny stream crossing the road, we stopped to look for a moment at the shoals of small fish§ swimming merrily in the still waters on either side of our path ; and tried to find an eel, but without success, though they are oftentimes to be seen there. A little further on we noted one of those gorgeous beetles‖ with green, gold and black elytra, but he luckily escaped

---

*Elaps lemniscatus and E. riisei.

†Stenogyra octona CHENM Mr Lechmere Guppy, to whom I am indebted for this determination, informs me this is the most widely distributed shell in the world.

‡Odontomachus hæmatodes.

§Rivulus miropus.

‖A Coptocycla.

the killing bottle just as he was being thrown into it and went winging his way through the sunshine, only to meet with a more ignominious death. Poor foolish beetle ! Instead of adorning a cabinet with his remains and being admired by scores of ento- mologists he allowed himself to be snapped up by a Qu'est-ce qu'il dit* which swooped down from its perch on a lofty bamboo and swallowed him with a single jerk. A curious sight here attracted our attention. We were turning back the leaves of the wayside plants, on the look out for the caterpillars of Micro- lepidoptera, scale insects, the smaller members of the cicada family, or anything else which we might find, when we noted a solitary ant,† running along the stem of the plant. Every now and again she stopped at the joint where the leaves sprouted and felt with her mandibles a little round excrescence growing there. "What is she doing?" I asked, and my companion, who knows far more about ants, and for matter of that, Natural History in general, than I do, told me that the little excrescences contained a honey sweet liquid which the ant was extracting. Near here my friend found a Queen-ant of *Camponotus artriceps,* which, having been fer- tilized, had, after the nuptial flight, retired to a secluded corner, where she was waiting for some workers to come along and find out her condition, when they would at once feed her, and build her a suitable home and so found a new Colony. What a wonderful and absorbingly interesting subject is ant-life. Every day fresh facts are becoming known about it ; ants have been proved to be Builders, Agriculturists, Cattle-herders, Slave- holders, Marauders—in fact there is hardly one of the primitive occupations of human life which is not in some way reproduced in the history of some species of ant. On the under surface of a Pandanus plant we saw an interesting wasp's nest. It consisted of a brown parchment like substance and in size and shape was not unlike a large pear, only the point, (where the fruit's stem would be) hung downwards, and it was there that the entrance was situated. With due precautions against "bites" (as the country Creoles call the sting) we got hold of the nest and rid it of its furious inmates—a feat which requires considerable coolness and dexterity. Being able to examine the nest at our leisure we could appreciate what great artists its builders were. In fact, all the "Jack Spaniards," as they are called, although they have no course of academical training are naturally great architects. The variety of form which their nests take must strike the most unobservant. Some species go in extensively for modelling in clay ; others make a special parchment of their own and a third

---

*Pitangus sulphuratus.
†Ectatomma tuberculatum, LATR.

kind construct their nurseries of sticks, particles of sand and leaves cleverly woven together. Some are miners, and excavate extensive burrows in the ground. All these elaborate constructions are nurseries, wherein they bring up their little ones and assiduously attend to them through their larval and pupal stages. The slight opening we made in the nest showed that the interior consisted of a series of cells in several stories, the straightness and symmetry of which would make many of our local architects blush, although they were built without the aid of rule or compass. Each cell contained a pupa or a larva, the former, being sepulchred in their narrow rooms by thin parchment caps.

One of the most curious facts in Natural History, is the wonderful way in which many animals are adapted to concealing themselves from their prey and from their enemies. I have read that that very showy animal—the caged Tiger—is most difficult to see in his native jungle owing to the striking similitude of his stripes to the long dried grass by which he is surrounded. A sleeping Howling Monkey* we are told looks like a heap of moss, and the Sloth† presents a similar appearance. When we get to the lower orders of creation we find this protective mimicry in its most prominent form. It is next to impossible to see a Whip Snake,‡ even in an almost leafless shrub, his long lithe body looking exactly like the trailing branches, his sharply pointed nose appearing to be a shoot. With his head and neck he frequently cleverly imitates the swaying of the branches of the bush on which he lies (he rarely coils) under a gentle breeze. There are spiders,§ which imitate ants; and insects‖ which seem to be fungus; grass hoppers,¶ which look like leaves, caterpillars which at first sight seem to be spiders; and beetles closely resembling bark. Therefore it is not surprising we should have found some remarkable cases of protective mimicry. Upon a shrub growing close to the ruin of an old house there were some small objects looking like green leaves just sprouting, but these same leaves, if we attempted to pluck them, had a trick of nimbly disappearing side-ways round the branch in a most un-leaf-like fashion. These creatures are commonly called "leaf hoppers."** The same plant yielded a still more marvellous illustration of mimicry. On the under side of a

---

*Mycetes seniculus.
†Cholœpus didætylus.
‡Oxybellis acuminata.
§Attidæ.
‖Orthesia and Tettigonidæ.
¶Steirodon dentatus St. and others.
**Tettigonidæ sp

leaf stalk was something which resembled a leaf rolled into a long spiral form ; it was veined like the leaves around it, its color was the same, and it had a stalk. We looked closely and still more closely and at last found it to be a green caterpillar* with a short tail (the aforesaid stalk) attaching itself to the real stalk by its hindmost pair of legs and extending the rest of its body in a perfectly motionless straight line at an angle of forty-five degrees from the rest of the leaf. Altogether it was as nearly like a rolled-up leaf, as it was possible for an insect to look. Near here there was a small bundle of dirty looking cotton, half the size of a walnut, caught in a web of gossamer, which turned out to be a large spider,† carrying its young ones in a bag of silky webbing. Further on we noted a dried flower stalk which had presumably fallen from the mango tree above, resting on a broad plantain leaf but flower sprigs do not move as this one did and closer scrutiny revealed the fact that it was an insect of the Mantis‡ family catching with its long barbed fore-legs at the flies incautiously skimming past it. In color it was black and the imitation it presented at a little distance, of a dried flower stalk, was simply perfect. In the pool at the foot of the hill we again observed the little fish *(Ciprinodontes)* which belong to the carp family, and also caught a glimpse of a cray-fish shooting backwards under a rocky ledge beneath the clear water. How the cray-fish so unerringly manages to shoot from the distance of at least a yard—tail first into its hole, is a matter for wonder when one first sees it perform the feat. It is only by long watching that we can find out how it is done. Canon Kingsley described it, in his inimitable language, only his remarks had reference to a lobster, but lobsters and cray-fish alike possess this peculiar faculty. Kingsley says : " If he wanted to go into a narrow " crack ten yards off, what do you think he did ? If he had gone " in head foremost of course he could not have turned round. " So he used to turn his tail to it, and lay his long horns which " carry his sixth sense in their tips (and nobody knows what that " sixth sense is), straight down his back and twists his eyes back " till they almost come out of their sockets and then made ready, " present, fire, snap !—away he went pop into the hole ; and " peeped out and twiddled his whiskers as much as to say ' you " could'nt do that !' "

We crossed over the stream by means of the stepping stones and ascended the steep incline on the other side. Here we met

---

*Belonging to the Geometridæ,
†Now being identified.
‡Genus Acanthops.

some barefooted labourers, who, as they approached a certain spot broke into a a short run ending with a long step over something in the path, calling out as they did so, *Fourmis chasseurs !* When we reached the place we found one of those most curious, of all curious facts, in Natural History. A column of hunting ants* four inches in width, was crossing the road on which they had formed a regular beaten track. They did not keep dressing, and distance, as do the units in civilized human armies—but each individual had plenty of room to march, in the fashion of barbarous nations on the warpath. We watched these ants for hours and the longer we did so the more forcibly were we reminded of the time when the Huns and the Goths, swept down in countless hordes upon the civilized nations of southern Europe, ransacking their cities, and destroying as ruthlessly as with fire, all which lay in their path. These ants, like their human prototypes, have no fixed abode, they build no elaborate struc- tures, nor do they excavate labyrinthine cities beneath the surface of the soil, as do many species. Hunting ants neither herd cattle (in the shape aphides) nor are they mushroom growers, or grain cultivators, or slave hunters as many other kinds are. They are merely great armies of savages whose only trade is war and pillage, in pursuit of which they march from place to place, carrying with them their queen and their young, resting only a few hours, or at most, a day or two, and then moving on again. Their mission is to destroy every living thing which cannot save itself by either speed or cunning. Even man himself retreats at the first indication of their approaching invasion and leaves them masters of his hearth and home. This column of ants had at intervals larger ants marching by themselves on its flanks. The jaws of these individuals were long and sickle-shaped, and they cut through the skin, drawing blood very quickly, as we found by actual experiment. Like the Goths, these ants had been on an expedition, and bore in their train the spoils they had captured. These comprised the legs, thighs, portions of the bodies of crickets, cockroaches, tree bugs, small cater- pillars entire, and portions of a centipede. At least one reptile had fallen a victim to them, for the scaly tail of a small lizard† still wriggled in the jaws of half-a-dozen hunters. Portions of spiders, ants' larvæ and pupæ of other species were also amongst the spoil. The main column had several small branch columns which investigated every nook and cranny, and dragged out and carried off, or cut up, according to their size, every lurking insect within crevices at a considerable distance from

---

* Eciton foreli, MAYR.
†Scolecosaurus cuvieri, FITZ.   Locally called *vipère.*

the line of march.   Once or twice we observed large pale-headed
fellows sitting on points of vantage, such as leaves, two pairs
of legs in air, mandibles widely distended, now and again touching with their rapidly vibrating antennæ those of some of the
passing ants.   They appeared to be giving directions, or
at any rate to be making some communication to the side
columns which were rejoining the main body.   We had previously noted the curious likeness of certain insects to their
surroundings and now we found some practical illustrations of
the utility of that likeness.   The hunters seemed to have very
feeble sight and to depend upon finding·out what was edible and
acceptable through their antennæ.   While we were watching the
column we thought a fragment of green twig moved, a second
glance and we were sure of it, it was one of those curious
"stick insects"*—*Cheval Bon Dieu*, as they are called by the
country people here.   He was about an inch and a half long
and he stood for minutes together motionless and apparently
inanimate, but withal well erect on his long legs, in the midst
of the passing column.   The hunters marched round him, under
him, over him even, and he did not move, sometimes one more
inquisitive than the others and probably sharper than the common
herd, touched him gently with her antennæ, or nibbled him with
her mandibles, but the cunning "stick," unlike the ancient
Roman Senator who rashly struck the Goth when the rude
barbarian gently stroked his beard, and so lost his own life and
those of his venerable companions—unlike him, the "stick"
remained immovable and the curious ant passed on.   There
were two more creatures within our view, who were equally wise
in pursuing this policy of death-like apathy.   One was a peculiar
looking spider† something like a little brown ball, with eight
abnormally thin and long legs.   He stood high above the crowd
and when an ant seemed about to seize one of his slender hair
like legs he raised it high, but if the ant seized it first he
remained as quiet as the "stick" did and waited until it was
released.   Sometimes the spider had three legs in air at once,
but he never attempted to get out of the line of march.   The
last example, was a tiny beetle, which had tucked his antennæ
and legs under his elytra and closely resembled the soil on which
he rested.   Curious to see the result, we dropped an ordinary
parasol worker ant‡ into the midst of the column.   The hunters,
save one, took no notice of her, but upon that incautious wight
the parasol ant fell angrily and avenged the insult severely.   We
had hitherto watched the steady stream of ants as it came down

---

*Bacillus sp.—belonging to the Phasmidæ,
†A Phalangidæ.
‡Atta cephalotes.

the hill-side, through the leaves and undergrowth, and crossed the road, and we now thought it worth while to follow it a little distance and find out where it flowed to.   We had not far to go, only to the hollow between the buttresses of a young silk cotton* tree on the other side of the path.   There the ants had formed their nomadic nest.   The hollow was nearly full of them, all clinging together, thousands upon thousands of thousands.   Bees swarming give a faint idea of how they looked.   The ants formed one solid mass, the proportions of which we could form no adequate idea of.   There appeared to be two or three large buckets full, and when we descended the bank a little way and looked up under it, we found that there were buckets full more of the seething mass of savage little creatures, clinging to each other by thousands of thousands and being joined every moment by more thousands.   Somewhere in the midst of that mass was to be found the queen and the winged males which my friend wished very much to obtain.   We tried to burn the nest with the view of finding the specimens we wanted smothered underneath the embers ; but this stratagem failed and only excited the ants to the greatest fury; they rushed hither and thither with widely opened menacing jaws and some of them managed to get up our trowser legs, so we beat a hasty retreat, defeated for the time being.   Having picked off all the ants we could find upon each other we proceeded on our way and following a narrow path to the left soon found ourselves on the banks of the brook at the bottom of the ravine.   Here the rain, which had been threatening for some time, came down in earnest, and we ran for shelter to a sort of shallow cave in the cliff on the opposite side, and to reach which we had to scramble up an almost perpendicular wall of rock.   In front of the entrance to this retreat hung a number of air roots belonging to the trees above, which had penetrated through the fissures and crannies of the soft stone.   Pendant to one of these there swung a lump of moss and grass, in the midst of which we detected the pretty eyes and sharp beak of a fly catcher of a species we were not able to determine, for the moment it saw it was discovered it forsook its aerial home for safer quarters.   The beautifully constructed nest contained three eggs, which we did not care to disturb.   The little mother anxious about her treasures, flew back again, but finding us still there, darted off, if possible, more scared than before.   Sitting at the entrance of our cave we watched the antics of a number of little frogs which jumped and scrambled on the rocks beneath, sometimes falling down terrific precipices, and picking themselves up again as if nothing had happened.   They were as busy as they could

---

*Eriodendron anfractuosum D. C.

be catching flies and mosquitoes with an alertness which was surprising. As they worked they sang, a shrill chirping little note, which was kept up without intermission. They were of two kinds, the most numerous being greenish brown, with an orange colored throat*; the others were perfectly black.† In length they were a little over an inch. It was most amusing to see how cleverly they stalked their prey and finally captured it with a sharp little jump and the throwing out of the tongue, which is a characteristic of the frogs. So long as we remained still, they would venture into the cave itself, hop on to our boots, all the while keeping up their incessant chirp, chirp, in the utterance of which, they swell out their throats to a considerable size. When the rain ceased, we ventured out. Upon one of the still pools in the river we found some insects which reminded us of many a sunny afternoon in England. The " water-skaters†" of Trinidad are not so large as those of Europe, but in shape, so far as we can remember, they are very similar. They do not live in the water, but on its surface, and it is as interesting as it is amusing, to see these insects skating, without skates and without ice. They frequently chased each other backward and forward over the sunlit surface, then two would meet, and after a short scuffle, the one who got the worst of it, would take to flight, in the course of which he would make a succession of most astonishing leaps. Others circled about in the leisurely manner of accomplished skaters, executing " eights" and other figures, on both "inside" and "outside edges" and with never a "spill" to break the monotony of the performance. Sometimes one would stand fast, his long legs stretched out, and his brown boat-like body perfectly still, as if he were glued to a sheet of glass, but a movement of our hands would send him off again at his topmost speed. But these gambols on the water are not for pleasure, they are the work by which the "skater" obtains his livelihood. He is ever on the look out for mosquitoes which come to the water to deposit their eggs and if they are not very quick about it they are at once caught by the skater who inserts a small beak into their bodies and sucks out their juices in less than no time, for the skater is a water-bug and is possessed of a capital appetite. They are not always doomed to skate about upon one particular pool however; when they become adults they develop wings and fly by night, so that should ever the water dry up, they are able to seek fresh lakes and rivers new. Another object of interest in this same pool was a half drowned mygale,‡ which had been probably washed down,

---

*Prostheraphis trinitatis, GARMAN.
†Now being identified.
‡Eurypelma virsicolor WALCK. Locally known as Tarantula.

by the heavy shower which had just fallen, from the branches of
the tree above, where he may have been straying from home in
search of bird's nests; he was rescued, however, after several
attempts at suicide, and placed on a tree where there was no
chance of his renewing his acquaintance with the stream. A
steel blue wasp* as large as a man's thumb and having yellow
antennæ excited our admiration, but our efforts to catch him, as he
cråwled on some moss grown rocks, he skilfully evaded. Not so
fortunate was a beautiful Epeira spider which hung in the centre
of a web a yard in diameter.

The river at this point is exquisitely pretty. It runs
down a succession of declivities into basin after basin, the rocks
or either side being some thirty or forty feet high. Up
the river the bamboos form a series of graceful arches. Lower
down the stream loses itself in a mass of green verdure every leaf
in which sparkles with drops of rain radiant with the sun-light
which penetrates the foliage above, but through which scarce a
glimpse of sky can be seen. One would hardly think it possible
that anything could live under the water where it rushes
with its greatest force over the smooth rocks in its precipitous
course, but even there there is life, and by putting an impedi-
ment in the water's way we discover underneath it, clinging to
the face of the rock, hundreds of dipterous larvæ. As we
ascended the path to the road above, we encountered a pretty little
snake† about nine inches long, probably on the look out for what
ever food the rain might have brought out. Directly he was caught,
he, after the manner of his kind, attempted to bite, but his little
teeth, strong enough for the ordinary purposes of daily life, were
not long enough to draw blood. If even they were, it would
have been of no consequence as he belonged to a very harmless
species feeding, when young, upon small lizards and frogs, and
when adult upon mice and birds; when full grown the country folk
call him "Machete Couesse"‡ the first word meaning cutlass, but
the interpretation of the latter, I have never been able to find out.
What he is called when young, it is impossible to say. His coat
showed a pretty white, brown and black pattern, but if he had
not come to an untimely end he would ultimately have donned a
uniform of a green brown color with two light longitudinal stripes
between side and back—one on either side of his spine. It will be
noticed I am particular in describing the position of these stripes
as being "between the side and back." I do so advisedly because if

*Pepsis sp
†Herpetodryas boddaerti or Coluber boddaerti.
‡Mr. de Suze informs me, since the above was written, that probably
Couesse means "grass" this may be so but is not a good description as
this species is found as frequently in trees as in grass.

one overcomes the natural repugnance felt to the ophidia, enough to enable him to examine them coolly, he will find that snakes are not, as many people imagine them, cylindrical in shape. Cross sections of many species will show that they are square, with rounded corners, others triangular, and some oval, but never cylindrical, a mistake which taxidermists frequently fall into. Further on we met a tremendous toad* in a bamboo clump, he was large enough to cover a soup plate. Toads are, as is well known, useful creatures, destroying many noxious insects but they are a great nuisance if they take up residence in the vicinity of bee-hives, as they will soon considerably lessen the number of bees. It is interesting and curious, however, to any one who can watch their proceedings philosophically and with equanimity for a few moments, to note the absurd gravity of the culprit as he squats near the entrance to the hive and the stolid air with which he projects his tongue and flicks up with it the bees as they settle before entering it. Every now and then he blinks his brilliant eyes as if he would say " an extra sweet tit-bit.' Why it is the bees do not sting the toad I cannot imagine ; if they do the toad most likely considers it in the light we regard hot sauce and mustard. Even the centipede falls a victim to his voracity, for I recollect seeing once, a toad (not very large) cramming into his mouth, with his fore feet, a particularly fine centipede. Toads should be encouraged in gardens troubled with parasol ants as they are fond of stationing themselves near the ants' line of march and catching the little creatures as they pass with their leafy burdens. Toads pick up quantities of dirt and stones when catching their food and it would be interesting to learn whether the stones and sand, which thus find their way into their stomachs aid digestion in the manner the pebbles swallowed by fowls do. The sun had come out brightly when we had again reached the road and as we walked along insects flew up from our feet, sometimes two or three at a time, and disappeared in the bush. At first sight they looked like flies, but closer examination led us to recognize an old acquaintance of our entomological rambles in Europe, namely the tiger beetle.† These beetles, however, are not so large, or nearly so pretty, as their European cousins, but they are just as fierce. My friend managed to catch one by quickly throwing a handkerchief over him, but his motto was *nil desperandum*, and he bit right and left with his sickle shaped mandibles, and kicked with might and main with his long slender legs, in his frantic efforts to get away from our grasp. If he was only a little larger, he would indeed

---

*Bufo marinus BOULENGER.
†Odontocheila bipunctata.

be a formidable fellow—he looked wild enough, small as he
was, with his prominent eyes, the marvellous mobility of his
parts and especially of his feelers.   In his larval stage he digs
little holes in which he waylays ants and other creatures.   In
his perfect state his character does not improve, for he is the
terror of all other insects of his own weight and size, while to the
smaller ones he must be a perfect ogre.   Our tiger beetle disposed of, we resumed our walk and found at the head of the
ravine that the stream crossed the road twice.   Hovering over
the water were some delicate dragon flies* which did not seem so
hungry as the larger species, which fly about in the evening and
whose sole object in life is eating.   Dragon flies are veritable
tyrants of the air.   Their youth is spent in water where
they work no less havoc among the other inhabitants.   To
do this they are furnished with a special apparatus; their
jaws are developed into a horrible pair of long arms which they
fold up over their faces, when they look like a mask.   These they
suddenly stretch out when they seize their helpless prey.   We
had now arrived at a cleanly kept cocoa estate, where we found
another species of " tic tack" in which we did not find any strong
points of interest.   This, however, could not be said of a pair of
small grey praying mantis* which moved rather faster than these
creatures do generally and like many other animals, much higher
in the scale of creation, they persistently placed the bough on which
they rested between ourselves and them.   Their colour afforded
one more instance of protective mimicry.   We also noted some
fine wood peckers busily searching the bark of the cocoa trees
for the grubs which do so much harm to the planters' crops.
We soon came upon the proprietor himself, a fine old gentleman
engaged in breaking cocoa pods, with whom we had a short chat
upon the prospects of the next crop, and cocoa pests in the shape of
rats and squirrels.   He said he was much troubled by these
animals and that he was careful not to kill snakes when he
found them, as he believed they ate the rats.   Such a sentiment
coming from such a source, greatly surprised us, as country people,
as a rule, are possessed of the idea that snakes should be killed
wherever met.   After we had rested, we started on our homeward journey, catching on the way a stick insect.   Stick insects
are not at all common, many people think, but at night, on a damp
bank, they may be found by hundreds.   When we reached the
hunting ants' nest, at the risk of some terrible bites my friend
thrust his hand into the midst of the mass of insects and captured
several hundreds, which, with many a vigorous stamp and hasty
exclamation, were soon imprisoned in a pocket handkerchief for

* Now being identified.

examination at some more convenient season. But the day was drawing to a close and we began to hurry home, having had a most enjoyable peep into the great book which Nature ever holds open to the mind which can enjoy the study of her manifold works.

AN AMATEUR NATURALIST.

(R. R. MOLE.)

A RAMBLE ON A COCOA ESTATE IN SANTA CRUZ.

HAVING made no arrangement for a special excursion on Whit-Monday, I decided to spend my holiday as I usually do, with some friends residing in the delightful and extensive valley of Santa Cruz, and, in order to show how much work can be done, even at our doors, I have prepared this account of my trip.

With this object in view, I started from my home at St. Anns about half past six o'clock on the morning of the 22nd of May, and hurried to catch the twenty minutes to seven tram car, so as to be able to get the first train going up the country.

Having succeeded in doing so, I was soon being rapidly carried eastward in the company of a friend, whom I had met at the Station, and who happened to be going to the same place as myself. Leaving Port-of-Spain behind us, we soon found ourselves rushing through the mangrove swamp to the east of the town, the noisy train startling, every now and then, a white heron or crabier. Rapidly passing drains in which a species of waterlily grew pretty thickly, fields of bulrushes, peasants' houses, and abandoned cane fields, in which other kinds of vegetation and cultivation are now springing up, we soon came to the sugar estates of Barataria and El Socorro ; and nearly opposite to the long avenue leading to the abandoned works of the latter, the train stopped beside a low roofed building, known as the San Juan Station, and the first on the line from Port-of-Spain.

Here our journey by rail ended, so we got out and prepared to do the rest of it on foot. Crossing the Station yard and the Eastern High Road, which lies at the back of it, we turned into the Saddle Road, which, starting opposite to the entrance of the Station yard, runs in a northerly direction throughout the entire length of the Santa Cruz valley, and crossing the low ridge at the north western end, from which it derives its name, carries the traveller back again to Port of-Spain via Maraval.

The morning had been a fine one, but now heavy clouds rolling from the east seemed to threaten rain, so we hurried forward, and at the junction of the old and new Santa Cruz Roads, we selected the latter, which is in many respects the better of the two. We passed on our way several pretty country villas, a straggling village, the inhabitants of which were chiefly East Indians and Creole Africans, more country residences, and a couple of bridges the first over a deep ravine and the second, known as Bagatelle Bridge, over the Santa Cruz River, which sweeps in a graceful curve, with many noisy cascades and whirling eddies, under it. After crossing this bridge, just beyond which there is on the right hand side of the road, a never failing spring of water, which renders this part of the road always wet, we did not take long to arrive at our journey's end, where, the usual greetings having been exchanged, we had a good rest and then breakfast.

After breakfast the rain began to fall, and continued to do so, at intervals, during the better part of the day. This unfortunately hampered my observations considerably, but taking advantage of every lull, I made several excursions into the cocoa plantation which adjoined the residence of my host.

Here I had fine opportunities of seeing the second staple product of the Colony in all its stages, and all the best methods of preparing it, but as Natural History was my hobby, I did not pay much attention to cacao, and merely noticed that the trees were putting out a fine lot of flowers and young fruit. On some of the latter I observed clusters of black ants of considerable size, and unpleasant smell when crushed. The labourers called them " cacao ants," and don't attempt to remove them from the pods. On some of the larger pods, there were *Coccidæ* or plant lice as well as ants, which probably is the cause of attraction of these ants.

On my first stroll, I went to the right of the premises, and into a well shaded patch of cocoa. Far above, the branches of the shade trees, sturdy trunks of which were easily distinguished among the more slender stems of the cacao, sheltered both the cultivation and myself from any strong sun light which there might have been, but which unfortunately, there was not, for even a shady cocoa piece is no shelter from a tropical shower. Besides cocoa, there were also some coffee trees planted between the rows, and here and there some pretty yellow banded wasps, apparently less ferocious than most of their kind, had built delicate paper-like, pear-shaped nests on the under side of the larger coffee leaves. These nests were ingeniously fastened to the leaves in such a way that although of considerable size they could not be seen by anyone viewing the leaves from above.

The trunks or stems of some of the cacao trees were covered with a coating of cob-web, made by a curious insect resembling an ear-wig, which resided under its delicate cover. While examining the trunk of a fallen tree, a flash of bright blue dazzled me, and looking up suddenly, I was lost in admiration of a glorious "Emperor" *(Morpho peleides)* which had just passed almost under my nose, and was now winging its way with leaping flight to the hills beyond the cultivation, giving me occasional glimpses of its glorious azure wings as it went. Digging into the trunk of the tree I unearthed an undeveloped millipede *(julus sp.)* or congori, and the larvæ of the ordinary boring beetle. In the trees above the shrill tones of the cicada rose and fell, I suppose in accordance with the insect's feelings, while on the cocoa tree to my right a pupa shell still clung, as it did when the insect emerged in its perfect state upon the great world.

Rain drops now began to fall thickly, so I was obliged to beat a hasty retreat to the house, where I waited with impatience for the rain to cease. Sallying forth as soon as the shower was over, I went on this occasion towards the river, which flows in front of the house, and strolled along the bank. Here, in the deep pools, were to be seen fish of various kinds. Shoals of sardines *(Hydrocyon sp.)* with silver sides and red tipped tails, strove against the current, and now and then one rose at a fly. Near the bottom of these pools the guabines *(Erythrinus sp.)* lay ever ready for their prey, as they are most voracious fish, and below them on the rocks and stones in the bed of the river, holding on by their sucker like appendages, were to be seen the curious and uninviting *mama cascadura*. In holes in the rocks on the sides of the river, lurked crayfish of various sizes, and inquisitive natures, ready to pounce on their unsuspecting prey. Close in shore, in the dead water of the reaches, might be seen small fry *(Cyprinodonta* and *Girardini)*. In the ooze of the pools water snails are lying and on the surface skims a curious spider-like insect. Under the rocks and stones in the uncovered portion of the river bed, small crabs *(Gecarcums sp.)* abound and now and then a frog or an ugly toad is disturbed in his day sleep. Over head and around the dragon flies cruised up and down, or rested motionless on twigs and leaves over the water.

Afternoon was drawing to a close when I turned homewards from this pleasant walk, and instead of retracing my steps, I selected a path which led through another bit of cocoa cultivation, in which on some shade tree a species of cicada known as the "razor grinder" was making the little dell resound with his evening song. A curious object, some distance off, mounting the trunk of a tree by a series of jerks, and resembling a squirrel, next attracted my attention. It turned out to be that most useful bird the *Charpentier* or *Carpenter* (a species of *Dendrocolaptes*) whose searching eye and powerful bill never fails to destroy the guilty insect which, perhaps, is sapping the life of the tree.

This bird was at one time shot down wherever found, by the cocoa planter, from whom it received the wrong

name of *Mangeur de Cacao*; for it was believed that it
destroyed the ripe pods and ate the sweet mucilage around the
beans. Observation of its habits, and experience, has since taught
the planter his error, for, though, it may attack pods into which
insects have already effected an entrance, it does not eat the bean,
and is recognised to day, by all sensible planters, as their best
friend, in destroying, as it does, the insects injurious to the tree
and its fruit.

Leaving the cocoa patch, the road traversed a guinea grass
piece, in which were planted a number of cocoanut palms, all of
which were in fine condition, then it led me through a fine
pasture surrounding the dwelling house, which latter I reached in
time for dinner. On a barren spot in the pasture, I could not
help noticing the performances of a number of handsome sand
wasps, which kept up a series of ærial dances around the spot, now
and then alighting to burrow a hole in the soft earth. Each had
its particular burrow, and when disturbed would perform a series
of ærial evolutions, and return to the spot it had selected to
resume its work. I tried to capture one of these wasps but not
having a proper net failed to do so.

On my return to the house the bad weather which char-
acterised the latter half of the day set in again, but the rain
cleared off in time to allow my friend and self to get the last
train to Port-of-Spain, where I left my companion, and, catching
a tram car to St. Anns, reached home just as the fire-flies were
illuminating mountain and valley with nature's night lights,
having spent a most pleasant day in spite of the inclement
weather.

SANTA CRUZ.

(T. I. POTTER.)

THE PROTECTION OF WILD BIRDS.

At the request of Mr. Syl. Devenish the Publication Committee
reproduce his report for the Ordinance for the Protection of
Wild Birds.

Port-of-Spain,

17th June, 1875.

THE HONOURABLE THE ATTORNEY-GENERAL, &c., &c.

Sir,—In compliance with the instructions received by me,
at the meeting of the Society of Arts held on the 9th March last,
I have the honour to forward to you, for the consideration of His
Excellency the Governor, the following Report on the practice

of shooting birds at all seasons of the year, a practice the results of which are becoming, every day, more apparent in the increasing number of grubs and all insects detrimental to our crops, although I believe the majority of our plumage birds are flowers sucking ones. I may mention, here, as an instance of such results, that when I was recently surveying in the Arima Ward Union, I saw on the estates rented in Oropouche by Mr. Ernest Rodulfo (a great place of resort for bird stuffers), considerable damage done to the crops by swarms of grass-hoppers, grubs, and insects, which were scarcely ever seen there, before, in any appreciable quantity.

1. Having for years past been almost constantly surveying in every district of the Colony, I can bear testimony to the relentless war waged, at all times, against our ornamental birds, though not by far so much against those for the table, and to the almost incredible absence now of days, of birds in our forests, which were formerly so much alive with them. In fact, the dearth of animal life in our high woods is sadly remarkable at the present time.

2. No attention, whatever, is paid by those who make a trade of bird stuffing, to the nesting, laying, or setting seasons, nor, in some cases, even to the moulting times of certain variety of birds, and, all the year round, in certain districts, particularly cocoa ones, there is a constant popping off which reaches its maximum in April and May.

3. It is worthy of remark that it must be much more remunerative to shoot birds for their skins than for their meat, as is every day proved by the very scanty supply of wild game birds in our markets, whilst thousands of dozens of stuffed birds, or skins, are constantly exported.

4. It is, perhaps, not generally known that 1 ℔ of powder and 5 ℔s of shot (No 12), and less than 250 caps, costing, altogether about $2. 20 will bring down an average of 25 dozen birds, generally sold to retailers, at 60 cents a dozen, or $15.00, for assorted birds, and at $1 per dozen for Rubis humming birds = $25.00 and that from 10 to 15 dozens a day can be shot and put in skins.

5. I have frequently met in distant parts of the Island, even in wild uninhabited forest, full grown able-men, eminently fit for more manly work, mostly Spaniards or of Spanish origin, at midday lazily lounging in their " *chinchorros*," smoking their pipes or long toms, and often singing and playing guitar, after having in the morning, shot and skinned their ten or twelve dozens birds.

6. After careful efforts made by me to endeavour and ascertain the exact period of nesting and laying of each sort of our wild birds, I regret to say that I do not believe that any satisfactory information can be gathered, without further investigation sytematically carried out.

7. The late Dr. Leotaud experienced so much difficulty in getting sufficient data on the subject, that in despair, as he repeatedly told me himself, when in the woods together, he had to remain almost silent on that interesting point, in his, otherwise, so complete and remarkable book on the birds of the Island.

8. Perhaps the Wardens may be asked to take an interest in the subject, and endeavour in their respective districts, to gather by themselves, as well as through their Ward Constables and others, much of the useful information required.

9. I have been myself, for years, at some pains to attempt it, and from the meagre knowledge I could procure, I have attempted something like a table of the nesting and laying times of some of our birds of the richest plumage, and beg to append it here, such as it is although I hope an absolute protection will be soon afforded most birds, for some years to come, during which time correct table may, no doubt, be prepared to regulate hereafter, the protective seasons.

10. With regard to our game, or wild table birds, there is certainly no such necessity for entire protection, and I believe they may still continue to be under the provisions of the " Wild Birds Protection Act" of 1873.

11. Our principal table land birds, are the ground doves (ortolans), the doves (tourterelles), two sorts of ramiers (wild pigeons) and a few others which are not so much troubled, (the ramiers, in fact, only appear in any quantity at certain times.) Yet I believe it would be advisable to prohibit their being pursued in March, April and May.

12. Our swamp, or water game birds, such as plovers (if they can be put in that category), several species of wild ducks, and others, are, with the exception of our water fowls and snipes, and a few lagoon birds, mostly migratory, though I have often found in our swamps nests and eggs of ducks, but they generally come over from the Main land, from July to December.

13. I must remark that I have adopted as being, really, the only ones by which they are generally known, the common, mostly French names, used in Dr. Leotaud's book and collection.

14. I might have swollen this report with many details respecting some of our birds, but I do not see of what use it would have been for the object in view, but, incomplete though it be, if it can, in any way, assist in framing the new Ordinance for the protection of wild birds, I shall feel more than repaid for the trouble it has given me.

15. Before concluding it, however, I wish to refer to one of our most exceptional set of birds, which I have omitted to mention here above, and which are considered one of the rarest delicacies in the Colony; I mean the Guacharo, or Diablotin (Steatornis

caripensis of Humbolt), and of which, in spite of the real difficulties of getting at them, as stated by Dr. Leotaud, the destruction
appears to me to be much more feared than he seems to believe.
I have myself, not many years ago, taken with my own hands,
in the caves at the top of the cabeceras of Oropouche, no less than
175 Guacharos, which, if I had not done so, would have no doubt
been caught next day by a party of greedy Spaniards whom I had
left at the foot of the mountain, preparing for an excursion to
those caves.

16. This quantity may give an idea of how the destruction
of those birds must inevitably be on its way, unless some measures
be taken to prevent it.

17. That the " Diablotins" are getting every year scarcer, is
proved by the very limited number of them to be, now, seen
offered for sale in the markets, where they are always much
" recherchés" as *tit bits* to be only procured once a year, and
where, some fifteen or twenty years ago, they were in their
season, abundantly found either fresh and alive, or " boucanés."

18. In fact, I am told that in some caves (cuevas) they have
already almost disappeared, whilst the numbers of those, in some
other caves which I have recently visited, both at Monos and
Huevos, as well as at Aripo and Oropouche, seemed to me to be
certainly considerably less than I saw before, only a few years
ago.

19. It would, therefore, be advisable to forbid the young
birds being taken away, for at least two or three years to come.

20. In concluding these very incomplete notes, I must crave
indulgence for the scanty information they may contain, but
having done my best to meet the object in view, it is no fault of
mine if I have not succeeded in furnishing more precise data
towards the framing of the new Ordinance for " the protection of
wild birds."

I have the honour be,

Sir,

Your obedient humble servant,

SYL. DEVENISH,

Sec. Corr. Com. Society of Arts.

P.S.—I enclose the schedules you have sent me, filled up at
your request with the popular names of the birds. in alphabetical
order.

IN MEMORIAM.

IT is our sad duty to record the death of Mr. George Vahl, a member of the Trinidad Field Naturalists' Club, at the early age of 32. Mr. Vahl was especially fond of boating and shooting and some six or seven weeks before his death he went on a shooting excursion up the Caroni where he contracted a severe cold of which he was unable to free himself. Having been for some time a victim of insomnia he was in the habit of using laudanum and on the 4th March, being unusually sleepless, he took a stronger dose than customary with the result that he was accidentally poisoned. He was taken to the Hospital and successfully roused from his insensible condition, but, as the week wore round, his strength, which had been seriously impaired by the cold before referred to, broke down and he gradually sank and passed away on Sunday evening, March 11th. The funeral obsequies, which were attended by the President and several members of the Club, were performed by the Rev. Russell Brown at Trinity Cathedral and the Cemetery. Mr. Vahl was elected a member of the Club on the 2nd September, 1892, and was an ardent Field Naturalist. He was particularly interested in the Hemiptera. He left amongst his papers a half completed account of one of his Caroni excursions, which he probably intended reading at a Club meeting. He was a native of Germany.

g the ne
best possible value for your money. *The*
squanders his means by paying above the market rate for art

BY SPECIAL APPOINTMENT TO

H.E. Sir F. Napier Broome, K.C.M.G.,

AND

Sir Charles Lees (Demerara.)

H. STRONG,

Piano Warehouse,

is supplied by every London Direct Steamer with

Pianos, Harmoniums,

AND

ORGANS,

all of which are specially made for this climate, and are
GUARANTEED FOR 20 YEARS, and may be had on the

New Hire Purchase System.

No Charge will be made for Hire if Purchased.

Mr. H. S. has a staff of competent men, he is therefore able
to comply with

All Orders to Tune and Repair

on the shortest notice possible.

Pianos, Harmoniums and Organs.

PIANO WAREHOUSE,

St. James Street, Port-of-Spain.

b.

o. 2.

onel
opy,
Syl.
ott,
ary.
res-
and
ead
g in
e a
ers,
ing
of
the
ted
nel
sell
op-
ees,
ct,
ted
m-
ns
2),

☞ All Communications and Exchanges intended for the Club should be addressed to the Honorary Secretary, Port-of-Spain, Trinidad, B.W.I.

JOURNAL

OF THE

Field Naturalists' Club.

Vol. II.   JUNE, 1894.   No. 2.

REPORT OF CLUB MEETINGS.

6TH APRIL, 1894.

PRESENT: Mr. Henry Caracciolo (President), Lieut.-Colonel Wilson, C.M.G., Sir John Goldney, Messrs. R. J. L. Guppy, T. I. Potter, C. J. Thavenot, T. W. Carr, J. R. Llanos, Syl. Devenish, M.A., J. Hoadley, J. Russell Murray, C. W. Scott, W. E. Broadway, R. R. Mole, and F. W. Urich, Hon. Secretary. The following elections took place : Rev. E. D. Wright, Corres. ponding Member ; Rev. E. J. Holt, Messrs. James Miller and Arthur Gaywood, Town Members.—Lieut.-Colonel Wilson read some "Notes on Camping Out, Fishing and Cariboo Shooting in New Brunswick," in the course of which the author gave a graphic account of birch bark canoeing upon Canadian rivers, lynx and bear hunting, trout and salmon fishing, beaver trapping and cariboo shooting. The notes also contained descriptions of the dress, the snow shoes and the sleighs of hunters in the Canadian wilds, and the severity of the cold was also commented upon. Sir John Goldney proposed a vote of thanks to Colonel Wilson, which was seconded by Mr. Devenish.—Mr. J. Russell Murray made some remarks upon the larvæ of some lepidop-terous insect he found destroying the bark of cocoa trees, but owing to an accident was unable to show the insect, which he had succeeded in breeding.—The President pointed out how the opinion expressed by the Club's Borer Com-mittee had been borne out by the subsequent investigations of Mr. Hopkins (published in *Insect Life*, Vol. vi., No. 2), and Mr. Massee of Kew.—Mr. Urich made some remarks about his recent stay at Caparo with Messrs. Brewster and Chapman.—A large number of bats as well as some fœti of various mammals were exhibited by the President.—Mr. Urich said in a recent excursion he managed to find two specimens of the only sweet water bivalve found in Trinidad, viz., *Anodon leotaudi*, Guppy, in a small brook in the far interior. Of this shell Mr. Lechmere Guppy wrote in a paper on the moluska of Trinidad, published some time ago, " I have never been able to obtain more than two or three specimens."—The meeting adjourned at 10.15 p.m.

4TH MAY, 1894.

PRESENT : Mr. H. Caracciolo (President), Sir John Goldney, Dr. Rake, Messrs. Syl. Devenish, M.A., T. I. Potter, Charles Libert, Arthur Gaywood, G. J. Reed, H. J. Baldamus, S. A. Cumberland, T. W. Carr, W. E. Broadway, and F. W. Urich, Hon. Secretary. Mr. William Brewster, of Cambridge, Mass., was elected an Honorary Member of the Club.—The President read a few " Notes on some Trinidad Bats " which contained much original observation on the habits of some of our species. A large number of stuffed specimens were shown to illustrate the paper. At its conclusion a hearty vote of thanks was accorded to Mr. Caracciolo and an animated discussion ensued in which all the members present took part.—Sir John Goldney gave a graphic account of the habits of the large fruit-eating bats of the Straits Settlements, which served for interesting comparisons with our species.—Mr. Cumberland read some notes on a large *Boa constrictor* he had poisoned. This reptile resisted a dose of prussic acid and it was a dose of strychnine that put an end to it. The specimen, which was shown, measured alive 10 feet more or less, but skinned it was 13 feet long.—Mr. Potter made some observations on Trinidad orchids, of which several species were exhibited.—Mr. Cumberland made some remarks upon a live specimen of the two toed sloth, *Cholœpus didactylus*, Linn., from Venezuela, which he had brought to the meeting. This mammal is in Mr. Oldfield Thomas's list as having been recorded in Trinidad by Ledru in 1810, but it certainly does not occur in the Island, and no doubt the specimen recorded came from Venezuela.—Mr. Potter drew attention to the nests of the caterpillars of a *Hydrias* sp. found at Monos.—The meeting closed at 9.30 p.m.

CLUB PAPERS.

THE SCHIZOMYCETES.

By Beaven Rake, M.D. (Lond.)

OUR knowledge of bacteria has advanced so rapidly during the last twenty years, that there is a danger that in this as in all good movements enthusiasts may claim too much, and that over-zeal on the part of some may bring bacteriology into disrepute. I refer to a tendency which exists to immediately discover a specific germ for every known disease. That this is premature will appear in the course of this paper, but there is no doubt that an intelligent study of bacteria is of the greatest importance to the physician, the surgeon, the farmer, the manufacturer, and indeed to all who are concerned with the lives of animals or vegetables. It therefore seemed that the subject was not unsuitable for the consideration of the Field Naturalists' Club.

The discovery of micro-organisms is no new thing. In 1683 the Dutch observer Leeuwenhoeck described organisms in saliva and putrid water, probably corresponding with what are now recognized as vibrio and leptothrix. For the next two centuries these organisms were the subject of much controversy. At one time they were regarded as animals, at another time as vegetables, while some observers doubted whether they were alive.

In 1773 Muller suggested a classification and others followed his example during the first half of this century. To Cohn, however, belongs the credit of first satisfactorily classifying the bacteria in 1872 and recognizing their position in the vegetable kingdom. But the twenty years which have elapsed since Cohn's researches have brought about a further change of opinion as to the position and classification of bacteria.

They are now regarded as belonging to the Schizomycetes or fission-fungi which come under the achlorophyllous division of the class protophyta of the group thallophytes. In other words bacteria are minute cellular plants, which increase chiefly by division and are without chlorophyll.

Owing to this absence of chlorophyll bacteria are unable to split up carbon dioxide into its elements, like the higher vegetable cells, but on the other hand they differ from animal cells by being able to obtain their nitrogen from ammonia compounds. Nencki has shown that active bacteria contain 83.42 per cent. of water, while in the dried state they consist of

A nitrogenous body called mycoprotein   ... 84.20
Fat...   ...   . 6.04
Ash 4.72
Undetermined substances 5.04

In structure bacteria consist of a cell wall which can be shown by the action of iodine to consist of cellulose, and cell contents which yield mycoprotein. In some species this is homogeneous, in others granular. Sometimes the power of retaining certain aniline stains is present, but not always. This is of great importance in determining the nature of some bacilli, for example, those of tuberculosis and leprosy.

In some bacteria the cell protoplasm contains starch granules as in *Clostridium butyricum*; in others, such as *Beggiatoa*, sulphur is present. In the latter organism the cells also contain a special pigment bacterio-purpurin. In other cases the pigment granules are outside the cells as in *Bacterium prodigiosum*, while in *Bacillus pyocyaneus* the pigment is partly within and partly outside the cells.

Many species of bacteria are provided with a gelatinous envelope, the result of either secretion from the cell or absorption of moisture and swelling of the outer layer of the cell wall. When this gelatinous material forms a matrix in which numbers of bacteria are congregated it is called a zooglœa. This zooglœa condition is a resting stage of the organism.

The forms of bacteria as will be shown presently are very various and have given rise to no little controversy.

With regard to the reproduction of bacteria, though, as implied by the name Schizomycetes, the characteristic mode of increase is by fission, there are other processes which resemble fructification. These are divided into two groups distinguished by the formation of endospores and arthrospores.

In endospore formation the protoplasm becomes granular, and at certain points in the chain of rods specks appear which gradually develop into highly refractive circular or oval bodies. After a time the cell walls and protoplasm disappear and the spores are set free. This is well seen in *Bacillus subtilis* and *Bacillus anthracis*.

In arthrospore formation as seen in *Leuconostoc mesenteroides* certain elements in the chain of cocci, apparently not differing from the rest, enlarge, and the walls become tougher and more refractive. The remaining cells die, and these enlarged cells are set free as spores capable of reproducing the growth in suitable soil.

Spore formation has not yet been observed in all bacteria, hence one of the chief difficulties in the way of a scientific classification.

Spores are invested with a thick membrane and are far more resistent to the action of chemical reagents and heat than the parent cells. They also retain their vitality when dried. This is a point of great practical importance. To destroy spores, infected articles such as clothes must be raised to a much higher temperature than that necessary to destroy bacteria. Again, in the preparation of sterilized media, the test tubes must be heated on several consecutive days in order to kill successive crops of spores.

It is probable that spore formation is not due to exhaustion of the nourishing soil and consequent provision for the perpetuation of the species as has been supposed by some. It is more likely that it is a result of free access of oxygen.

Anthrax bacilli have been found in the soil where bodies of animals dead of anthrax have been buried, and thus a real danger to the community arises. Klein, however, has pointed out that if mice and guinea pigs which have died of anthrax are kept unopened, the bacilli degenerate and disappear. Infection of a burial ground must therefore result from the interment of bodies of animals in which necropsies have been made, or in which the hides have been soiled with excretions or blood. The practical lesson to be learned is the same in either case, that all animals dead of any suspicious disease should be at once cremated.

I may state here that a series of observations which I made on the earth over the graves of lepers or where pieces of leprous tissue had been interred gave negative results.

The classification of bacteria like that of many other groups of plants and animals is still in a transition state, but provisionally the classification suggested by Zopf may be accepted as a good one for working purposes.

He divides the Schizomycetes as follows :

GROUP I. — *Coccaceæ.* — Cocci and thread-forms resulting from juxtaposition of cocci. Fiss.on in one or more directions.

    Genus 1. Streptococcus.
        2. Micrococcus.
        3. Merismopedia.
        4. Sarcina.
        5. Ascococcus.

GROUP II. — *Bacteriaceæ.* — Cocci, rods and thread-forms. No distinction between base and apex in latter. Fission in one direction.

    Genus 1. Bacterium.
        2. Spirillum.
        3. Vibrio.
        4. Leuconostoc.
        5. Bacillus.
        6. Clostridium.

GROUP III. — *Leptotricheæ.* — Cocci, rods and thread forms. Distinction between base and apex in latter.

    GENUS 1. Leptothrix.
        2. Beggiatoa.
        3. Crenothrix.
        4. Phragmidiothrix.

GROUP IV. — *Cladstrichccæ.* — Cocci, rods, spirals and thread forms. The latter with false branchings.

    Genus 1. Cladsthrix.

Before noticing the genera mentioned in this table a few remarks may be made on some practical points connected with bacteria.

One of the most interesting and important of these is the question of pleomorphism.

In Cohn's first classification mentioned above he divided bacteria into four groups :

1. Sphæro-bacteria ; globules *(micrococcus.)*
2. Micro-bacteria ; short rods *(bacterium.)*
3. Desmo-bacteria ; long rods *(bacillus* and *vibrio.)*
4. Spiro-bacteria ; spirals *(spirochæte* and *spirillum.)*

Billroth in 1874 opposed this division and went to the other extreme. He regarded all Cohn's forms as developmental forms of one micro-organism which he called *Coccobacteria septica.*

In 1875 Cohn answered Billroth, still maintaining that distinct genera and species existed. He produced a second classification, placing the bacteria under a new group, the Schizophytes, and adding several genera.

Our present knowledge admits a mean between the extreme views of Cohn and Billroth. We now accept the doctrine of pleomorphism which was to some extent anticipated by Lister in 1873 after observations on a bacterium in milk, but was first definitely formulated by Lankester in the same year after his researches on beggiatoa to be referred to later. He says : " The existence of true species of bacteria must be characterized, not by the simple form features used by Cohn, but by the *ensemble* of their morphological and physiological properties, as exhibited in their complete life-histories."

This theory of pleomorphism has since been abundantly confirmed by the researches of Neelsen and Cienkowski on the bacillus of blue milk, of Zopf on beggiatoa and cladsthrix, of Van Tieghem on clostridium, and of others. It has the merit of being simp'e and at the same time rational. We should not think of separating a fern into two species because the young fronds differ in shape from those of the fully grown plant, nor should we suppose that the common holly contains two species, because the leaves at the base are prickly for purposes of defence, while those at the top of the tree are plain. Neither would anyone imagine that the e irly cotyledons of a plant just emerging from the soil represent a species distinct from the fully developed plant, because the latter shows totally different leaves and stem. And yet, to compare small things with great, those who made the above gross mistakes would not be much more ridiculous than those who tr'ed to divide bacteria by absolute morphological characters. Science is always slow to give names, and provisionally we must accept Zopf's classification which is based on the theory of pleomorphism.

We must, however, clearly recognise that while a bacterium can change its form it cannot change its nature, and be converted from a harmless into a pathogenic form, as was asserted by Buchner. This theory has been disproved by Koch and Klein.

The most important practical question which arises with reference to any micro-organism is whether it is pathogenic or capable of invading and multiplying in living tissues, thus producing disease in man or animals, or whether it is non-pathogenic or saprophytic, i.e., only able to develop on dead tissue, and therefore found in sloughs and discharges. To determine this point Koch has laid down the following four postulates :—

1. The micro-organism must be found in the blood, lymph or tissues of the diseased man or animal.

2. The micro-organisms must be isolated from the blood, lymph or tissues, and cultivated on suitable media outside the body. These cultivations must be carried on through successive generations.

3. A pure cultivation thus obtained must when introduced into the body of a healthy animal produce the disease in question.

4. In the inoculated animal the same micro-organism must again be found.

These postulates have been fulfilled in the case of certain diseases which attack animals, such as anthrax, glanders, tuberculosis and malignant œdema. In human diseases there is an obvious difficulty in the last two postulates which involve inoculation of a healthy human subject. In such cases we have to be content with experiments on animals, but here we are faced by the difficulty that certain diseases such as leprosy are not found in the lower animals.

Lastly a few words may be said on the distribution of bacteria outside the animal body. They are found in earth, air, and water.

Portions of garden mould taken from the surface and dropped into sterilized culture fluid always infect it, producing a culture of cocci and bacilli. In winter Koch has found that all organisms are absent at a depth of one metre, in soil which has not been recently disturbed, which does not consist largely of decomposing material, and into which there is no unusual soakage of water.

Recent researches have shown that bacteria play an important part in the nitrification of soil. As these bacteria are chiefly found near the surface, the necessity at once becomes apparent, in agricultural operations, of frequently forking over the soil round the roots of trees or plants.

In air bacteria usually occur attached to particles of dust.

Perfectly still air becomes pure by subsidence of these particles. Bacteriological examination of air over the surface of the liquid in main sewers has shown an absence of bacteria.

All water except from great depths contains bacteria. Rain water becomes contaminated from the dust which it carries down, while surface water is infected from the ground through which it soaks. The risk of contamination of well water and river water by bacteria derived from sewage is obvious. Even sea-water is not free. **Dr.** Bassenge of the *Stosch* shewed me cultivations which he had made from water taken at various depths in the Atlantic Ocean.

From the above considerations it is evident that the bodies of man and the lower animals are always exposed to the danger of infection, and that many bacteria enter every day in food or water, or attached to dust are swallowed, inhaled, or brought in contact with open wounds or sores on the surface of the body. The aim of preventive medicine therefore should be not only to destroy bacteria whenever possible, but also by attention to the ordinary rules of health, to diminish the receptivity of the individual. For we cannot but believe that bacteria will continue to exist in spite of all our efforts to exterminate them.

I propose now very briefly to examine the various groups and genera set forth in the classification of Zopf. I have for the most part followed Crookshank's descriptions.

Group I.—*Coccaceæ.*

This possesses only cocci and thread-forms resulting from the juxtaposition of cocci. Fission may occur in one or more directions. Five genera are included in this group :—

1.—*Streptococcus.*—These are known as chain-cocci, because the individuals always or occasionally remain united together to form chains. Division is in one direction only.

Numerous species of streptococcus are described, many of which are associated with disease in man or animals, though to what extent this relation is pathogenic has yet to be determined. In this family occur the many forms of streptococcus pyogenes so frequently found in the pus of abscesses. Many of them are readily recognizable by the naked eye from the resemblance of the colonies which grow in glycerine agar to drops or streaks of oil paint of various colours. Other species have been described in connection with erysipelas, diphtheria, puerperal fever, endocarditis and several other human diseases, and also in many diseases of the lower animals such as swine erysipelas, cattle-plague, foot and mouth disease, disease of grey parrots, and the "flacherie," of silk worms

2. *Merismopedia.* These are also called plate cocci from the fact that they divide in two directions forming lamellæ or plates.

Most of the members of this genus are saprophytic. One species, *Micrococcus tetragonus,* found in tubercular cavities and sputum is saprophytic in man, but pathogenic in the lower animals, causing death in guinea pigs and mice in from two to ten days.

3. *Sarcina.* The genus Sarcina is characterized by division in three directions forming colonies in cubes or packets. Hence these are called packet-cocci.

They are all unassociated with disease. The best known is the *Sarcina ventriculi* described by Goodsir of Edinburgh in 1842. These cocci occur in groups of four or multiples of four producing packets with rounded off corners like bales of cotton. They are found in the stomach of man and animals in health and disease.

4. *Micrococcus.* This genus is characterized by division in one direction. The cocci after division may remain aggregated in irregular heaps, and are therefore called mass cocci. They never form chains.

There are several species of micrococcus and some of them are found associated with scarlet fever, measles, whooping cough and other diseases in man, but none of them have ever been proved to be the cause of these diseases. Koch however, has shown that certain micrococci produce septicæmia, pyæmia and progressive suppuration in rabbits, and Klein has described a coccus which gives rise to pyæmia in mice.

5. *Ascococcus.* The individual cocci are like micrococci, but they grow in characteristic gelatinous pellicles, and are hence called pellicle-cocci.

Only one species is known, *Ascococcus billrothii.* On the surface of nutrient media these cocci form a cream-like skin, made up of an enormous number of globular or oval families, each of which is surrounded by a thick capsule of cartilaginous consistency.

GROUP II. *Bacteriaceæ.*

This group possesses mostly cocci, straight or bent rods, and straight or spiral thread-forms. The cocci may be absent. The thread-forms present no difference between base and apex. Fission occurs only in one direction.

There are six genera in the group,

1. *Bacterium.* This genus is characterized by cocci and reds or only rods, which are joined together to form threads. Spore formation is absent or unknown. The absence of spore formation is now used to distinguish bacteria from bacilli. The arbitrary

division according to the length of the rods was found to be impossible, for a rod at one stage of its growth or under certain conditions might be a bacterium, while at another time it might be long enough to be considered a bacillus.

There are many species of bacterium associated with disease, but not one has been absolutely proved to be pathogenic in man. The bacterium of pneumonia discovered by Friedländer has been considered by some to be the cause of pneumonia, but inoculations of animals with cultivations of the bacteria have only succeeded in some cases.

In animals many pathogenic bacteria have been described, such as the bacterium of diphtheria of pigeons, the bacterium of fowl-cholera, the bacterium of septicæmia in rabbits, and the panhistophytum ovatum which has been proved to be the cause of pébrine in silk worms.

Other bacteria are chromogenic, producing yellow, red, brown, violet and other colours. The best known of these is bacterium prodigiosum which occurs on bread, boiled rice, and starch paste, and has been known as the bleeding host.

In this genus also occur the zymogenic saprophyte which gives rise to the ammoniacal fermentation of urine, and also that which oxidises alcohol in wine and other fruit-juices into vinegar.

2. *Spirillum.* This genus contains threads which are in the form of screws, and are made up of long or short rods only, or of rods and cocci. Spore formation is absent or unknown.

One of the best known examples is the spirillum of relapsing fever which was described in 1873 by Obermeier. It occurs in the blood of patients suffering from relapsing fever. The spirilla are absent from the blood during the non-febrile intervals. Carter in Bombay inoculated monkeys successfully from the human subject.

Another member of this genus which has attracted much attention of late years is the comma bacillus which Koch claims as the cause of cholera. His opponents, however, say that comma bacilli are not found in all cases of cholera. There are also possible sources of error in the inoculations of animals which have been carried out.

Since writing the above I see in the *Lancet* of February 10, 1894, that Prof. Cunningham of Calcutta has had certain distinct forms of comma bacilli obtained from cases of cholera, in continuous cultivation for nearly four years, all being under precisely the same conditions of environment. These bacilli have continued to present well marked differences, and Prof. Cunningham regards them as distinct species. If this is so, cholera cannot be regarded as due to the presence of any particular species of comma bacillus in the intestine,

Other comma bacilli are found in cholera nostras, in saliva and in old cheese. These were formerly believed to be identical with Koch's cholera bacillus, but many now regard them as distinct species.

3. *Leuconostoc.* This genus consists of cocci and rods. Spore formation is present in the cocci.

One species is described, *Leuconostoc mesenteroides*, well known as frog-spawn fungus or gomme de sucrerie, a formidable enemy to sugar manufacturers. So rapid is the growth of this organism that forty-nine hectolitres of molasses, containing ten per cent. of sugar, were converted within twelve hours into a gelatinous mass.

The cells occur singly, in chains, and in zoogloea, surrounded by a thick gelatinous envelope, so firm that sections can be cut with a razor. In addition to increase by division, leuconostoc is also reproduced by the formation of arthrospores as already described.

4. *Bacillus.* This is a genus containing cocci and rods or rods only, forming straight or twisted threads. Spore formation is present either in the rods or cocci.

This genus contains several most important species. Foremost among these is the *Bacillus anthracis*. This is the cause of splenic fever in animals, and of charbon, malignant pustule or woolsorter's disease in man. The morphology and life history of this bacillus have been very fully worked out, and as it completely fulfils the four postulates of Koch the bacillus anthracis is accepted as a type from which to study the various bacteriological methods.

As was stated earlier in this paper anthrax bacilli have been found in soil in which animals which have died of the disease have been buried. A more common mode of dissemination, however, is by the excreta, and blood and discharges from the nose and mouth of moribund animals. The bacilli find a nourishing soil in decaying animal and vegetable matter ; spore formation goes on rapidly, there being free access of oxygen, and the grass is extensively contaminated. Floods may carry these spores over adjacent meadows.

Animals may be directly infected through wounds in the mouth caused by siliceous sedges or grasses, or by wounds of insects. It is also probable that they may be infected through the intestinal and pulmonary mucous membranes.

In man the disease is met with in those who have to handle the wool or hides of animals which have died of anthrax. Infection takes place either through a wound or abrasion of the skin, or by inhalation of the spores attached to dust, which then develop in the respiratory and alimentary tracts. The disease most frequently occurs in certain manufacturing centres where the hides of animals are treated after being imported from abroad, or brought up from the country, such as Bermondsey in the south of London, and Bradford in Yorkshire. It is commonly known as woolsorter's disease. The stringent regulations now in force in England, as to cattle infected with splenic fever will doubtless diminish the frequency of the disease, but there is still a serious danger to be apprehended from hides imported from abroad. In the *British Medical Journal* of January 20, 1894, I see that a case of malignant pustule occurred in Paris recently, the source of infection being goat-skins imported from China. The skins were found to contain the dermestes vulpinus in two stages of development. The bodies of these insects and their excrement contained the bacillus anthracis, and guinea pigs inoculated with these insects or their excrement reduced to powder, died of charbon.

Pasteur by cultivating the anthrax bacillus succeeded in producing a protective vaccine for animals. This protection, however, only lasts for a time.

If the bacillus is passed through different species of animals its virulence is altered. Bacilli from sheep or cattle are fatal if re-inoculated into these animals, but if inoculated into mice, the bacilli thus obtained lose their virulence for sheep or cattle, and the latter are protected for a time against virulent anthrax. Also a culture that is vaccine for sheep kills a guinea pig and then yields bacilli fatal to sheep.

The *Bacillus tuberculosis* is another member of this genus which has been constantly in evidence since its discovery by Koch in 1882. Much attention has been given to its life history, though not with the same success as in the case of the anthrax bacillus, for spore formation though described by some observers has not been satisfactorily made out. Koch's four postulates are fulfilled in the case of animals, though it is obvious that the last experiment of the chain cannot be applied in the case of man. There is, however, little doubt that tubercle bacilli blown about with dust constitute a real danger to the community. In the *British Medical Journal* of January 13, 1894, Dr. Miller records the discovery of tubercle bacilli in dust from a house in which three people had died of phthisis at varying intervals, and two others had shown physical signs of the disease.

The *Bacillus lepræ* is very closely allied to the bacillus tuberculosis in size and form, but especially in its behaviour to staining reagents. These two bacilli are separated from all others by the fact that they retain certain aniline dyes when treated with nitric or sulphuric acid, while other bacilli are decolorized. We have, however, advanced very little in our knowledge of the life history of the bacillus lepræ, for it is by no means certain that it has yet been cultivated, though many have claimed success. Inoculations of animals have failed, though the bacillus is very resistant. In a fowl which I inoculated with a piece of leprous nodule, I found leprosy bacilli more than two years after inoculation, though there had been no development of the disease.

Time will not allow me to do more than mention a few of the many other interesting members of this genus. The *Bacillus mallei* has been fully worked out, and is now generally accepted as the cause of glanders. The *Bacillus alvei* has been shown to produce foul-brood in bee-hives. The - bacillus of malignant œdema, of septicæmia of mice, of swine typhoid, of swine erysipelas are also accepted as pathogenic in animals.

It has been shown that the inoculation of garden earth in mice and rabbits induces fatal tetanus. At the point of inoculation abscesses form and from these abscesses bacilli mixed with other organisms have been obtained and cultivated. A culture of this mixture produces the same disease. Recent researches by Kitasato and others have added much to our knowledge of the bacillus of tetanus.

Among the saprophytic bacilli I can only refer to the *Bacillus subtilis* or hay bacillus, a very common and widely distributed organism, and a frequent intruder in other cultivations.

5. *Vibrio.* This is a a genus containing screw shaped threads in long or short links. Spore formation is present.

One species is described, *Vibrio rugula.* It occurs in vegetable infusions causing fermentation of cellulose. The rods are curved or with one shallow spiral. They form swarms and grow out into threads curved in a screw like manner. Endospores are formed at one extremity, which then becomes enlarged giving the rod the appearance of a pin.

6. *Clostridium.* This genus resembles bacillus except that spore formation takes place in characteristically enlarged rods One species is pathogenic, causing a disease in cattle which is variously known as blackleg, quarter-evil, rausch-brand or symptomatic anthrax.

A common species is *Clostridium butyricum* which converts lactic acid in milk into butyric acid and produces the ripening of cheese.

GROUP III.—*Leptotricheœ.*

This group possesses cocci, rods and thread-forms which shew a distinction between base and apex. The threads are straight or spiral.

The group is divided into four genera.

1. *Crenothrix.* In this genus the threads are articulated ; the cells are without sulphur ; the habitat is water.

One species is known, *Crenothrix kuhniana*. It occurs as little whitish or brownish tufts in wells and drain-pipes. It fouls the water and may stop up the narrower pipes.

2. *Beggiatoa*. In this genus the threads are not articulated ; the cells contain sulphur granules ; the habitat is water.

Three species are described, the best known of which is *Beggiatoa roseo-persicina*, which has been already mentioned as furnishing the material from a study of which Lankester formulated his doctrine of pleomorphism.

The cocci form at first solid families bound together by gelatinous substance. Then they become larger, ovoid or globular and hollow, containing watery fluid. These hollow families are often perforated, presenting a reticulated appearance. Finally they may become broken up into irregular structures. In the cocci of the older vesicles dark granules of sulphur occur. These micro-organisms occur sometimes in such quantity that whole marshes or ponds may be coloured red by them.

3. *Phragmidiothrix*. In this genus the threads are without joints ; successive sub-division of cells is continuous ; the cells are without sulphur ; the habitat is water.

One species is known, *Phragmidiothrix multiseptata*. It consists of cocci and threads. The latter are separated by transverse partitions into discs. Repeated transverse and longitudinal division takes place in these discs, resulting in the formation of cocci. These cocci develop *in situ* into slender threads.

Phragmidiothrix is especially characterized by this constant sub-division, but it also differs from beggiatoa in the absence of sulphur and from crenothrix by the absence of a sheath. It occurs in sea water attached to crabs.

4. *Leptothrix*. In this genus the threads are articulated or not articulated ; successive sub-division of cells is not continuous ; the cells are without sulphur.

The threads are composed of long rods, short rods, and cocci. The threads may break up into spiral, vibrio, and spirochæte forms. They occur in large numbers in the mouth in man and animals, and are believed to be connected with dental caries. Two species are described.

GROUP IV.—*Cladstricheæ*.

This group possesses cocci, rods, threads and spirals. The thread forms have false branchings.

The group contains only one genus.

1. *Cladsthrix*. This has already been mentioned as the micro-organism on which Zopf worked when establishing his classification of bacteria founded on the theory of pleomorphism. Its development can be traced from cocci to rods and threads.

These threads form false branches by single rods turning aside. Thus in Cladsthrix we find every foim of the Schizcmycetes represented.

Two species are described. One is very common in still and running water ; the other occurs in the lachrymal canals of the human eye.

In the above short account of the Schizomycetes I have endeavoured to approach the subject mainly from the standpoint of the naturalist. Time would fail me were I to notice the many other points connected with this group of organisms. I must be content to hope that the few remarks I have made may not be wholly uninteresting to this Society.

March 2, 1894.

A DAY'S INSECT HUNTING AT CAPARO.

BY R. R. MOLE.

A SWEET April morning, with a dewy softness about the atmosphere, the very feel of which brings back forcibly to the memory a showery spring morning in old England. Above, a cloudy sky with here and there streaks of azure blue ; a heavy bank of white cloud lying over to the Westward above the lofty line of verdant forest, which stretches mile upon mile, into the interior. But it is only the freshness of the morning and the clouds which remind us of home, for look where we will we have evidence that we are in the tropics. The palm leaf* roofs of one or two houses peep here and there out of the surrounding gardens, which are full of gigantic broad-leafed plantain and banana plants, resplendent with the crystals of the showery night which has just passed away. Here are orange and mango trees— none of any size, however, for we are in quite a new locality, where four years ago stately balatas† and majestic ceibas‡ reared their haughty heads among a myriad other, lesser, but no less beautiful forest trees, the aerial play-grounds of long-bearded howling monkeys,§ who morning and evening roared out their welcomes and farewells to the sun with the regularity with which he rose and set. Graceful little capuchins‖—or matchins as they are called here—sprang from bough to branch, or swung by their

---

* Carat—Sabal glaucescens.
† Mimusops dissecta, R. Br.
‡ Eriodendron anfractuosum.
§ Mycetes seniculus.
‖ Cebus sp.

prehensile tails from the arms of trees at giddy heights, playing the while, elfish tricks upon each other, accompanied by many a queer grimace and frolicsome antic.   The clanging clamour of the toucan* and busy tap-tapping of the woodpeckers† were to be heard all round, while birds of every kind sang anthems of thankfulness to the Great Creator for having bestowed on them the Gift of Life.   Here, where we stand, looking forth upon well-cultivated cocoa, kola and plantain, roamed by day bands of fierce little quenks,‡ grunting and digging up succulent roots and rubbing their hairy backs and flanks against the trees, ever and anon disturbing the gentle, great-eyed Gouazoupita deer§ with its timid fawn from the covert which it had chosen wherein to ruminate upon the memories of the preceding evening's raid on the sugar-cane or corn patches of the small holder living on the verge of the forest, and driving them away to more secluded retreats. The timid agouti‖ with nervous step stole under the forest trees and picked up his living with now and again a joyous frisk and curious antic, for the agouti, though surrounded by many foes, is in a quiet way a merry little fellow, especially when gun and hound are far away.   His great cousin, the lappe or paca,¶ in glossy spotted coat, foraged here and roamed about at will, ready to take to flight at a second's notice and seek the refuge of the friendly but sluggish stream which meanders between its deep banks only a stone's throw off.   Then when night fell and the great fire beetles** flitted through the trees and the "moping owl did to the moon complain," strange uncouth forms stole out of hollow trees, nooks and crannies in boughs and branches, or burrows in the earth.   The matapel,†† with his long preternaturally solemn nose, small ears, rough fawn-coloured coat, with his pre-hensile tail curled round some stick or branch as a *point d'appui*, with his enormously powerful claws began tearing up rotting wood and inserting into the orifices so made his worm-like tongue, drawing in, each time he retracted it, scores of the white ants or termites which constitute his *summum bonum*.   Near him too one might have seen sundry unrecognizable shapes, busy scratching and rooting, which if they would have allowed us to have taken a closer look we should have found to be armadillos,‡‡ or, to use the creole name, tatous, busy with earth-worms and other crea-

---

* Ramphastos vitellinus, Licht.
† Dendrornis susurrans, Jard.
‡ Dicotyles tajacu, L.—Collared Peccary.
§ Cariacus nemorivagus.
‖ Dasyprocta aguti, L.
¶ Cœlogenys paca, L.
** Elateridæ.
†† Tamandua tetradactyla.
‡‡ Tatusia novemcincta, L.

tures, the catching and devouring of which constitute their
nightly occupation. Above in the trees, dancing and flinging
himself about in the queerest attitudes, his presence perhaps
invisible to the eye, yet making itself known by a certain subtile,
all-penetrating, mean, disgusting, horribly effusive scent, would
have been the porcupine.* There, on yonder rising ground,
cautiously dragging his heavy body about in the shade, ever and
anon throwing himself into his position of defence, and sharply
tapping on the dried leaves with rapid strokes of his horn-tipped
tail, might have been found—he is found there yet—the deadly
Mapepire z'annana,†—the Silent Death of the Black Night as
someone has rendered it, the worst member of the *Crotalidœ*, the
Mapepire, on the look-out for rats and the other small deer which
constitute his sustenance. Having struck and gorged his victim
he frequently chose—as he does, even now—some spot where the
rising sun would warm his rough scaly coils before he retreated to
his hole, often too the sanctuary of hard-hunted agouti or armadillo, to sleep off the effects of his midnight repast. Over there,
in yonder thicket too is often met his congener—Mapepire-barcin‡
either on a convenient spot close to the water's edge, or else coiled
in the forks of some low bush, ready to strike his death-dealing
fangs into the trembling body of the unlucky bird or small mammal which incautiously ventures within his reach. But now the
scene has changed ; the woods have been felled and cocoa, coffee,
kola, plantains, and bananas have been planted in their place
and most of these wild creatures have been driven from this spot ;
but nightly they pursue their gambols, hunt and are hunted, live
their lives and perpetuate their species within a quarter of a
mile of where we stand. There are others which have not so retreated before the encroachments of man. They are the frisky
little squirrels§ and the field rats‖ and mice which levy tribute
on the cocoa pods, the feathered tribes, including the gorgeous
parrots¶ and gaily dressed cornbirds** the flashing humming

---

* Synetheres prehensilis.

† The Creole name (meaning pine apple Mapepire) for Lachesis muta
Dr. de Verteuil thinks Mapepire is a Carib or Gallibi word.

‡ I have seen it spelt *barcin* and *valsin* and have been told these
words meant "striped," and "dancing" But both Mapepires are striped
and all vipers dance, *i e.* circle round, with the head ready for launching at
an expected aggressor. Dr. de Verteuil informs me that "barcin" is
derived from the Guarahaon and means faint or blurred stripes. The
scientific name of this snake is Bothrops atrox.

§ Sciurus æstuans, L.

‖ Nectomys palmipes, All. & Chpm. ; Heteromys anomalus, Thompson ;
Echimys trinitatis, All. & Chpm. (Piloree).

¶ Pionus menstruus, L.

** Cassicus persicus, L.

birds.* Gay butterflies seem to exult in the open space provided for them in which to display their gaudy colours. At night the opossum, or manicou† as she is commonly called, often laden with seven or eight clinging little ones, revenges herself for the invasion of her domains by raiding the fowl roosts or stealing plantains. Heavy boas creep forth from drains and thickets on the watch for any living thing which can be stowed away in their capacious stomachs. But although these creatures have not been driven from their old haunts and have profited rather than lost by man's invasion and destruction of their sylvan fanes, there are others who wage persistent, unrelenting, never ceasing warfare upon the products of his labour and who organize great invading armies into his territory, in token of their never relinquished claim to that which he has wrested from them by right of superior force, a right to which all other wild things have acquiesced. These stern but tiny opponents are members of the insect world, and it is to the parasol ants and ants and insects in general that we intend to devote ourselves this fine but showery morning

With this object then we three, members of this Club—Mr. Albert Carr, the owner of the plantation *Itcarrdonum,* Mr. F. W. Urich and myself took our cutlasses, knives, tweezers, a plentiful supply of pill boxes, tobacco tins, and small bottles, chloroform, &c., and sallied forth through the little pasture and across the felled tree which constitutes the somewhat slippery bridge over the sluggish stream, reduced to almost its lowest point, but after heavy rains a boiling rushing river which undermines banks and sweeps away all before it. Then through the well-kept promising young cocoa amongst which we noted many kola plants ; then skirting the bank of the river again, keeping a wary look-out for the snake so well described, if its name applies only to his poisonous powers " the atrocious one "—but perhaps it is a misnomer—this lovely morning disposes us to regard kindly even the Mapepire barcin, for he only uses his dread powers in obtaining his livelihood, or in self-defence. Fortunately, we did not tread on one—unfortunately we did not see one, although they are to be met here sometimes. Mr. Carr pointed out a place where an iguana‡ had been seen scooping out a hole in the bank in which to lay her eggs ; also where not so very long ago he and his brothers, who are the mighty Nimrods of the quarter, had surprised a school—one hardly knows what to call a litter of baby alligators§—fierce little things ready to fight to the last.

---

* Of humming birds Iere possesses no less than seventeen species ; probably there are several more.

† Didelphys marsupialis, L.

‡ Probably Iguana tuberculata.

§ Alligator sclerops.

From thence we reached a spot where some enterprizing parasol ants had started a nest. The nest was soon dug out of its recess in the earth near a.drain and looked not unlike a very mouldy dutch cheese in size and shape. It was broken up and one or two specimens taken including the big-headed fellows, who with fiercely snapping jaws always come to the front when their home is attacked. One of them cut through a piece of stick presented to it, which in size was at least half the thickness of its own body. But this nest did not contain what Mr. Carr desired so much to shew us, and after duly destroying it and scattering the fungus beds and digging in the soil around it, we proceeded to another. Here we were more successful, and Mr. Carr, at the risk of the severe bites of the enraged insects, boldly took up the nest with his bare hands. In its centre was a big helpless-looking ant, a very Goliath amongst the others, rather larger than a bee, which so far as we could see was covered with parasol ants of the very smallest size, and whose red bodies contrasted strongly with the dark brown one on which they looked almost like parasites. The insect's helpless appearance and its sur-roundings decided the question in our minds that this was undoubtedly the queen of *Atta cephalotes* and not *Atta fervens*. In this nest we observed several ants engaged in busily removing the white shroud in which the young ones pass their chrysalis stage, and which was so ably described in the *Journal* recently by our greatest local authority on parasol ants—Mr. Tanner. Several other small nests were visited—never a large one, for Mr. Carr takes care to destroy them as fast as they form, and several more of the large helpless ants were found—each one with smaller ones clinging to it—and only one in each nest. The trees were also examined, and several species taken—among them a reddish-looking fellow* with a buckler-like shield over the head and thorax. This ant was found principally upon a Bucare-Immortelle, although they were also present on several other trees. Mr. Urich, who was the entomologist of the party, identified these ants as belonging to the cryptocerus genus. After awhile we came to another part of the river, which here presented a dark and unhealthy appear-ance. A closer investigation shewed numbers of fish on the sur-face near the banks gasping faintly and many of them looking as if they would like to desert their natural element. Mr. Carr surmised the water had been poisoned and certainly it looked like it for some of the fish were quite dead. Curious to relate, however, the poison which was so disastrous to the finny inhabi-tants of the stream, had no visible effect on a large species of

---

* Cryptocerus clypeatus, Fabr.

water snail* many of which were larger than a man's fist.   Of
these we gathered a dozen or so and Mr. Carr informed us that
country folk call these huge freshwater molusks "congs," and
cook and eat them.   He could not tell us, however, what their
gastronomic merits were and we did not feel brave enough or
sufficiently enthusiastic to ascertain them for ourselves.   Of the
fish we collected several species, among them the ubiquitous
"guabin,"† with his formidable teeth ; the " cascarab," a pretty
little fellow rather deep in the body with a dark spot on either
side near the tail, reminding one of the marks on the John Dory
of European seas, piously attributed to the finger and thumb of
St. Peter in some quaint old legend almost now forgotten.   The
river catfish with their flat heads and long feelers were also picked
up in numbers, and one of these which we opened contained a
dozen and a half of small fry the size of a minnow, which are
known in Caparo as sardines.   The poison, it was sagely supposed,
had affected the small fish first and the catfish seeing their
stupid, half-drunken condition, had gorged himself with them,
later to fall a victim himself to what he doubtless at first considered
a piece of rare good luck.   Then we got a queer little fellow
which is called a " Maman Cascarado," which holds on to the
stones in the river bed, by the sucking action of its specially
adapted mouth.   Several other kinds were noted, duly taken and
docketed and consigned to spirits for the examination of wiser
heads than ours.   The sun, which had shone brightly for the past
hour, was now obscured by black clouds and a few large drops of
rain, the precursors of a heavy tropic shower, drove us to find
refuge in a half ruined hut of which only the carat thatch, the
corner posts and southern wall (also of palm leaves) were standing.   But although so ruinous the house was far from uninhabited.   In one corner the wood ants‡ had constructed a mighty
nest undisturbed.   Spiders of many kinds had suspended their
snares from the roof and no doubt had we examined the thatch
we should have found much to interest us.   As it was in its apex
we saw the ancient slough or cast skin of a large "tigre,"§ or
as he is called at the British Museum the " South American rat
snake." Apropos of this skin, the owner of which we were told
was caught months previously, we expressed our doubt of the
tigre eating rats at all but Mr. Carr assured us that it was so.
One Sunday afternoon as he lay in his hammock he saw a tigre
come out of the long grass a little way off, cross the road and
make for the house.   It ascended one of the supports of the

---

* Ampullaria urceus.
† Pronounced " Wawbin "
‡ Termes sp.
§ Coluber variabilis.

gallery in which he was taking his siesta and so gained the roof where it disappeared in the thatch. It had not been there long before sundry squeals and rustlings betrayed the fact that the tigre had good reasons for visiting the house. He had secured a rat. After swallowing it he descended to the ground again and made off to his retreat in the long grass, or what is more probable, disappeared in it *en route* to his home in some tree near-by, for the tigre is essentially a tree snake. The shower was by this time over and we again sallied forth and presently found another parasol ants' nest. We began digging with our cutlasses and the second clod of earth removed revealed a prize which was worth coming all the way to Caparo for, even if we got nothing else, a peripatus,* about two and a half inches long. Unfortunately it had been slightly bruised but nevertheless was exceedingly acceptable. A closer search gave us a smaller one which was also bagged amid considerable excitement. The discovery of these interesting creatures was quite unexpected in such a locality as the purlieus of an ants' nest and though we had turned over logs by the dozen that morning without success we never imagined for a moment we should get peripati in such a situation. Many persons perhaps do not know what a peripatus is. It is an animal in shape and in the number of its legs very much like a millepede, or, as it is popularly known here " conger-ee," but it is destitute of the ring-like armour which is so striking a characteristic of the " conger-ee." Its brown, naked, skin is often adorned with a bright stripe or a pretty mottled pattern. It is able to contract or extend its length in a wonderfully elastic fashion. From its head it puts out a couple of feelers, like a snail does its " horns." From ducts at the base of these feelers it exudes, at will, a glutinous substance and woe be to the unfortunate wood-louse or other creature which permits the peripatus to touch him with it. He is bound as firmly as ever insect was in spider's web, and the peripatus devours him forthwith. The habits of these creatures are very little known and they secrete themselves so well that they are rarely found. They are often passed over as being slugs. One or two species are known to be indigenous to Trinidad, but either of them is a *rara avis*, albeit certain members of this Club and outsiders too, are constantly looking for them. Another curious point about certain species of peripatus is that they are supposed to be hermaphrodite and are able to individually and independently perpetuate their kind. A prolonged examination did not give us any more specimens, but resulted in the finding of another large ant of the kind previously described. Wandering on we came to the extreme boundary of the cultivated

---

* Peripatus trinidadensis, Sedgw.

portion of the property. The forest trees on a small hill had been felled and afterwards burnt, preparatory to the land being taken into cultivation. We ascended the rising ground and found it commanded a pretty extensive view of the cultivated portions of the valley, which, from this point, looks as if it is enclosed, as in a ring fence, by dense forest. Near here we found a cocoa beetle* on the bark of a wild chataigne or chesnut, and shortly afterwards the larva of another one was taken from the burrow he had made in the stem of a young cocoa tree. Amongst the half-burnt logs we noticed several skinks† or slippery-backed lizards with pretty, inoffensive-looking heads, sparkling eyes, and smooth and sleek closely-lapped scales. They were far too wary to allow themselves to be caught, however, had we attempted it, and so we left them to their own devices, which apparently is sunning themselves on fallen logs and waging incessant warfare upon insects and smaller saurians, for they are incorrigible cannibals, these same pretty innocent-looking things. A look at a hole from whence an armadillo had been recently taken and which Mr. Carr informed us bore unmistakeable signs of having been also the home of a snake—perhaps a deadly mapepire, or perhaps an inoffensive Jack,‡ who knew, for both love holes in the earth, though the former is fonder of hills than the latter, who usually makes his habitat in damp localities. But it was time to return as breakfast was waiting us. We soon got back and after refreshing ourselves devoted the next hour or two to the proper and safe disposal of the many specimens we had collected. After resting we determined to return to the river to secure some of the fish for the table, for although the fish are poisoned, or, perhaps, more correctly speaking, made stupid, yet they are perfectly good for edible purposes, which, indeed, is the object of the perpetrators of this vile mode of obtaining them. The water on the other hand is polluted and rendered unfit for drinking purposes for days after, besides being made additionally objectionable by the decomposing bodies of the fish which die and are not collected by the poacher. Our kind host was much annoyed at this occurrence, as he depended for his water supply on the river and he therefore made numbers of enquiries about the place as to who had been guilty of the unneighbourly offence of poisoning the water. This time we were accompanied by Mr. Arthur Carr and several contractors, with a strong following of boys, to whom, perhaps, the news was not an altogether unmixed evil. In a short time a large basket full of fish was collected, and we then proceeded to find out what was the agent used in

* Steirostoma depressum.
† Mabuia agilis.
‡ Epicrates çenchris.

their destruction. After clambering some way along the banks of the river we came to some large stones which had been evidently used as a rough mortar upon which to pound some fibrous substance, which was unanimously declared to be " balbac "* liane. Wading into the river, long pieces of the bruised liane were secured by the boys and a large portion was also discovered on the bank. A little further on the remains of the plant from which it had been obtained, still hanging from a forest tree was to be seen. During a rest we now took on the bank Mr. Carr said there were many lianes used for the purpose of securing fish, but balbac was the most common one. With regard to the method of application he said there were many superstitions, and amongst them were these : When a man wishes to obtain a fish dinner with as little trouble as possible to himself, which is by the aid of balbac, he is particularly careful to avoid spitting in the water, for he believes if he commits such an indiscretion the liane will have no effect on the fish. Neither must he for the same reason touch the water with a cutlass, because if he does "he cuts the poison." In accordance with this belief, when he has got a sufficient supply of fish, he chops the water with his cutlass to prevent more being poisoned than are necessary, and also to avoid doing harm to persons who may happen to drink of the river lower down. With reference to this fish poisoning it may be interesting to note that Bates, in his admirable book, " The Naturalist on the River Amazon," 2nd ed., p. 242, describes a similar custom in Brazil : " There was " a mode of taking fish here which I had not before seen " employed, but found afterwards to be very common on the " Tapajos. This is by using a poisonous liane called Timbó " *(Paullinia pinnata.)* It will act only in the still waters of " creeks and pools, A few rods, a yard in length, are mashed " and soaked in the water which quickly becomes discolored with " the milky deleterious juice of the plant. In about half an " hour all the smaller fishes, over a rather wide space around the " spot, rise to the surface, floating on their sides, and with the " gills wide open. The poison acts evidently by suffocating the " fishes ; it spreads slowly in the water and a very slight mixture " seems to stupefy them. I was surprised, on beating the water " in places where no fishes were visible in the clear depths for " many yards round, to find sooner or later, sometimes twenty- " four hours afterwards, a considerable number floating dead on " the surface." [Mr. Broadway here exhibited mounted leaves of the balbac.] Having secured more water snails, we made a short excursion through the forest ere returning, and saw a

---

* Paullinia sp.

rough scaffolding which had been erected between a couple of trees from whence it is the practice to shoot agoutis which are enticed within range by sundry tempting heads of corn or bananas. Once upon a time a new hand at this sport was duly installed upon such a perch as this, and after waiting for a considerable period his patience was rewarded. But the sportsman was so taken up with the merry jumps and friskings of the lively little beast, the curious way in which it ate and washed face and ears with its fore feet, that he forgot to shoot. At length an unusually comical gesture elicited from him a loud peal of laughter, at which the agouti, surprised no doubt, took himself off so quickly that the gun was presented too late. The hero of this adventure is as much twitted about his agouti shooting as that angler, who, failing to catch trout by the ordinary means, was discovered purchasing enough fish at the fishmonger's to fill his basket, before going home. Our friends had shewn us the poisonous *balbac* liane, and now they pointed out to us a liane which is a veritable travellers' friend. A huge piece of *croc-chien** was cut down, from which dripped the clearest and most palatable of water, which we tasted in an impromptu cup made of the broad leaves of the *terete*.† To get a plentiful supply it was necessary to cut the liane clean through in one place and with the least possible delay to sever it again six feet further on, so as to prevent the water running away.

I have reached the end of my notes of this, to me, delightful excursion. When I had written this far I turned back and found the heading to this paper was "Insect Hunting" and I was at once reminded by a guilty conscience that I was about obtain your kind and indulgent attention by false pretence. My only plea in palliation of my grave offence is the absorbing interest which surrounds everything in Nature. When one has escaped from the dusty streets, the daily routine of office and can throw aside the cares and worries of every-day life for a while— when one is free to fly from the congested centres of civilization— is free to go into the sweetly solemn forests, to hear the rich swelling and falling of the perpetual anthem which ever ascends to highest heaven from bird and insect choristers, rustling leaves and murmuring brooks, as they sing their endless song of praise to the Great Father who gave them being—when one inhales the fragant incense with which tree and flower laden the fresh breezes and so contribute to the continual service of praise to Nature's God, which began when time was not and will only cease when it has passed away. When the eye revels in the glories of light and shade, the golden sunlight, the sombre storm clouds, the

---

* Belonging to the Dilleniaceæ.
† Ichnosiphon, sp.

magical effects wrought by wind and rain upon the quivering greens, browns, and yellows, and countless tints of the forest—when one notes the infinitude of forms in which life is present and the perfection of design with which each form is adapted to its own peculiar circumstances and requirements and the amazing dissimilarity in detail, yet no less wondrous unison of the whole, the mind thirsts and pants to drink in deeply of the intoxicating and exhilarating cup presented to it, and, attempting it aims at too much. Though there may have been previously a settled plan of action decided upon and a determination to examine and to inquire only into one particular detail, yet the very fact that the beautiful work of Nature lies all round one, sets the senses at war with each other in their efforts to grasp all the deep meaning and entrancing beauty of the scene,—preconceived schemes are forgotten and the mind receives a number of beautiful photographic impressions forming a series of pictures which are the merest fragments representing only a very small part of the whole. In Nature the intelligent mind always finds something to instruct, edify and amuse—she is so great, so grand, so harmonious, "age cannot wither her charms, nor custom stale her infinite variety."

May 5th, 1893.

NOTES ON A TRINIDAD BUTTERFLY.

By the Rev. E. D. Wright.

HYPOLIMNAS MISIPPUS inhabits the Old World and is also found in the West Indies. It is remarkable on account of the dissimilarity of the sexes and of the female mimicking in appearance and manner of flight certain butterflies of the genus *Danais*, belonging to a different family, the *Danaïdæ*. These latter are distasteful to birds &c., and hence enjoy an immunity from attack—they have a slow, heavy flight and are easily recognisable, so the advantage to the female *misippus* of resembling them is obvious.

The male *misippus* is black with a large round white patch in the middle of each of the wings—and two smaller white marks at apex of the forewings. Hence it is popularly called the Four Continents in Trinidad, where it is not uncommon. The female has the wings brown and the lower half of the forewing black in which is an oblique bar of very clear white. Thus it bears a very close resemblance to *Danais chrysippus* in the east. There is also another form of the female in the east in which the clear white bar is absent ; this has been named *H. inaria* and mimics another species of *Danais* (*D. dorippus* ?) The female misippus in the collection at the Institute is of the first form with the white bars. This insect also inhabits Jamaica where it seems to be rare—and a female in the possession of Mr. C. B. Taylor, of Kingston, is of the second or *inaria* form, has no clear white bar across the forewing and bears a very close to resemblance to *Danais jamaicensis*, which I believe is peculiar to Jamaica. There are therefore both in the Old and New Worlds two forms of the female, one corresponding to a *Danais* with a white bar across the forewing and one corresponding to a Danais without this white bar.

It has been said that this insect is an introduction into the West Indies ; but the facts which I have given seem to be against this—but on the other hand if it is indigenous it affords a very remarkable case of distribution being common (with both its forms of the female) to the Eastern and Western Worlds. Further information about this insect and its larva, food plant in Trinidad &c., is very desirable.

March 10th, 1894.

TELEPHONE No. 128.   ESTABLISHED A.

REMEMBER

That when purchasing the necessaries or luxuries of life, it is to
interest to secure the best possible value for your money. *The man wi
squanders his means* by paying above the market rate for articles in daily use
is inviting ruin, which will not be slow in answering the invitation. For
this reason sensible people who desire to die solvent, patronize when
making their purchases

THE

Popular Establishment owned and managed by Mr. P. A. Ramsey, situated
at the south-eastern corner of Upper Prince Street and Brunswick Square.
The Proprietor of this well-known place of business has made it his duty
to see that his customers are supplied with *the best and freshest Drugs,
Chemicals and Medicines,* the stock of which is replenished by every steamer
arriving from Europe and the United States, thus affording the

CREOLE

population of the country an opportunity of procuring the same quality of
drugs as would be supplied to them in London, Paris or New York, and in
order to secure their being accurately dispensed, the Proprietor has secured
the services of *a Staff of Skilled Druggists* who perform their duties under
his personal supervision, hence affording a double guarantee to those who
favour him with their prescriptions. And in order to further popularize
this favourite

PHARMACY

the Proprietor keeps on hand a large and varied stock of the best known
Patent Medicines, Toilet Requisites, Perfumery, and other articles usually kept
in stock in a Pharmacy. In order further to encourage Cash Customers,
a quantity of *Pictures, Fans, Toys,* and other

ACCEPTABLE SOUVENIRS

Are daily *given away gratis* to ready money purchasers.

Civility and Attention to all !!!

A NIGHT BELL

Is attached to the premises and all calls are promptly attended to.

Communications and Exchanges intended for the Club should be addressed to the Honorary Secretary, Port-of-Spain, Trinidad, B.W.I.

Price 6d.   Annual Subscription, 3/.

Vol. 2.   AGUUST, 1894.   No. 3.

J'engage donc tous à éviter dans leurs écrits toute personnalité, toute allusion dépassant les limites de la discussion la plus sincère et la plus courtoise.—LABOULÈENE

Trinidad·Field·Naturalists'·Club.

NATURA MAXIME MIRANDA IN MINIMIS.

Publication Committee :

H. CARACCIOLO, F.E.S., *President.*
P. CARMODY, F.I.C., F.C.S. ; SYL. DEVENISH, M.A.
B. N. RAKE, M.D,   R. R. MOLE,
F. W. URICH, F.E.S., *Hon. Secretary.*

CONTENTS :—

Report of Club Meetings—

    June 51

    July 53

Notes on Camping Out, Fishing and Cariboo hunting in New Brunswick 55

Preliminary list of Trinidad Reptiles and Batrachians, including— 77

    Description of a new Trinidad Lizard 80

    A rare and interesting Snake ... 85

    Description of a new Trinidad Frog ... 88

☞ All Communications and Exchanges intended for the Club should be addressed to the Honorary Secretary, Port-of-Spain, Trinidad, B.W.I.

JOURNAL

OF THE

Field Naturalists' Club.

VOL. II.   AUGUST, 1894.   NO. 3

REPORT OF CLUB MEETINGS.

1ST JUNE, 1894.

PRESENT : Mr. H. Caracciolo (President), Professor Carmody, Dr. Rake, Messrs. Syl. Devenish M.A., J. R. Llaños, C. J. Thavenot, Henry Tate, T. W. Carr, H. J. Baldamus, Charles Libert, W. E. Broadway, J. Hoadley, S. A. Cumberland, R. R. Mole and F. W. Urich, Hon. Secretary. The visitors were Messrs. Alick Hamlyn, and Sokowiski, (Columbia), and Master Carmody. Dr. J. R. Dickson was elected a Town Member. A letter was read from Professor C. V. Riley (Washington), thanking the Club for his election as an Honorary Member. A communication was read from the Club's taxidermist upon some animals which he was preparing for the collection. A letter was also read from the Librarian of the British Museum with reference to numbers of the journal required to complete the first volume in the Museum Library. The Secretary read a communication from Mr. A. B. Carr upon the cocoa beetle, *(Steirostoma depressum)* which contained much original, interesting and useful information, on the habits of this cocoa pest, with some practical suggestions as to methods of dealing with the same. Specimens of the larvæ and the beetle were exhibited. Mr. Cumberland read a short paper by the same gentleman upon the " Poor me one " *(Nyctibius jamaicensis)* No 32 of Dr. Leotaud's collection. The song of this goat sucker is remarkably mournful and is described by the peasantry as " Poor me one," but is attributed by them to the little ant-eater *(Cyclothurus didactylus)*,—

commonly and erroneously known as the sloth. Mr. Carr
was the first person to discover that the song was that of a bird
and not the cry of a mammal. Mr. Cumberland, who had also
made some observations upon the bird in question, gave a remarkably clear and characteristic illustration of the song and
further read some original notes which fully bore out Mr. Carr's
remarks. The value and interest of the paper was enhanced, by a
stuffed specimen shot by Mr. Carr, and ably prepared by
Mr. Cumberland being handed round for examination. An
animated discussion followed in which Mr. Devenish drew the
attention of members to the resemblance which he said existed
between the bird in question and the Guacharo (*Steatornis caripensis* Humb.) Dr. Rake read some original observations upon
the nesting habits of the small night jar *(Nyctidromus albicollis*
Gm.*)* and compared it with those of the common English
species. Eggs of the Trinidad bird, which is popularly
known as "Watchman of the Road," were handed round for
examination and further discussion ensued.—The President described how he had suffered from the bites of Vampire bats in a
recent visit to Gasparillo Island, and stated that the wounds were
triangular in shape, the pieces being bitten right out. Fowls,
said the President, suffer severely from the bites of these animals
which attack them between the toes and in one night, during
his stay, two fowls died from loss of blood.—The President drew the
attention of the Club to a beautiful example of Gorgonia (belonging to the Coral order) presented by Mr. W. Scott Kernahan who
obtained it in the Second Bocas at a depth of eighteen fathoms.
To the stem of this remarkable plant-like animal were attached a
number of bivalves and a star-fish, which the Club's conchological
member, Mr. R. J. Lechmere Guppy, C.M.Z.S., had kindly determined as *Avicula vitræ* and *Asteroporpa annulata.* Dr. Rake
placed on the table a fine specimen of *Hyla crepitans* (tree frog)
which, he said, was in life a beautiful white. The President drew
attention to an albino agouti presented by Mr. P. Gonzales.
Mr. Broadway exhibited four species of *Castnia* moths (Cane
suckers) amongst which was a beautiful white one recently taken.
Mr. Henry Tate shewed two fine butterflies, one a species of
Zeonia? and the other a species of *Catogramma*, which he had
secured at Maraval, and which are not at all common, especially
in such perfect condition. Mr. R. R. Mole exhibited a living
coral-snake *(Elaps lemniscatus.)* 46 inches long, caught when in
the act of swallowing another snake, by Mr. Watkins of Maraval.
The specimen was in perfect condition and attracted considerable attention. Mr. Mole also drew attention to a fine snake*

---

* See Preliminary List of Reptiles published in this number—page 85.

(Phrynonax eutropis) which was recently captured by Messrs. Hoadley and Hamlyn at Maraval, and which differed from any snake which Mr. O'Reilly had in his Reptilium or existed in Dr. Court's collection, or which he or Mr. Urich had seen during the period they had been engaged in studying the ophidia of Trinidad. It was a singular fact that since this snake was caught two more had been received from Caparo and another from Santa Cruz. For the purposes of comparison he drew attention to other species, *(Herpetodryas macrophtalmus* and *Coluber baddaerti)* confined in the same box. The snake he thought would turn out to be a species which had never previously been recorded from the island.—Mr. Mole placed on the table two living specimens and one dead one of the rare Trinidad bivalve *Anodon leotaudi* and said that they were found, with nine others by Mr. Carr and himself, during a recent visit to the High Woods. He also exhibited some large water snails *Ampullaria urceus* which were found caked in the dry mud of a ravine which had been empty for months. In the orifice of each specimen he found some sixty or seventy young ones of the same species which were also on view.—Mr. C. J. Thavenot laid on the table a good example of the long spine from the tail of a gigantic ray or devil fish.—The proceedings concluded with an unsuccessful attempt to get a fine young specimen of the Anaconda or Huillia, *Eunectes murinus,* (brought to the meeting by Mr. Mole) to feed. The lights and unwonted number of spectators had evidently roused the reptile's modesty and he refused to gratify the members with an exhibition of the methods by which he captures his prey. The meeting closed at 9.45.

6TH JULY, 1894

PRESENT: Dr. Rake, Messrs R. J. L. Guppy, John Goode, Henry Tate, C. Libert, A. B. Carr, H. J. Baldamus, T. I. Potter, W. E. Broadway, Lechmere Guppy jr., T. W. Carr, J. R. Llaños, W. S. Tucker, J .Russell Murray, and F. W. Urich, Hon. Secretary. In the absence of the President Mr. R. J. L. Guppy took the chair. Votes of condolence were passed to the President and to Mr. R. R. Mole, the former on the loss of his child and the latter on the death of his brother, and the Hon. Secretary was directed to write letters expressing the Club's sympathy with these gentlemen. The following gentlemen were elected Members of the Club : Mr. L. O. Howard (Washington) Honorary Member and Mr. John Thomson, Town Member. Mr. Lechmere Guppy jr., read an interesting paper entitled " Notes on some Trinidad Butterflies." The biology and habits of several species of butterflies were ably described and the author's capital drawings of the insects in all stages of development and his carefully prepared specimens made the paper most valuable. On its conclusion the Chairman made some remarks as to the importance of the members of the Club studying the life history of our fauna. A beautiful *Papilio* from the woods of Oropouche was shown by Mr Guppy, jr.—The Hon. Secretary drew attention to the second instalment of Trinidad Mammals received from the Taxidermist ; they were : 1 Collared Peccary (*Dicotyles tajacu*, Linn) 1 Ocelot *(Felis pardalis)* 1 Mangrove dog, *(Procyon cancrivorus, Cuv)* 1 Ant-eater *(Tamandua tetradactyla)* 1 Little Ant-eater (*Cyclothurus didactylus*) 1 Squirrel, (*Sciurus aestuans)* 2 Piloree rats *(Echimys trinitatis* All and Chpm) 1 Spiny rat, (*Loncheres guianae*, Thos) and 1 Manicou Gros Yeux (*Didelphys trinitatis* Thos n. sp.) The Secretary pointed out that the last named mammal was a co-type of a new species recently described by Mr. Oldfield Thomas of the British Museum. Mr. Chapman's remarks about the tendency of the Pilorees to lose their tails were read from his report on Trinidad Mammals and gave rise to an interesting discussion as to this peculiarity.—Mr. A. B. Carr exhibited a skin of *Vampyrus spectrum* which, when killed, measured 2 feet $7\frac{3}{4}$ inches across the wings and said that the specimen was taken on his estate at Caparo. The stomach, said Mr. Carr, contained vegetable matter although the animal boasted such an array of formidable looking teeth. Mr. Broadway showed a case of some rare Trinidad butterflies amongst which was conspicuous a specimen of *Siderone marthesia*, Cram., of which the British Museum authorities are so anxious to acquire specimens. Other interesting species were *Catonephele numilia*, Cram., *Telegonus midas*, Cram., *Phyciodes ianthe*, Fab., *Symmachia menetas*, Drury.—The Secretary on behalf of the President, laid on the table an interesting wasps' nest from Venezuela. Mr. Henry Tate invited notice to two very fine *Hesperidœ* taken at Maraval, which none of the members present had seen before.—Mr. Murray produced a double tailed banana flower from Siparia which he said was the first he had ever seen and which he could only regard as an extraordinary freak of nature.—The Secretary, on behalf of Mr. R. R. Mole, invited the members to inspect an interesting family, consisting of a snake *(Epicrates cenchris)* and her eighteen young ones which were born on the 1st July. He further said that the mother had been five months in captivity and had refused all food during that period. He pointed out the wonderful difference in the markings of the old and young snakes. These creatures, he added, belonged to the constrictors and were perfectly harmless, while they waged unceasing war upon the enemies of the cocoa planter, viz the rats and squirrels and the birth of such a litter upon any cocoa estate in the island would be far more efficacious in the destruction of vermin than a year's shooting while it would be absolutely impossible for them to damage the trees, which gunning often did.—Mr. Albert Carr mentioned that these snakes were frequently found in ants' nests but though fat bearing plentiful marks of the ants' mand bles. —Mr. Urich thought it highly improbable that the snakes lived upon either the ants or their larvæ. They probably found the ants' nest a suitable retreat.—Dr. Rake said, as the Members present all probably knew, Mr. Broadway had been appointed Curator of the Botanic Gardens, Grenada, and as this was the last meeting before his departure, he would take the opportunity of proposing a vote of thanks to Mr. Broadway for all the valuable services he had rendered the Club since its foundation and wishing him every success in his new position. Mr. Broadway, he continued, always contributed to the success of the meetings by exhibiting interesting specimens,—the ease of butterflies he had shewn that night was a good example—and, he (the speaker) was sure Mr. Broadway would be much missed at future meetings. The motion was seconded by Mr. Potter and unanimously carried. Mr. Broadway, who was loudly cheered, thanked the Members in suitable terms, after which the meeting adjourned at 9.40 p.m.

CLUB PAPERS.

NOTES ON "CAMPING-OUT," FISHING AND CARIBOO HUNTING IN NEW BRUNSWICK.

By the Hon. Lieut.-Colonel. D. Wilson, C.M.G.

IN a moment of enthusiasm, after listening to an interesting and instructive paper read by one of the young members of the Field Naturalists' Club, I was weak enough to remark " I wish " in my younger days, when I frequently spent weeks and months " at a time in hunting, fishing and shooting, in the Backwoods of " New Brunswick, I had"—as was so strongly advised by, I think, the President of this Club—" *made a note of everything* and really studied Natural History in every detail." By that innocent remark I gave myself away, for an energetic member at once collared me, and said, " you must give us a paper about " hunting and the backwoods," and I half promised to look up some of my very rough notes and do so. They were notes, not written by a " Field Naturalist," but merely by a very ordinary sportsman as a short diary of events during some hunting and shooting trips in the far north. Even a half-promise ought to be kept *if possible* as the Canny Scot would say—and that is my excuse for being before you to-night.

I must in the first place ask you to disabuse your minds of any such idea as that I am going to read a learned paper on Natural History in any sense or branch. I am going to do nothing of the sort. I am merely going to give you a few personal reminiscences and recollections of very happy days, spent chiefly in the company of the "red man," the North American Indian, in the remote backwoods of that part of Canada, which formerly, and in my day, constituted the separate Colony of New Brunswick and now forms one of the United Provinces of Canada. I may shortly state how I came to be "gadding about" in these backwoods.

I went out there in 18——better leave it blank, for I am beginning to be no chicken—as Private Secretary to the Hon. Arthur Gordon, now Lord Stanmore, then Governor of that Colony, afterwards one of Trinidad's most able and distinguished Governors, and a Governor, I may remark, who did more perhaps than any other, towards opening up the magnificent resources of this most fertile Island.

Besides accompanying the Governor in the exploring expeditions into the remote parts of his Government which he always undertook during the summer months, and besides spending some three or four months one season in visiting, as Commissioner on Salmon Fisheries, every river in the Province from its source to its mouth, I used every winter to get a few weeks' leave, in order to go out "cariboo hunting and trapping."

In summer our expeditions beyond civilization were almost always performed in canoes—generally in the fragile, but most serviceable birch-bark canoes of the Indians. Sometimes, but much more rarely, in the "dug-outs" used chiefly by lumbermen.

These birch-bark canoes are perfect marvels of lightness, of capacity, and buoyancy, and in a certain way—though so fragile—of strength. They are usually "worked" by either one or two men, with light single bladed paddles in deep and smooth water and with the current, and with long poles in shallow water, rapids, and especially against a rapid current.

If one man only is "working" the canoe, he kneels or stands in the stern and places his passengers or freight sufficiently far forward to trim the canoe. If two men are paddling or poleing they take their places at the bow and stern, passengers and freight in the middle.

I do not know a more beautiful sight than to see two men, whether Indians or white men, who know how to do it, pole a birch-bark up or down a difficult rapid. You have no doubt many of you, often admired the precision and "time" of a man-of-war's boat, or of a good racing crew. Well it is nothing to the precision, time, and nicety of touch required under such difficulty of circumstances, when you often have your pole snapped, by being caught between rocks &c., and have to pick up your spare pole, without losing your balance, and in time to keep swing, and perhaps to fend off from some dangerous rock or snagg.

As a general rule you work your poles on the same side of the canoe, by a delicate touch drawing the canoe towards the "hold" you get in the bottom, or edging it away from it, as occasion requires.

In the same way in paddling you steer and direct your canoe by slightly turning, or rather "cutting" outwards or inwards the blade of your paddle. By this means when paddling alone and on *one* side of the canoe only, you send the canoe straight ahead, or even turn it towards the side you are paddling on.

There were two kinds of birch-bark canoes which we used, according to the tribe of Indians we were with. The Micmac, belonging to the tribe of that name, were used chiefly on the Northern rivers—the Restigouche, the Metapedia, (along the valley of which the Canadian Pacific Railway now runs, but which was then little known) the Nepisiquit, and the Miramichi all falling into the Bay of Chaleur and Gulf of St. Lawrence. These canoes were larger, heavier and better sea boats, but slower and not so handy in a rapid river as the other.

The other the Milicete was used by the Milicete Indians of the St. John river and its tributaries, and was extremely light, fast, and handy.

My own pet canoe (Milicete) was made of one single sheet of birchbark without a single patch, about 18 feet long, was so light that I could take it up by the centre bar, swing it on to my shoulders, and carry it bottom upwards, across a " portage "; and yet was rooomy enough to carry four men and some baggage.

In these are stowed guns, fishing rods, baggage and cooking things, and the provisions for the trip—generally consisting of little else than flour, salt-pork, tea, (in large quantity) and sugar and molasses or syrup, for we depended almost entirely on our fishing-rods and guns for everything else. One thing the Back Woodsman or Indian Hunter never forgets and never goes without and that is his *axe*. The *axe* there takes the place, and more than the place, of the *cutlass* of the Trinidad hunter.

With it the North American Woodsman cuts down the trees for the night's fire and the poles for the camp, and with the aid of the piece of wood (speedily cut out on the spot for the purpose) peels the large sheets of bark generally used to cover the camp with and make canoes of, &c.

Thus equipped we used to start off, it might be for a week, it might be for a month or two, to follow the course of some of these magnificent rivers for hundreds of miles, generally ascend-

ing one, "portaging" over to the head waters of another which
we would in like manner descend, camping where and when we
felt inclined, and generally travelling from 12 to 50 miles a day,
stopping for a couple of days at any spot where the fishing or
shooting was unusually good ; or to explore some side stream in
order to trap or shoot, beaver, musquash, lucifee, &c.

I am afraid I was too keen a sportsman to be a good Field
Naturalist, and my great ambition was—I do'nt know that I
need be ashamed of it for I was young, and I was cram-full of
Fenimore Cooper's novels—to be a great and mighty hunter, able
to follow a trail, read signs, shoot, and paddle a canoe &c., &c.,
accordingly.

I do'nt know that I was ever more proud in my life than
when on one occasion we met a party of Micmac Indians, and I
overheard them questioning our own head Indian, old Gabe as
to who and "what sort" we were individually. It came to my
turn, and old Gabe removing his pipe from his mouth, grunted
out in the quiet Indian way "That, that is Kawanick, 'most as
good as an Indian." Coming from old Gabe, who was a great
hunter, I thought it the highest praise that could possibly be
given to be thought "most as good as an Indian." Kawanick
was my universal Indian name. It means the 'otter,' and was
bestowed upon me, I believe, because I was rather good at
swimming and diving and particularly good at catching fish, having
fished our Scottish rivers from my earliest boyhood. We all
had our Indian names. The Governor was the "Sagum," meaning
the "Great Chief." Young C. who frequently accompanied us,
was "Loks," the panther or painter, because he swaggered occa--
sionally in a gorgeous head piece of a most brilliant red colour,
and so on.

The charm of these canoe voyages in the way of shooting
was that you never knew what you might at any moment come
across. Large game, ducks, partridges, &c., &c., &c.

Game was even then getting scarce. You seldom or never
got much ; but you had the pleasing excitement that at any
moment you might sweep quietly round a bend of a river and
get a chance shot at a moose, a deer or a cariboo or a lucifee
(lynx) or even a bear. The two latter are particularly difficult
to get a shot at, they are so cunning.

Poleing up the Nepisiquit one day I saw what turned out to
be a lucifee swimming across some distance ahead, but out of
range. We poled up opposite the spot as quietly as possible,
hoping to get a shot at him on the bank. I may remark that a
canoe is a capital thing to "stalk" in, because, not only does it
make no noise, but you can get down in the bottom of the canoe,
and canting it slightly to one side, and putting your hand with

paddle or pole over the further side, gradually edge your canoe towards the game, concealed by the higher side of the canoe. I saw two little tips of ears over the stem of a large fallen tree. I stood up to try to get a " sight " of even half an inch of the head, but no, as I slowly straightened myself up, slowly down went the tips. I went quietly down again till I was nearly flat in the canoe in hopes that he would take a good peep over, but no, as quietly did the two little tips rise and fall again. It was my only chance and I fired knocking the dust off the edge of the tree, but never did I see an inch of that lucifee again.

The lucifee, though so shy and difficult to get near on ordinary occasions, is, however, a very ferocious little beast if you, so to speak, put him in a corner : and I shall never forget the fury with which one, which I caught in a trap during a winter hunting expedition, rushed at and attacked my Indian and myself, when we tried to get near enough to brain him with an axe. He was caught by a leg in a powerful steel trap attached by a chain to a tree, and he rushed trap and all at us to the full extent of the chain, very much as a cat protecting a litter of kittens will do at a small cur. We brained him at last and got a beautiful skin.

The lucifee, I have used the usual local name, is sometimes called the painter or panther. The proper name ·I believe is lynx *(Felis canadensis.)* Colour brown or grey, spotted with a darker shade. Stout, very powerful limbs, short stumpy tail, and upright ears with a little tuft at the points.

Bears are extremely shy—have a very keen sense of smell, and of hearing, and have a marvellous aptitude for disappearing and never again showing, and are most difficult to get a shot at.

On one occasion I crossed a " portage * " between the head waters of the Grand River (running into the St. John) and the Restigouche alone about dusk. There were lots of tracks of bears on the soft ground and mud, and I was eagerly looking out for a shot. I retraced my own track, after a short interval of some 10 or 15 minutes. There, on the top of my own foot tracks, was the fresh trail of a bear, which had apparently followed me along for a considerable distance, but though I had crept stealthily along, watching for a chance, I never got a glimpse of him.

On another occasion on the Nepisiquit River, waking up in

---

* I ought perhaps to describe what a " Portage " means. It is the path or track that passes from the head waters of one river to another, from one Lake to another, or which leads past a high fall or rapid, up or down which you cannot pass in a canoe, and over which " portage " you must carry your canoe and baggage. It is a word constantly in use in the Backwoods.

camp one fine morning we saw on the top of a bare hill, some distance off, a dark brown spot, moving about. By the aid of our field glasses we discovered it was a large, brown bear, feeding on the blue berries, or as we call them in Scotland, *blae berries*, which were plentiful. I at once started off with one of our party to try and stalk him. After a long climb we sighted him feeding quietly some 400 to 500 yards off. I was greatly tempted to have a shot even at that uncertain range; but there was a most tempting dry ravine in front of me, which, if I could only negotiate it silently and successfully, would bring me within 150 yards, and the wind was the right direction. I tried it, and my companion stopped and lay down in the bush to watch. I crept quietly through the ravine after taking my bearings (no pun meant!) carefully avoiding every little branch that could crackle under my foot (mocassins of course and no boots) but when I reached the brow of the hill, no bear was to be seen. My companion joined me and we got on the bear's tracks and found he had gone clean away, and pretty fast too; and my companion told me it was simply marvellous the way in which he had disappeared. He suddenly raised his head and looked for a second in our direction, slipped behind a little bush, and never showed again, though my companion was steadily watching, and, you would have thought, there was not sufficient cover on the hill to screen a rabbit.

Many a time I have tracked and been equally close to a bear, and thought myself quite certain of getting at least a shot, but I never did, and I have always regretted that I did not take that 400 to 500 yards chance.

I am not quite sure whether these Bears are the *Ursus arctos* or the *Ursus americanus*, but I think the latter is correct.

We got a couple of them when quite small cubs, and kept them chained up at Government House, Fredericton, but after a time, as they grew up, they became quite dangerous, and we had to shoot them. They were most comical little scamps—with as much humour as a monkey—but their play turned by degrees into "horse play" and they would suddenly and like lightning make a grab at your neck with their powerful fore paws, which was anything but fun.

As I have before said, in these canoe voyages we used to camp in one place for two or three days sometimes, to trap, fish and shoot. Trapping to me was most interesting work—you had to learn to track and "read signs" as only an Indian can; and to learn all the ways and habits of the various animals which you were going to trap, and which were chiefly beaver, lucifee, musquash, mink, fox, marten, erimine, otter &c.

We always carried a few steel traps, something like rat traps, but of various sizes and very strong; but where we stopped long enough to lay out some lines of traps, and especially in winter hunting "dead-falls" were also used—heavy logs of wood so arranged with two little cross-pieces of stick, that on the bait or trigger being touched they fall on, and *squash*, the animal touching it.

You have probably all heard of the wonderful instinct and ingenuity of the beaver. They are certainly great waterworks engineers, much more learned and clever in erecting dams and reservoirs than many human engineers. I have had many opportunities of seeing and examining their dams, and some of them were really wonderful constructions. They are by no means alike, but vary according to the locality, the lay of the ground, to the size of the river or stream to the force of current, &c., &c. In fact the beaver studies "local circumstances" and human engineers would do well to do the same.

Let me try to describe one dam which we broke up and hunted out on a small stream running into the upper waters of Nepisiquit. The place was well selected, a low spur of hard ground running across the valley on either side, but still leaving a wide space between, and not a narrow gully in which the force of current (when the stream was in flood) would be too great for the dam to resist. Two or three large trees (pines I think), had been felled by the beaver on each of these spurs so as fall across the stream, tending slightly downwards.

Now it requires a considerable amount of "backwoods" knowledge or ingenuity to make a large tree fall exactly where you wish. A good axe-man knows well how to do it, possibly the G.O.M.—who is said to be great in that as in some other things —knows how to do it, though there are some who think that he attempts to *throw things down* without knowing where they will fall, or what the result will be. You must not cut a notch all round the tree, leaving a small pin in the centre which finally breaks and the tree falls in any direction as may happen. You must cut *in* an even level notch on two sides only—the side you want the tree to fall *to*, and the side you want it to fall *from ;* and the former should have an even surface a little lower in the stem, say $\frac{1}{2}$ an inch or an inch, than the similar notch with similar level surface on the latter ; the first notch, the side you wish the tree to fall to, should be deeper than the other and pass the centre of the stem.

Well on this principle does the beaver fell his trees which are often of great size. He nibbles with his sharp gouge-like teeth, a little all round the stem it is true, but, as you will always find, much deeper and slightly lower down the trunk on the side which the tree falls to, and is meant to fall to.

To go on with the description of this particular dam, up-right stakes, most of them cut and carried by the beaver, some floated down and then "placed," had been put in nearly per-pendicular where there was little or no strong current, but with a very considerable batten or slant, on either side where the current was strong. Into this barricade branches, drift wood and other stakes had been stuck and woven, and the whole then made solid with stones and mud, so that it formed a solid wall perhaps 8 or 9 feet wide at the bottom, and tapering up to a width of about a foot at the top, with a slight semi-circular sweep down-wards, and a carefully made lip for the surplus water to run over. Where the current of the stream is very rapid dams are made with a similar sweep upwards—convex to the current.

This dam had created a small lake several acres in extent, and of very unequal depth, dotted all about with the beaver houses, with their smooth, round, well plastered roofs rising about a foot or more above the level of the water. The entrance to these houses is always under water, and the houses contain several stories or rooms, communicating with each other.

The Indians say, and from my own observation I believe it is true, that in addition to their favourite food a plant called *Nuphar luteum*, which grows at the bottom of lakes and rivers, and is something like a cabbage stalk, the beaver stores a winter supply by sticking into the mud in the bottom of the deeper parts of its dam, the tender branches of birch, poplar, and willow trees on the bark of which it also feeds.

A great deal of "savvy" and care is required to successfully trap beaver. The hunter generally goes to the spot in the stream or lake by his canoe, and carefully from his canoe looks along the bank until he finds a trail by which the beaver are in the habit of leaving and returning to the water. The hunter must not, if he can possibly help it, get upon the bank. His scent would betray him and no beaver would come near that spot. He sets the trap (steel trap) in the water at the end of the trail. It must be a certain depth. Not too shallow, because if it is, the beaver gets caught by a *fore-leg ;* in which case he invariably (no doubt after the most careful consideration of whether there is any other possible manner of getting away) bites off his *fore-leg*, leaves it in the trap and scuttles away with *three legs*. If, as is often the case, the hunter has to fasten the chain of the trap round a sapling or tree above water, he carefully afterwards laves water over the chain, the tree, the bank, or anything that he may have touched to take away the scent, he then sticks some tender saplings or slips of birch, poplar &c., (cut somewhere else) after carefully washing them, here and there about the trap, handling them below water. Finally he sprinkles from a bottle, which he carries, some "castoreum" or other lure on the bank.

One remarkable instance of the "knowingness" of beaver I remember, and I think it occurred at the dam which I have tried to describe,

I went up the stream to the dam one evening with our chief Indian "Gabe" and set one or two traps. Next morning we paddled up to look at them.

We found one of the traps gone; the chain had been fastened round a small birch tree about 5 or 6 inches in diameter. On examining the spot we found the birch tree was cut down by beaver, and on more carefully examining the stump, and the tree lying alongside, we found that a "chunk" of the tree was missing. Beaver had cut down the tree *below* the chain, and then (finding it would not come away) had cut the tree again *above* the chain, and gone off, beaver, trap, and chunk of tree!! We felt rather sore, or as the Yankees would say, "mean."

But on proceeding in the canoe up the dam or lake, we saw a commotion in the water, and paddling rapidly up, we found it was the "chunk" of wood trying to float on the top of the water, attached to the "chunk" was the chain, attached to the chain was the trap, and to our great delight in the trap was a very large beaver, which we carried back in triumph; he was caught by a hind foot.

After the sober reflection of many years, I think I should have let that ingenious beaver go; but I didn't. As I said before I was *young*. I carried that particular "chunk," with the neat cutting of the beaver's teeth at either end, home to Scotland with other "relics" and I am not sure that it is not there to this day, as the story goes, with various other hunting trophies.

The fishing—salmon and trout—which we had on some of these canoe voyages, was just about the best one could get in the world.

I remember one day our passing down two or three long, deep stretches of the Nepisiquit River, beautiful-looking water for trout, and much to my surprise I could not for a long time get a rise, though I fished carefully every here and there. Our Indians (who belonged to and knew the river) at once explained it. They said the trout in these deep stretches of water are very large and always go in "schools." "You will fall in with a school somewhere, *and when you do you will have to take off one of those* TWO *flies you fish with.*" I was fishing with the usual cast of two or three trout flies.

Sure enough we soon did fall in with a school, I at once hooked a three or four pounder, and long before I could pull him in, another fish of equal weight had seized my second fly. I forget whether I got out both of them, but I immediately discarded my second fly, and with one fly (of any colour) pulled in trout after trout, one every cast, for a couple of hours, till I had filled up half my canoe. They were all from two to six pounds weight. The curious thing was, the more splashing a fish made when being landed, the more the others used to collect and congregate round the place, and immediately when your fly again touched the water, there was a general " rush " for it.

I never before saw such fishing, but it was probably the first time almost that the upper Nepisiquit had ever been fished with a rod.

Lower down the same river, below the " Falls," which are 30 feet high, clear fall, besides the rapid above—and which salmon cannot pass—the salmon fishing is excellent, and I have got as many as seven or eight salmon in a morning before breakfast.

Of course we generally stopped for a day or two when there was particularly good fishing, but during the trip some of our rods were always kept ready fixed along the gunwale of the canoe, ready for a cast in any likely pool, and we lived largely on these excellent trout which you never get tired of, as you very soon do of salmon, if you get it every meal.

I will give a few short extracts from one of my old note books (which I was fortunate enough to find) written during a trip up the Tobique River across the Watershed (a 3 mile portage) and down the Nepisiquit in 1863. The party consisted of the " Sagum, " " Loks " and myself and the faithful Gabe— and we had a party of Tobique Indians (Milicetes) with their canoes to take us up the Tobique. And a party of Micmac Indians, with their canoes, came up the Nepisiquit to meet us at the " portage " and take us down the Nepisiquit to the Gulf of St. Lawrence.

" *Monday, 3rd August.*—Camp 2 miles above the Tobique " Rapids. Started before 5 a.m. Reached Hutchison's (the " furthest outlying settlement or farm) at 11 a.m. The Sagum " received and answered the usual address—we wrote and " despatched letters, etc., and then poled on, and camped about " 24 miles up at the mouth of the Wampskeyhegan River, where . " I caught a good supply of trout. An awfully hot day. Black- " flies and mosquitoes very bad in the evening, but a cool night."

" *Tuesday, 4th August.*—Started rather late and proceeded " up about 10 miles. Stopped for lunch and caught many trout, " and a good many more after lunch on the way up, chiefly at " Banks Bradoon " and the ' Oxbow.' Camped at ' Deep " Hole,' 24 miles up, where I caught some more trout. Saw " two salmon rise, but could not raise one myself. The Indians " say salmon do not take fly on the Tobique. (They spear

" plenty.) I don't believe it. When they (the salmon) reach
" the upper waters—the spawning ground—they are sure to rise."

[This I found was the case, for after we passed the " Forks "
of the Tobique where the several branches meet, I caught two or
three salmon nearly every day on our way.]

I skip from 4th to 17th August, as a continuous record is
interesting only to the man to whose memory it recalls every
incident of every happy and often exciting day ; and go on to
" *Tuesday* 18*th Aug., On Nepisiquit.* Started after breakfast and
" went down to the Forks where we had splendid fishing ; caught in
" two hours with two rods five or six dozen trout, none of which were
" under 1 lb. some over 4 lbs. Dined there ; paddled on a few miles
" and camped for the night at a place where we got more fine trout.
" In the evening I went up a beaver stream opposite with Gabe,
" and set some beaver traps, near a dam."

" *Wednesday* 19*th Aug.*—I went out very early with Gabe to
" look at the traps. Found a large beaver in my first trap, caught
" by a hind toe. In one of the other " traps was a quite little one.
" After breakfast we decided to break up and hunt out the dam
" before leaving. So we all canoed up, and drained it, and then
" set to work on the houses. After a while a beaver bolted out,
" I was just going to fire, when I saw it was a young one, so I
" called out to Gabe and the others and we pursued it and caught
" it in a landing net. Soon after we started another little one
" and caught it in the same way alive. We could not start any
" more and returned with our two new pets, which we put in an
" empty biscuit barrel. It was very rainy in the afternoon so we
" remained in the same camp and caught 3 or 4 dozen more trout
" many of them over 4 lbs. weight."

Such was the sort of daily record. I may mention that the
two little beaver pets were not only safely conveyed to Government House, but after a winter there were taken home by the
Governor and presented to the Zoological Gardens where I saw
them some years after.

So much for " Camping out " in summer, but to my mind
" Camping out " in winter in order to hunt cariboo or moose, and
to trap, was far better fun ; though it is tremendously hard
work, if you go thoroughly in for it, and share the work with
your Indian Hunter, which every true " back-woods " hunter
should do. I used by the generosity of Governor Gordon,
generally to be able to get three weeks or a month's leave in the
middle of winter, and I invariably went out for the whole
time to camp, generally alone with my Indian Hunter, Peter Saul,
(and accompanied sometimes by his *boy*), in order that I might learn
to track, trap and hunt " like an Indian " as Gabe described it.
It must be difficult for any one who has never been in the far

north to realize what hunting and camping out in midwinter there is. You must try to realize the thermometer going down to 32° below Zero (64 degrees of frost), the rivers not only frozen over but in some places frozen *solid*, the ground covered with snow in some places 12 or 15 feet deep or more, and days of what in Scotland we used to call "blin drift."

You leave your comfortable and warm house and burning fires, to go off in the woods 40 or 50 miles back from the furthest settlement of any kind, you make your own camp, you cook your own food, and you may see no human face but your Indian's for the whole month. Yet it is a glorious work if you can stand it.

I used to generally go out in the early winter when game is in better condition. Of course you walk on snow shoes, you are supposed to have learned that accomplishment before you think of going out winter hunting, and you must haul the whole of your worldly goods, provisions, guns, buffalo robes, blankets &c., from the furthest settlement or lumber road in taboggans to the place where you mean to camp.

As to equipment you wear mocassins put on over at least two and often three pairs of very thick woollen socks or hose and sometimes a piece of blanket wrapped round besides. Mocassins are of several kinds—the two most in use are made of moose hide, one kind of soft hide tanned or cured by the Indians, and like *very very* thick strong chamois skin—the other is the skin of the hock of the moose, cured with the hair on, drawn on like a boot. They are most comfortable, giving free circulation to the blood, a matter of absolute necessity in such severe cold, but they would hardly compare in point of elegance with the sharp pointed, exquisite little "monstrosities" of the 'mashers' of the Ball Room which seem year by year to become more like the foot gear of the typical Chinese lady. Thick woollen under garments, strong knickerbockers, a thick lined flannel shooting jacket, and, in great cold, a blanket overcoat with a hood which will pull right over your head, and a fur cap which will turn down over your ears and neck, completes your dress. Of course with a belt in which is slung your axe, hunting knife, ammunition, &c.

I ought not to omit the covering for the hands, an important thing, for you have got to remember that if you touch the steel barrel of your rifle, in an *extreme* cold frost, with an uncovered finger, you may burn the bit out—not of your rifle but of your finger—and you will find that if you get your fingers frost bitten there is an end to your shooting for the time. I found the best thing was a pair of thinish leather or kid gloves, thinly lined with wool or flannel; and with which you could with ease handle your rifle and touch a trigger, and over them to wear what in Scotland are called "hum'le mittens," which you can draw off (from your right hand at any rate) in a second when you want to fire your rifle.

My Scottish friends who are present will know what "hum'le mittens" are; to the uninitiated I may explain that they are warm worsted woven gloves, a sort of bifurcated bag; in one compartment of which you put your thumb, in the other your *four* "fingers in a bunch" and no better covering could you have in extreme cold.

Taboggans are long narrow sledges or sleighs about two feet wide. They are of several kinds. One kind is made of a very thin, wide board of hard wood, the whole width of the tabogan and indeed forming its *sole,* the front of which is bent upwards in a sort of circle, and attached to very slight stiffening runners. This is the kind which is chiefly used for the fashionable amusement of "tabogganing" down hills, &c.

The other kind and which I used to prefer for hunting expeditions, is made with two strong stiff runners joined together by cross bars and shod either with thin plates of a very hard wood (I forget the name) which takes a good polish on the snow or ice, or with thin plates of steel about four or five inches wide which are also bent or turned up in front, so as to mount over fallen trees, branches, inequalities of ground, &c., instead of catching in them.

Taboggans are drawn by a strong rope or hide belt which you put across your shoulders and under your arms, so as to leave your hands and arms free when necessary.

I had myself a nice light steel shod taboggan as my powers of hauling on a long day's tramp on snow shoes were far inferior to those of Peter who used to take a load twice as heavy as mine.

In snow shoeing through the woods, whether hauling taboggans or not, you always go in Indian file.

The leader has much the hardest work, as not only has he to cut away any tree or branch, but he has to "break path" as we used to call it, which in soft deep snow is very hard work. Late in the winter when the snow is well settled, and a crust almost ice, has formed on the top of it, you do not sink at all, but it is different in the early winter when the snow is soft and light. Generally speaking every one steps exactly in the foot— or rather snow shoe—steps of the leader, but in hauling, the two first men while following exactly in the same track, sort of cross step each other, which breaks down and hardens a track just wide enough for the taboggans to run in.

Hauling in your things from the last settlement to where

you mean to camp, building your camp (if it is to be used for any time) and at the end of your trip hauling out your cariboo if you have any, to the nearest lumber road, is very hard work indeed.

Peter was a splendid Indian, a first rate hunter, and very devoted to me. But I nearly got into sad trouble the day I first won his affections. It was at the same time so comical that I think I must relate the circumstance to you.

It was my first winter hunt alone with Peter. We sleighed out, our empty taboggans hitched on behind the sleigh, to the last settler's; and slept the night side by side, wrapped comfortably in our buffalo robes, on the floor of the *one* room of the small shanty with a big fire at our feet.

Early next morning we started to do our 15 to 20 miles "haul" to where we meant to camp. It was unusually hard work, and a warmer day than usual at that season, and by mid-day we were both "pumped." To relieve Peter who had been "breaking path" all the way, I suggested that we should exchange taboggans, and that he should take my smaller and lighter taboggan and load. He did this with alacrity, but with the lighter load he soon shot ahead of me and left me; and at last getting completely *done* with my heavy load, I left the taboggan and doubled after him, and in three or four miles caught him up; I was in no amiable mood, and pitched into him, pretty roughly I am afraid, for "leaving me alone in the woods." We had sat down in the meantime to "blow," and he like me also lost his temper and rounded on me. "Call that leaving you," he said, "I'll show you what leaving you means. What will you do now if I go off through the wood there and leave you altogether ?"

Were such a thing to happen I knew I was "done for."

So I quietly dropped my rifle across my knees, click went the lock to full cock, and I said "I will tell you what I should do— as you reach that tree," and I pointed to one some 80 yards off, "I should fire at you, I *will not be left* alone."

Dead silence, and no one moved, for hours. Peter gravely pulled out his pipe and began to smoke. I pulled out mine and did ditto. He re-filled and lit up again, I did the same. For the life of me I dare not say another word, for if I did I knew I should lose all future influence over him, but it was a miserable hour or more to me, for I was thinking what a "corner" I had got into. I *must* keep my word, or I was done for as to getting on with him. If he did go, and I had to keep it, what then ? Of course I meant to wing him only, but even so, what a risk !

At last I saw him glance quickly up at me from the corner of his eye, and I began to feel that the day was my own, but I did not move a muscle. Another glance up, and then a grunt, and "Sure, I didn't mean that I was going to leave ycu, I was joking."

Click—back to half cock went my rifle, and I quietly said: " Glad to hear it, Peter, we are going to be chums and comrades " in these woods for the next three weeks at any rate, and I will " rely upon you, if you rely upon me, but you must begin by " believing that what I say I'll do, I do."

" I believe you," he replied, "suppose we go on, I will soon " get the taboggan and catch you up—you rest or go on slowly " in that direction."

And he did catch me up in an incredibly short time, and from that moment we were sworn chums, and I think each placed the fullest reliance in the other for ever afterwards.

We took it in turns for the rest of the day to " break path " and reached our proposed camp about dusk after a very hard day's work.

The winter camp you make varies, according to whether you mean to occupy it for a night, or two; or whether you mean to occupy it for a few weeks, and hunt the country all round; which is what we generally did.

A scratch camp for the night is generally made something like a summer camp—open at one side where you make a huge fire.

A strong pole or cross piece, between two trees or forked posts, and a " lean-to " roof, made by slanting poles with cross pieces, covered thickly with spruce fir branches, and then thickly covered with snow, which you shovel up with your snow shoes, and then batten down hard and smooth with your snow shocs and hands, so that it is air proof. The ends of the camp being filled up in the same way.

Snow, leaves, and rubbish are cleared away from the inside and the ground covered thickly with spruce fir branches, with which you also make your bed. A most comfortable bed it is, if well made. There is a great art in making it well. You get a quantity of the points of the feathery spruce fir branches about a foot long, and regularly *thatch* the ground thickly from the foot to the head of the bed, by sticking the *stick* end of the branches in layers into the ground—very much as natives here thatch a roof with carat—only that the branches are stuck in the ground instead of being tied to the rafters of a roof. The bed is beautifully springy and the smell of the fir leaves or branches is delightful. You cover this with a buffalo robe, and cover yourself with a blanket or two and another buffalo robe.

Then comes the most important part of the proceedings— the night's fire. You fell four or five enormous trees, generally pitch pine, and cut them in lengths or logs, eight or ten feet long (about the length of your camp) and haul them on the taboggans to the camp, where you place them in a pile lengthways in front of the open side of the camp—very much like a pile of cordwood, only that the logs are put lengthways—you put in some stakes of very green hardwood which will not easily burn, to keep them a little in position, and you light your fire inside (on the camp side) of the huge pile, and soon have a fire fit to roast an ox.   As one log is burnt through you roll down another and tumble the ends of the ones burnt through, into the fire, which is thus kept up all night.

I have slept in a *very* scratch camp like this, in deep snow, when the thermometer was 32° below Zero—*64* below freezing— and enjoyed it, but to do so, you must do a fair share of the work, wood-cutting and camp-making, &c., if only to keep yourself warm.

A "permanent camp," as we called it, was generally regularly built as a log hut.   You fell long logs, a foot or more in diameter, for the sides, shorter ones for the ends, notch and cross the ends at the corners of your hut, and slope the roof down towards the two ends, so as to leave a big open hole for the smoke to go out in the centre, stuff up the cracks between the logs with moss and snow, or build up snow firmly against the outside.

Your fire, which need not be quite so huge as in an open camp, you pile up across the middle of the hut, you make one small door, made of bark or spruce branches closely matted, and no window.

Such a camp can be made wonderfully snug, and we used often to return to a well built camp of this sort for several successive winters.

It only required a few repairs, and perhaps a new roof, and fresh branches for bed, etc., and there was no bother about estimates, passing requisitions, pre-audit &c., you immediately set about it and did it.

Of course, it is of great importance to select a good site for your camp—well sheltered in the first place—plenty of trees suitable for your fire within easy haulage—and, if possible, a water spring.   You can, of course, always have water by melting snow, but *spring* water is better ; and the Indians are wonder- fully clever at finding out a spring even when it is covered over with snow.

The game hunted in winter was chiefly cariboo (the American reindeer) and moose (elk).   Sometimes in the early winter you might still find a bear out on the prowl.

Moose used formerly to be killed, or rather butchered, in large quantities late in the winter, when the snow is very deep and has a hard crust, sometimes regular ice, on the top. The moose cannot run or sometimes even walk through this, as he is an enormous size and weight and has a small sharp hoof, and sinks right down through the crust and is caught and cut by it.

As the winter gets on and the snow gets very deep, the moose "yards" as it is called. That is, two or three or more get together, they select a tract of ground with plenty of food for the rest of the winter, and they keep to their "yard" trampling it down continually for the rest of the winter, and never leaving it.

When driven out of this yard in the late winter (the cows are then heavy in calf) they can easily be run down and I am told used often to be butchered with axes.

It was against the law to kill them after (I think) 1st Feb. ; it was no sport and I am glad to say I never hunted them in the *late* winter. In the early winter they are very difficult to stalk.

The cariboo is a much smaller animal though yet about the size of a cow. Their hoofs are very different from those of the Moose and spread out, (something I am told, like a camel's) in such a way as to prevent their sinking much in the snow at any time ; and to enable them to travel without breaking through at all where there is a hard crust or thin ice. Consequently you can never get a good shot at them except by downright good tracking and stalking. Still they should in my opinion only be killed in the early winter, when they are fat, and the breeding season is not close. Their senses of smell and hearing are extremely acute, and great skill is required in stalking them.

They are found generally in herds of from two or three up to 30 or 40, and I have been told even 100.

Cariboo on the march, from one part of the forest or from one feeding ground to another, always go in Indian file—each stepping exactly in the footsteps of the leader—so that if you come upon a track where perhaps 30 have passed, you would, until you have been *educated*, think that only one cariboo had passed along.

The Indian hunter however knows better, and from the depth of the foot mark, and numerous other signs will tell you with wonderful accuracy, how many there are in the herd, the pace they were going, the number of minutes, hours, or days since they passed &c., &c.

When the herd begins to feed they of course scatter all about, and if it is in thick wood, it is sometimes very difficult then to track them up till you can get a shot, without some one of the herd "winding" you. '

I was generally very lucky in my cariboo hunts—but it is very uncertain work, and I have often seen a party come back with none,

I remember one winter I went out to our favourite district with Peter. The second day we got on the tracks of two and I shot one of them, fortunately for our pot, for we did not see another fresh track, far less get a shot, for three weeks.

It was a Sunday morning and our sleigh was to come out to the woods for us on Tuesday. I decided (as we never or hardly ever hunted on Sunday) to stay in camp, tidy up and cook and sent Peter out with a taboggan to haul in the carcase of the *one* solitary cariboo we had killed. He returned soon after breathless, saying " quick, quick, there's, a herd of cariboo on the way."

I jumped up, shoved a handful of cartridges in my pocket, seized my rifle and was off. As we hurried along on Peter's back tracks, he said he was afraid he had spoilt the chance, that he had left the taboggan in the middle of a large barren, without thinking when he saw them, and as they were feeding up in the direction of it, he was afraid they would "wind " it, and be off. We reached the big barren and sighted the herd—five cariboo, some 1,000 yards off. Peter said they were getting close to the taboggan, and almost immediately they "winded " it, collected and started off at a trot, quite 700 yards from us. It seemed our only chance and I fired at the only one with good horns, and had the mortification to see my bullet knock up the snow just under him—and they entered some thick high wood.

Peter said we had better follow them a bit to see what direction they took, as we might be able to come across their tracks next day as we went out to the settlement.

So we followed on to the highwoods, floundering along, as we had left our snow shoes, in order to stalk more quietly, when we first saw the herd. After following the track into the thick wood for a bit, Peter's face suddenly glowed with animation, he started to run and said " come along we will get a shot yet, for they have not seen us and are not frightened, they are only *walking*," this of course he *read* from the tracks.

We started to run, slippery work it was too, over trees &c., without snow shoes and I had several tumbles in some of which I must, as I afterwards found, have dropped some of my cartridges and we kept it up for two or three miles.

At last we sighted them far ahead on a barren, and bending away to the left, almost at a right angle, towards another belt of thick wood, and to our great delight still walking quietly. "Now's our chance " said Peter, "there's an open streak of " barren down the middle of that wood which they must cross ; if " we can get there before them you are sure of a shot." It was easy said, but I was " well pumped." However I put on all I knew and we did reach before them. I whispered to Peter "don't mind though I let one, two, or even three pass without "firing, I *must* have the one I fired at, the one with the horns," as I spoke the herd appeared crossing the open 200 yards off. As soon as the horned one came in sight I fired and dropped him, the rest of the herd galloping off.

When you drop one out of a herd of cariboo like that, the rest after galloping for 1,000 yards or so, almost invariably stop in a bunch for a moment and look back for the lost one. I was too "blown" to run more, so I shoved another cartridge in my rifle, gave it to Peter saying "run and shoot" while I went to "bleed" the fallen cariboo, for I have known one dropped like that to get up, (while you run on after the others) and be no more seen.

As I was giving the *coup de grace* I heard—bang—bang; and ran on to Peter—"I hit two" he said "come on and you'll get another chance." We started again I loaded as I ran and caught sight of them and again they stopped to look back. I fired both barrels, killed two and wounded one. We again pursued, I got another chance and killed another; the fifth, the last was badly hit, and limped about some distance off while we were cleaning those killed. To my regret I found I had not another cartridge to put him out of misery, but he must have been badly wounded for we found him frozen dead next morning. So we got the whole herd, five of them.

There is this advantage in winter hunting, your game keeps good for any time, and there is no extra charge for freezing it when the thermometer is 20° to 30° below zero! You bleed and clean your cariboo or moose; and then fix or stake it out to freeze in a shape which will be easily lashed on a taboggan— drawing the head out straight to the front; doubling the fore legs close up to the body and pinning or staking them there; and stretching and pinning the hind legs straight out behind, so that they will drag or trail behind your taboggan. You leave him so, being careful to mark the place and putting up a piece of rag or something to scare foxes or wolves over the carcase, and a day or two before the end of your trip, you go out and haul in the several carcases to the nearest spot where you can get a team and sleigh in. On the occasion which I have mentioned after cleaning and "fixing" the four cariboo (which were killed 8 or 10 miles from our camp) we returned to camp picking up our snow shoes, the taboggan and the solitary cariboo killed three weeks before on our way. Our sleigh from home was to come out the next day but one, and so we had to get these five cariboo hauled out to the nearest lumber road in *one* day. To make matters worse it snowed very heavily all night, and next morning there was some two to three feet deep of *new* snow on the ground.

Peter the boy, and I started at daylight with the two empty

taboggans and such hard work was it breaking path that it was 1 p.m., before we reached the cariboo. The carcases were completely covered with the new snow and it was sometime before we could find them. We hauled out *three* to our camp by 4 o'clock, so much easier is it to haul on a broken and beaten track —and the weather having quite cleared up during the day and there being good moonlight, after a good dinner and rest Peter and the boy went back in the evening and hauled in the other two cariboo getting back to camp about 10 p.m.

Next morning we were lucky enough to get a lumber camp team or "span" of horses and sled to haul them and us out to the settlement. By this stroke of luck though we had been three weeks without seeing a fresh track, we were able to go home with a good sleigh load of six fine cariboo, besides other skins (trapped) and birds.

Referring to the severe cold in winter in New Brunswick, Providence has mercifully ordained that when the cold is very. intense, there is hardly ever any wind. To face a wind in the open, when the thermometer is down to 30° or 40° below zero, would be impossible.

There was one day, long after spoken of in the Province as *the Cold Friday*, when there was a strong wind with an intense cold. It was in fact what has been called in later years a "blizzard," and a fierce one too. In many places the settlers did not dare to leave their houses even to run across the yard to their barns to feed their cattle.

Of course back in the woods, if you are in a thickly wooded country you are well sheltered, and do not feel such a wind so much.

On that particular Friday I was camping out with Peter, and we went out to hunt cariboo as usual, merely wearing our big thick blanket over-coats which we did not generally do when hunting unless it was unusually cold, and selecting a line of country where we would be in thick forest all the time. We did not feel it very much, and had no idea, until after we returned, what a fearful day it had been out in the open settlements.

Still, coming in thick high wood to a long narrow strip of barren, certainly not 100 yards wide, which we must cross, Peter suddenly, when still in the thick bush, stopped, and said " now we must make a rush for it, you must run all you can " across that little bit of open or we shall be frozen." He made me put my hood right over my head, and tighten it right round so that only my eyes and nose and little bit of cheek was exposed, and did the same thing himself. We "sloped" our rifles, and ran across all we knew. It was only a matter of a minute or two, yet on reaching the shelter on the other side my face was frost-bitten, and had to be revived in the usual way by gently rubbing it with snow. Going on, however, in the thick wood we felt no great inconvenience.

It must be remembered that though so cold it is extremely *dry*, and the light snow shakes off you like dust, leaving you perfectly dry.

At the request of my friend Mr. Devenish, to whom I remember telling the story many years ago, when we were camping out together on the Ortoire River, I will tell you what he calls the "last match" story. It is nothing very wonderful but it was a very *wet* night, for it can rain in New Brunswick nearly as heavily as in Trinidad.

It was in one of our summer canoe voyages. We had gone up the upper waters of the right hand branch of the Tobique, and wished to get to two large almost unknown Lakes—Long Lake and Tobique Lake—out of which the tributaries of this branch of the Tobique take their rise. At a place called the Forks, we found the streams so low that the canoes, loaded, could go no further, and we camped.

Next morning (Saturday, 8th August) leaving our canoes and baggage and most of the men and telling Gabe and another Indian to start in an empty canoe at day-break next day to try to get up to Tobique Lake and meet us, the "Sagum," "Loks" and I started with three Indians by land for Long Lake. The portage was said to be eight miles, but it was a long ten, and no track or path of any kind. On reaching Long Lake we set the Indians to work to make a "scratch" spruce bark canoe.

Next day (Sunday 9th August) the canoe was finished about 12 and we embarked with Noel, one of the Indians, and proceeded up the Lake, telling the other Indians to wait where we had camped for the night till towards evening, in case we should have to return. If we did not, they were to recross the portage to the depôt camp.

After paddling several miles up the Lake, we found a portage track and crossed by it to Tobique Lake, where we found Gabe and Lola. They said they had got up with very great difficulty, there being hardly any water in the river and that it would be impossible to take more than two in the canoe.

So the "Sagum" and "Loks" went with them, and I returned the way we came with Noel, across the portage to Long Lake, with great difficulty, however, for the wind was rising strong against us and it was a very frail and temporary canoe and was already showing signs of coming to grief. We hoped to catch the two men at our last night's camp, for they had "grub" and we had none.

To our great disgust they were gone, and we only found
the dying embers of the fire. No food, getting dark, and the
rain beginning to come down in torrents. Hungry as I was
and anxious to join the others I was inclined to take the deserted
camp for the night, but I asked Noel what an Indian would
do if he were alone and unencumbered with a white man. "Oh!"
he said, "if I were alone I could easily find my way and recross
the long portage to the (depôt) camp to-night."

It was some nine or ten miles as I have said, not a vestige
of a track, and it was getting pitch dark and pouring rain, so
I did not think he could. But I was a little nettled and thought
I would like to see him try, So I said, "all right you carry
my gun and axe, so that I may travel lighter, off you go, I
will keep up with you somehow, if I cannot I will shout.

So off we went, and a terrific pace did that fellow take me
at through the bush for a while, but I soon saw that he was going
anything but direct, and could not possibly find his way on
such a night; and at last to my secret relief he stopped, and
said it was no use he could not find the way, it was impossible.

We set to work, peeled some bark and made a camp and
were proceeding to make a fire, when I found (I was soaked to
the skin) that the matches were a mass of wet pulp. I remembered that I had seen two straggling matches which had somehow or other got into my purse or pocket book which I carried
in a small waterproof bag in an inside pocket. You can imagine
with what care I got them out and proceeded to strike them
under cover of the bark roof. The first went out. One damp
match between us and a cold, comfortless, hungry, wet night!
I never struck a match so carefully in my life, and I succeeded
and the dry inner birch bark, which we had carefully procured
for the purpose blazed up and we had a fire.

It was immediately rewarded by two good results. The
first was that as the blaze sprang up there was a flutter overhead
and looking up I saw on a tall tree a partridge craning his neck
and bewildered by the sudden light. I did not give him long to
consider it—crack went my gun and down he came. We had at
any rate something for supper. But a still better result followed
for we then heard shouts from a hill about a mile off, which we
soon recognised as coming from the two Indians, who had set off
to cross the long portage a little before our return down the Lake,
and who like ourselves, had been caught by the rain and darkness and been obliged to camp midway.

They had heard my gun and then seen the blaze of our fire,
though we could not see their's and shouted to us; as they had
the "food" there was little doubt what to do, and guided by their
shouts for a time and then by the light of their fire, we soon

reached their camp, and enjoyed a hearty supper, and with a good fire at our feet, wet as we were we put in a very good night or the remainder of it, joining the rest of the party at the depôt camp next day. I should like to give you some account of " Moose Calling " in the autumn months, and wild goose and duck shooting, as well as some further account of winter hunting &c., but I am afraid I have trespassed upon your kindness a great deal too long already and must conclude.

" KAWANICK."

6th April, 1894.

A PRELIMINARY LIST OF THE REPTILES AND BATRACHIANS OF THE ISLAND OF TRINIDAD.

By R. R. Mole and F. W. Urich.

WITH DESCRIPTIONS OF TWO NEW SPECIES
By Professor Dr. O. Boettger.

THE present list has been prepared with the intention which actuated Mr. Oldfield Thomas when he wrote the Preliminary List of Mammals viz : to bring together all the records of the Reptiles and Batrachians which have been published as occurring in Trinidad and which are scattered through various books and papers, many of which are difficult of access, and to so form a basis on which a complete scientific list may be formed. During the last four years we have collected as much material as possible, so that with very few exceptions all the species here mentioned have been examined by Professor Dr. Boettger and Mr. G. A. Boulenger of the British Museum (the latter through the kindness of the Zoological Society of London) and to these gentlemen we are indebted for a great deal of assistance. Professor Dr. Boettger, especially, has helped us considerably, not only in examining specimens, but also in looking up records in literature which is not available to us.

In his book on Trinidad Dr. de Verteuil gives a list of 26 reptiles, but the species of most of these are not to be recognised, except by the local names. This excellent work contains some biological notes of interest, to which we would call attention. It was our purpose to have added some similar notes to this paper, but subsequent consideration decided us to make it merely a Preliminary List.   We intend, however, on some future occasion to publish some notes on the life history of the reptiles and batrachians mentioned herein   We have already embodied in a paper, which was read before the Zoological Society of London in June, Notes on some 22 species of Snakes.   This paper will be published by the Society at an early date.

The present list shows a total of 76 species :—

Tortoises ...	...	...	...	6
Lizards ...				25
Snakes ...				33
Batrachians	...	...		12
				——
				76

Of these species 21 are recorded for the first time from the Island and two species are new to science.   The specimens thus recorded for the first time by us are marked with an asterisk before the number.

REPTILIA.

ORDER: CHELONIA.

Family : TESTUDINIDÆ.

1. *Nicoria punctularia*, Daud.

Boulenger, Cat. Chelon., Rynch. and Crocod., Brit. Mus., 1889 page 124.

Emys sp. of de Verteuil's List.

Recorded in Trinidad by Boulenger "Female and young from Trinidad living in the Gardens of the Zoological Society, now in the British Museum."

2. *Testudo tabulata*, Walb.

Boulenger, Cat. page 157.

Testudo carbonaria, Spix.

First recorded in Trinidad by de Verteuil.

Family : CHELONIDÆ.

3. *Chelone mydas*, Linn. (Green turtle.)
Boulenger, Cat. page 180.

4. *Chelone imbricata*, Linn. (Hawksbill turtle.)
Boulenger, Cat., page 183.

5. *Thalassochelys caretta*, Linn. (Loggerhead turtle.)
Boulenger, Cat., page 184.

Family : CHELYDIDÆ.

6. *Hydraspis gibba*, Schweigg.

Boulenger, Cat., page 224.

> H. gordoni, Gray.   Proc. Zool. Soc., 1868, page
> 563, pl. 42.
> Emys sp. of de Verteuil.

First recorded in Trinidad from Mount Tamana.

ORDER : EMYDOSAURIA.

s. Crocodilia.

Family : CROCODILIDÆ.

*7. *Caiman sclerops*, Schneid.

Boulenger, Cat., page 294.

> Alligator sclerops of de Verteuil.

New to the Island. Common in Central and South America.

ORDER : LACERTILIA.

Family : GECKONIDÆ.

8. *Gonatodes vittatus*, Licht.

Boulenger, Cat. Lizards, Brit. Mus., vol. i., page 60.

Recorded by S. Garman, Bull. Essex. Inst., vol. xix., 1887, page 17, from Port-of-Spain. This is the common little lizard found on walls and fences about town. The female remained unknown until 1887, when it was described by Garman. Found also in Dominica, Grenada, Curaçao.

*9. *Gonatodes ocellatus*, Gray.

Boulenger, l.c. i., page 60, pl. 5, fig. 1.

New to the Island. Male and female. Last mentioned new to science.

10. *Gonatodes ferrugineus*, Cope.

Cope, Proc. Acad. Nat. Sc., Philadelphia, 1863, page 102.

First discovery Prof. Theo. Gill. (E. D. Cope teste.)

11. *Thecadactylus rapicaudus*, Houtt.

Boulenger, l.c. i., page 111.

> Platydactylus theconyx of de Verteuil.

"Found by W. J. Cooper, Esq , on Anguilla,* a rock near Trinidad" (Boulenger.)   Recorded in Trinidad by Garman l.c. page 17.

---

* So far as we know there is no rock near Trinidad called Anguilla. One of the Leeward Islands is known by that name.

12. *Hemidactylus mabuia,* Mor. de Joun.

Boulenger, l.c. i., page 122.

First recorded in Trinidad by Boettger. Kat. Reptilien, page 28.

*13. *Sphœrodactylus molei,* Bttgr., n. sp.

First discovery R. R. Mole.

Char. Maxime affinis *Sph. glauco,* Cope, sed supra- et infralabialibus solum quaternis, cauda inferne serie mediana squamarum dilatatarum carente et colore diversus.—Snout pointed, as long as the distance between the eye and the ear-opening, once and a half the diameter of the orbit; ear-opening very small, roundish, rather smaller than a digital pallet. Rostral small, without longitudinal cleft above; nostril pierced between the rostral, the first labial, a supranasal and two postnasals; four upper labials, median smallest; four lower labials, first very large, fourth very small; mental moderately large, truncate posteriorly, followed by polygonal rather large scales passing gradually into the granular gular scales. Upper eyelid with a slightly greater but not spine-like scale above the middle of the eye. Upper surfaces covered with equal, very small, smooth, convex, juxtaposed granular scales, largest on the snout; abdominal scales much larger, imbricate. Tail, longer than head and body, cylindrical, tapering, covered above with small, smooth scales arranged in verticils, inferiorly with larger irregular scales, without a median series of transversely dilated plates.—Light brownish grey above, with two rather obsolete paler dorsolateral longitudinal streaks; a dark streak on the side of the head, beginning from the nostril and passing through the eye; tail with four to five white black-edged transversal spots in regular intervals; limbs dotted with whitish, digits annulated with blackish and whitish; lower surfaces whitish, the throat and the præanal region with or without a few very fine brownish dots.

Hab. Caparo, Trinidad.

Total length	51.5		m.m.	Fore Limb	7.5 8	m.m.
Head	6.25	7	,,	Hind Limb	9.5 10	''
Width of head	4.25	4.25	,,	Tail	25 (injured)	
Body	20.25	20	,,			

Of the 18 known species of this genus 5 live in S. Domingo, 4 in Jamaica, in Cuba and in Central America, 2 in St. Croix, and in Venezuela, one in Martinique, Antigua, St. Lucia, St. Thomas, in the Bahamas, in South U. S. America, and in U. S. of Columbia.

Family : IGUANIDÆ.

14. *Anolis alligator,* Dum. Bibr.

Boulenger, l.c., vol. ii., page 31.

A. trinitatis, Reinh. & Lütt.

First recorded in Trinidad by Reinhardt and Luetken, Videnskab Meddel, 1862, page 269. Also recorded by S. Garman, Boulenger, etc. Common in the West Indies.

15. *Anolis biporcatus,* Wgm.

Boulenger, l.c., vol. ii., page 89.

First recorded in Trinidad by Boulenger from specimens presented by C. Taylor.

*16. *Anolis chrysolepis*, Dum. Bibr.

Boulenger, l.c., vol. ii,, page 89.

New to the Island.

17. *Polychrus marmoratus*, Linn.

Boulenger, l.c., vol. ii., page 98.

First recorded from Trinidad by de Verteuil.

18. *Liocephalus herminieri*, Dum. Bibr.

Boulenger, l.c., vol. ii., page 166.

Recorded from Trinidad by Duméril and Bibron.

19. *Uraniscodon plica*, Linn.

Boulenger, l.c., vol. ii., page 181.

Hypsibatus agamoides of de Verteuil.

Common in all tropical South America. The largest known specimen is in the British Museum and measures 394 mm.

20. *Iguana tuberculata*, Laur.

Boulenger, l.c., vol. ii., page 189.

First recorded from Trinidad by Boulenger, also by Garman.

Family : TEJIDÆ.

21. *Tupinambis nigropunctatus*, Spix.

Boulenger, l.c., vol. ii., page 337.

De Verteuil records *Salvator merinœ*, which is a synonym of *T. teguixin*, L., but no doubt he means the above species. Prof. Dr. Boettger writes regarding this species : " Although *T. teguixin* is supposed to occur in the " West Indies, I doubt its occurrence in Trinidad very much. I have " never seen West Indian specimens of this species ; there are neither any " specimens in the British Museum from the British West Indian " Colonies." First recorded in Trinidad by Garman.

*22. *Centropyx striatus*, Daud.

Boulenger, l.c., vol. ii., page 340.

New to the Island. A female from Caroni new to Science. Recorded only from North South America.

23. *Ameiva surinamensis*, Laur.

Boulenger, l.c., vol. ii., page 352.

Garman, on West Indian Reptiles, etc., l.c., page 2, as *Ameiva atrigularis*, var. nov.

Ameiva vulgaris of de Verteuil.

First recorded in Trinidad by Garman.

24. *Ameiva punctata,* Gray.
Boulenger, l.c., vol, ii., page 360.

Ameiva major of de Verteuil.

25. *Cnemidophorus murinus,* Laur.
Boulenger, l.c., vol. ii., page 362.

26. *Cnemidophorus lemniscatus,* Daud.
Boulenger, l.c., vol. ii., page 364.

Recorded by Boulenger from specimens presented by C. Taylor

*27. *Scolecosaurus cuvieri,* Fitz.
New to the Island. Recorded in Grenada by Garman.

Family : AMPHISBÆNIDÆ.

28. *Amphisbæna fuliginosa,* Linn.
Boulenger, l.c., vol. ii., page 437.

29. *Amphisbæna alba,* Linn.

Family : SCINCIDÆ.

30. *Mabuia aurata,* Schneid.
Boulenger, l.c., vol. iii., page 189.

First recorded in Trinidad by Garman as *M. ænea,* Gray l.c. page 2 9 .

31. *Mabuia agilis,* Raddi.
Boulenger, l.c., vol. iii., page 190.

De Verteuil records *Eumeces spixi* from Trinidad. This species is synonymous both of *M. aurata* and of *M. agilis* and either may be meant.

ORDER : OPHIDIA.

Family : TYPHLOPIDÆ.

32. *Typhlops reticulatus,* Linn.
Boulenger, Cat. Snakes, Brit. Mus., 1893, vol. i., page 27.
Recorded in Trinidad by Boettger.

Family : GLAUCONIIDÆ.

33. *Glauconia albifrons,* Wagl.
Boulenger, l.c., vol. i., page 63.

First recorded from Trinidad by Boulenger from specimens presented by Sir A. Smith. Also found by Garman.

Family : BOIDÆ.

34. *Epicrates cenchris,* Linn.

Boulenger, l.c , vol. i., page 95.

First recorded in Trinidad by Boulenger from specimens presented by C. Taylor.

*35. *Epicrates cenchris,*

var. *fusca,* Gray.

Boulenger, l.c., vol. i., page 96.

New to the Island; previously recorded from Venezuela.

Squ 49 G. 10/9 V 237 A 1 Sc 55.

36. *Corallus cookii,* Gray,

var. *melanea,* Gray.

Boulenger, l.c., page 100.

First recorded in Trinidad by Boulenger from a specimen presented by J. H. Hart.

*37. *Corallus cookii,*

var. , *ruschenbergi,* Cope.

Boulenger, l.c., 101.

Xiphosoma hortulanum, Mole & Urich, Proc. Zool. Soc., 1891, page 447.

New to the Island. This is a common snake on the banks of the Caroni. It is also found in the interior, but generally near streams.

Squ 45 G 14/15 V 258 A 1 Sc 108.

38. *Eunectes murinus,* Linn.

Boulenger, l.c., page 115.

Boa murina of de Verteuil.

First recorded in Trinidad by de Verteuil.

39. *Boa constrictor,* Linn.

Boulenger, l.c., 117.

Squ. 89 G 19/20 V 240 A 1 Sc 53.

40. *Boa diviniloqua,* Laur.

Boulenger, l.c., page 118.

Boulenger records this snake from Trinidad. So far as our experience goes it is not found here, being confined to St. Lucia and Dominica. We have seen a specimen from the former place where it seems to be tolerably common and is called " Tête chien " by the peasantry.

Family : COLUBRIDÆ.

a. Colubrinæ.

*41. *Streptophorus atratus,* Hallow.

Boulenger, l.c., page 293.

New to the Island. Previously recorded from West Ecuador and Venezuela.

Squ 19 G 1/1 V 143 A 1 Sc 48/48 & 1

42. *Geophis lineatus,* Dum. Bibr.

Duméril & Bibron, Erp. gen. vii., page 105 (Rhabdosoma,)

This is the common little-snake found under stones and rubbish in gardens in Port-cf-Spain.

Squ 15 15 15 15 G 3 3 2 3 V 142 143 150 145 A 1 1 1 1 Sc 12/12 & 1,
12/12 & 1, 13/13 & 1, 13/13 & 1

43. *Liophis cobella,* Linn.

Jan, Icon. Ophid., part 16, pl. 5, fig. 1.

First recorded in Trinidad by Garman.

Squ 17, G 2/2, V 157, A 1/1, Sc 61/61 & 1

*44. *Liophis reginæ,* Linn.

Jan, Icon. Ophid., part 16, pl. 6, fig. 1.

New to the Island.

Squ 17 17 G 2 & 2/2 3/3 V 146 138 A 1/1 1/1 Sc 80/80 & 1, 79/79 & 1.

45. *Liophis melanotus,* Shaw.

Jan, Icon. Ophid., part 18, pl. 3, fig. 4.

Recorded also from Grenada, Tobago, Venezuela and United States of Columbia.

Squ 17 G 3/2 V 148 A 1/1 Sc 57/57 & 1.

46. *Coluber corais,* Cuvier.

Jan, l.c., part 48, pl. 4, fig. 6, pl. 5, fig. 1 (Spilotes.)

Recorded also from N. Brazil, Guiana and Venezuela.

Squ 17 G 2/2 V 210 A 1 Sc 72/72 & 1

47. *Coluber variabilis,* Wied.

Schlegel, Essai s.l. phys. d. Serp., pl. 6, fig. 1-2 ; Günther, Cat. Colubr. Snakes, page 99 (Spilotes.)

First recorded in Trinidad by Mole and Urich P.Z.S., 1891 page 448

Squ 17 G 1/1 V 224 A 1 Sc 111/111 & 1

*48. *Coluber poecilostoma,* Wied.

Jan, l.c., part 48, pl. 5, fig. 3, 4 ; Günther l.c., page 100 (Spilotes.)

First found in Trinidad by Mr. A. B. Carr from whom we received a specimen. New to the Island.

49. *Coluber boddaerti*, Seetz.

Jan, l.c., part 31, pl. 4, fig. 3 (var. bilineata.)

First recorded in Trinidad by Garman.

Squ 17 G 2/2 V 183 A 1/1 Sc 64/64 & ?

*50. *Phrynonax eutropis*, Blgr.

Boulenger, Cat. Snakes, Brit. Mus., vol. ii., 1894, page 22, pl. 1, fig. 1.

First found at Caparo by Mr. A. B. Carr, also found at Maraval by Mr. John Hoadley and Mr. Alex. Hamlyn. Professor Dr. Boettger writes : " This rare and very interesting snake is known by a single specimen in the " British Museum from unknown origin. The only difference is that the " type in the British Museum (Squ 25 G 2 V 191 A 1 Sc 126/126 & 1) has " 25 scales instead of 23 a difference which is undoubtedly individual. This " species is nearly allied to *Spilotes fasciatus*, Peters (Mon-Ber-Akad Berlin " 1869, page 443) *Ahætulla polylepis*, Peters (l.c. 1877, page 709), the scales of " which in 23 rows, are only very slightly keeled.'

" The habitat Trinidad is quite new for this snake."

Squ 23 G 2/2 V 199 A 1 Sc 134/134 & 1

51. *Herpetodryas carinatus*, Linn.

Boulenger, Proc. Zool. Soc., 1891, page 355.

Squ 12 G 2/2 V 165 A 1/1 Sc 93/93 & 1.

*52. *Herpetodryas macrophthalmus*, Jan ?

Jan, l.c., part 31, pl. 2, fig. 2 (carinatus var.)

First found in Trinidad by Mr. A. B. Carr, from whom we received specimens. Prof. Dr. Boettger, who determined this species from a specimen we sent him, could not give a definite opinion as to this snake as the material was not sufficient. New to the Island.

Squ 12 G 2/3 V 173 A 1/1 Sc 171/171 & 1.

53. *Leptophis liocercus*, Wied.

Jan, l.c., part 49, pl. 6, fig. 1.

Dendrophis liocercus of de Verteuil.

Also recorded from N. Brazil, Guiana, Ecuador, St. Lucia and Tobago.

Squ 15 G 2/2 V 163 A 1/1 Sc 131/131 & ?

b. Dipsadinæ.

*54. *Homalocranium melanocephalum*, Linn.

Jan, l.c., part 15, pl. 2, fig. 4.

First found in Trinidad by Mr. W. E. Broadway from whom we received specimens. New to the Island.

Squ 15 G 3/3 V 149 A 1/1 Sc 63/63 & 1

55. *Leptodira annulata*, Linn.

Günther, Cat. Colubr. Snakes, page 166.

First recorded in Trinidad by Boettger and Garman.   Found all over tropical and central America.

Squ 21 G 1/1 & 3 V 190 A 1/1 Sc 89/89 & 1

56. *Dipsas cenchoa*, Linn.

Jan, l.c., part 38, pl. 2, fig. 1 (Himantodes.)

First recorded in Trinidad by Garman.

57. *Oxybelis acuminatus*, Wied.

Jan, l.c., part 33, pl. 4, fig. 2 (Oxybelis aheneus.)

Dendrophis aurata of de Verteuil.

First recorded in Trinidad by Garman.   Known from tropical Peru, N. Brazil, Guiana, Venezuela, Central America to tropical Mexico and Testigos Islands.

Squ 17 G 3/3 V 184 A 1/1 Sc 170/170 & 1.

58. *Scytale coronatum*, Schneid.

Jan, l.c., part 34, pl. 5, fig. 3.

First recorded in Trinidad by Gunther.

Squ 19 G 2/2 V 189 A 1 Sc 85.

59. *Oxyrrhopus plumbeus*, Wied.

Recorded in Trinidad by Boulenger.   We have the head and tail of a specimen from Caparo from Mr. A. B. Carr.

Squ 19 G 3/2 V? A 1 Sc 86/86 & 1.

c. Elapinæ.

60. *Elaps lemniscatus*, Linn.

Jan, l.c., part 42, pl. 5, fig. 1.

First recorded in Trinidad by Gunther and Garman.

Squ 15 G 4 V 222 A 1/1 Sc 32/32 & 1

61. *Elaps riisei*, Jan.

Jan, l.c., part 48, pl. 6, fig. 3.

Elaps corallinus, Gthr. partim, non Linn.

Elaps corallinus of de Verteuil.

First recorded from Trinidad by Garman.   Also found in St. Thomas.

Squ 15 G 3 V 183 A 1/1 Sc 45/45 & 1.

Family : AMBLYCEPHALIDÆ.

*62. *Leptognathus nebulatus,* Linn.

Jan, l.c., part 37, pl. 5, fig. 3.

New to the Island. Recorded in N. Brazil, Guiana, Venezuela and West Indies.

Squ 15 G 1/1 V 184 A 1 Sc 91/91 & 1

Family : VIPERIDÆ.

a. Crotalinæ.

*63. *Bothrops atrox,* Linn.

Jan, l.c., part 47, pl. 2, fig. 2, 3.

Trigonocephalus jacaraca of de Verteuil.

First found in Trinidad by Mr. A. B. Carr from whom we have specimens. New to the Island. Also found in Mexico, tropical Central America, and N. South America to Ecuador and North Brazil.

*64. *Lachesis muta,* Linn.

Schlegel, Essai. Phys. Serp., pl, 20, fig. 19-20 (Crotalus); Duméril & Bibron, Erp., gén. vii., page 1485.

Crotalus mutus of de Verteuil.

New to the Island.

Squ 37 G 98 V 225 A 1 Sc 44/44.

BATRACHIA.

BATRACHIA ANURA.

Family : RANIDÆ.

1. *Prostherapis trinitatis,* Garman.

Phyllobates trinitatis, Garm., Bull. Essex Institute, vol. 19, 1887, page 13.

Prostherapis herminæ, Boettger, Ber. Senck. Nat. Ges., 1893, page 37.

First discovery in Trinidad by Garman. Also found in Venezuela.

Family: ENGYSTOMATIDÆ.

*2. *Engystoma ovale*, Schneid.

Boulenger, Cat. Batr. Sal., Brit. Mus., 1882, page 163.

New to the Island. Recorded in United States of Columbia and Venezuela.

Family: CYSTIGNATHIDÆ.

*3. *Hylodes urichi*, Bttger, n. sp.

First discovery F. W. Urich.

Char. Statura et colore aff. *H. cerasino*, Cope, sed minor, lingua latiore, tympano multo majore, discis membranorum non truncatis.— Habit of *Prostherapis trinitatis* (Garm), but more slender. Head a little broader than the body. Tongue a large oval, entire, extensively free behind. Vomerine teeth in two small oblique groups behind the choanæ. Snout conical, a little longer than the greatest orbital diameter, canthus rostralis distinct ; nostrils close to the tip of the snout ; tympanum distinct, 2/5 the size of the eye. Fingers moderate, first shorter than second ; toes moderate quite free, disks small, much smaller than the tympanum, not truncate ; subarticular tubercles well developed, two metatarsal tubercles. The hind limb being carried forward along the body, the tibio-tarsal articulation reaches the tip of the snout, or beyond. Skin smooth above ; a small tubercle on the upper eye-lid and a curved fold above the tympanum ; sides indistinctly, belly distinctly, granular.

Blackish or grey above, uniformed or marbled and with darker markings on the head and sides, occasionally a triangular lighter blotch across the interorbital space, lower surfaces whitish, clouded and vermiculated with blackish ; limbs indistinctly cross barred ; groin, upper anterior and lower posterior face of femur and inner face of tibia *carmine-red.*

Length of head and body	...	...	20½	20	m.m.
„ „ „	...	...	8	8	"
Breadth of head	...	...	8½	8½	"
Diameter of orbita	...	...	2½	2½	"
Tympanum	...	...	1	1	"
Anter. members	...	...	13	13½	"
Poster "	...	...	35½	37	"
Femur	...	...:	10½	11.	"
Tibia ...	...	...	12½	13	"
Disk of 4th toe ...	...	...	5/8	5/8	"

Habitat : Trinidad, on the banks of streams, under stones, and in the woods, under the leaves in damp places.

This small species is distinguished by the bright, but somewhat changing colouration and marking with grayish white, blackish gray and carmine red. Among the *Hylodes* species with a granular belly and with vomerine teeth, situated in small groups behind the chaonæ, it is very much like *H. cerasinus*, Cope, from Costa Rica, but nearly half the size, the tongue broad oval, nearly round, the tympanum between 1-3rd and ½ the size of the eye, the fingers much shorter, and the disks broad oval, but not truncate in front. *H. whymperi*, from Ecuador also shows a similar blood red colour on the groin and on the hind limbs, but is otherwise entirely different.

4. *Leptodactylus longirostris*, Blgr.

Boulenger, l.c., page 240, pl. 16, fig. 3.

*5. *Leptodactylus pentadactylus*, Laur.

Boulenger, l.c., page 241.

New to the Island. Recorded also in Ecuador, tropical Brazil, Guiana, Venezuela, Dominica and St. Lucia.

6. *Leptodactylus typhonius*, Daud.

Boulenger l.c., page 246.

First recorded in Trinidad by S. Garman who erroneously ascribed it to *L. longirostris*, Blgr. Also found in Venezuela, Guiana, North Brazil.

7. *Leptodactylus caliginosus*, Girard.

Boulenger, l.c., page 247.

First recorded in Trinidad by J. H. Hart. Report of Botanical Gardens, 1890.

Family : BUFONIDÆ.

8. *Eupemphix trinitatis*, Blgr.

Ann. Mag. Nat. His., 1889, page 307.

Engystomops pustulosus, Bttgr., non Cope.
Kat. Batr. Sammlung, page 33.

Eupemphix pustulosus, Bttgr., non Cope.
Ber. Senck. Nat. Ges. 1893, page 40.

First discovery J. H. Hart, in the Botanical Gardens, Trinidad.

9. *Bufo marinus*, Linn.

Boulenger, Cat. Batr. Sal., Brit. Mus., 1882, page 316.

First recorded in Trinidad by Boulenger from specimens received from Sir A. Smith. Also recorded by Garman.

Family : HYLIDÆ.

10. *Hyla crepitans*, Wied.

Boulenger, l.c., page 352.

Common in Trinidad and Venezuela.

*11. *Hyla coriacea*, Pts.

Boulenger, l.c., page 367.

> H. pardalis, Garman, non Spix (Garman, Bull.
> Essex Inst., vol. xix., 1887, page 16, from
> Trinidad.)

Recorded also from Surinam.

*12. *Phyllomedusa burmeisteri*, Blgr.

Boulenger, l.c., page 428.

New to the Island. This frog does not lay its eggs in water but constructs a nest by glueing together, with a natural secretion, leaves of growing plants over-hanging water.

2nd Feb., 1894.

Communications and Exchanges intended for the Club should be addressed to the Honorary Secretary, Port-of-Spain, Trinidad, B.W.I.

Price 6d.   Annual Subscription, 3/.

Vol. 2.   OCTOBER, 1894.   No. 4.

J'engage donc tous à éviter dans leurs écrits toute personnalité, toute allusion dépassant les limites de la discussion la plus sincère et la plus courtoise.—LABOULBÈNE

Trinidad·Field·Naturalists'·Club.

NATURA MAXIME MIRANDA IN MINIMIS.

Publication Committee:

President.

H. CARACCIOLO, F.E.S., *V.P.*:   P. CARMODY, F.I.C., F.C.S.,

SYL. DEVENISH, M.A.,   R. R. MOLE,

F. W. URICH, F.E.S., *Hon. Secretary.*

CONTENTS :—

Report of August Meeting	91
Secretary's Report	91
List of Members...	97
Financial Statement	99
Publication Committee's Report	100
Report of September Meeting...	102
New List of Trinidad Butterflies and Moths ...	102
Tucuche (Mount) Ascent of	104
Night Jar—Note on the Breeding of...	109
Cacao Beetle—Note on	110
Trinidad P...	

☞ All Communications and Exchanges intended for the Club should be addressed to the Honorary Secretary, Port-of-Spain, Trinidad, B.W.I.

JOURNAL

OF THE

Field Naturalists' Club.

VOL. II.   OCTOBER, 1894.   No. 4.

REPORT OF CLUB MEETINGS.

3RD AUGUST, 1894.

THE Annual Business Meeting of the Club took place at the Victoria Institute on the 3rd August. Present Mr. H. Caracciolo, F. E. S., President ; the Hon. Lieut. Colonel Wilson, C.M.G., Dr. Rake, Messrs. Syl. Devenish, M.A., J. G. Taylor, J. R. Llaños, R. J. L. Guppy, C.M.Z.S., W. S. Tucker, J. T. Rousseau, B.A., Henry Tate, B.A., Alfred Taitt, John Thomson, C. W. Scott, J. Hoadley, L. Guppy, jr., T. W. Carr, T. I. Potter, F. V. H. Spooner, Arthur Gaywood, R. R. Mole and F. W. Urich, F.E.S., Hon. Secretary & Treasurer. The minutes of the last monthly meeting having been read and confirmed the Secretary presented his Annual Report upon the work of the Club during the past year. It read as follows :—

SECRETARY'S REPORT.

ACCORDING to Rule 6 I beg to submit to the Members of the Club the Third Annual Report of the Club's transactions for the year 1893-4.

In the course of the past year I was away in the country for a month, during which time Mr. R. R. Mole kindly performed the duties of Secretary.

The number of members on the list at the close of the year 1892-3 was 88. There have been elected during the one just closed 5 Honorary, 2 Corresponding, 34 Town and 4 Country members, making a total of 124 members. The losses through death, resignation and other causes have been nine.

Amongst the working members the Club has lost, specia mention must be made of two : Mr. G. W. Hewlett, who was one of the first members and who recently went to South Africa and Mr. W. E. Broadway, who has been appointed Curator of the Botanic Gardens, Grenada. The latter is one of the founders of the Club and since its formation has never missed a meeting of any kind. He devoted considerable time to its affairs and his interesting exhibits at the monthly meetings will be much missed in the future. I am sure all the members of the Club will join me in wishing Mr. Broadway every success in his new sphere and thanking him for the services he has rendered this Society. A list of members is appended.

Although the balance in hand is only $26.60 as compared with $103.54 of last year I may state that the Club is on a sound financial basis. There are advertisements to be collected which will bring up the receipts from the journal to very near $240.00 besides the next half-year's members' subscriptions amounting to about $175.00 both of which may be looked upon as valuable assets. Our collections are absorbing much of our funds and will do so for the next year or two, but when once these are completed, we shall have funds to improve the *Journal* and increase the Library which, at present, depends only on donations and exchanges. The usual statement of finances is appended.

During the year there have been twelve Ordinary Meetings and one Ladies' Evening. The Second Annual Meeting was held, at our usual place of meeting, the Victoria Institute on the 18th of August 1893. The Ladies' Evening was held on the 6th December 1893 and, thanks to the patronage of Lady Broome, who takes a lively interest in the Club and its doings, was a success in every way. The Ladies' Evening was more of a social nature and on that occasion the accommodation at the Victoria Institute proved inadequate for the company who did us the honour of being present. The increased attendance at the Ordinary Monthly Meetings mentioned in my last Report has been fully maintained during the past year, and the papers and communications read have quite equalled if not surpassed those of previous years. The exhibition of specimens of interest has been fully kept up and the Club's thanks are due to the gentlemen who thus helped to make the meetings attractive.

The *Journal* is published as usual ; for details I refer members to the Report of the Publication Committee. Most of the numbers of Volume I are now out of print.

Through the generosity of Sir John Goldney the nucleus of a Library has been formed and on behalf of the Club I tender sincere thanks to Sir John for his handsome present. The following is a list of the additions to the Library during the year 1893-4.

PRESENTED BY SIR JOHN TANKERVILLE GOLDNEY.

Kirby,—Text Book of Entomology.
Bates,—Naturalist on the Amazon.
Wood,—Homes without Hands.
Hudson,—Naturalist in La Plata.
Cassel's Natural History—6 Volumes.
Beale—On the Microscope.

PRESENTED BY DR. B. N. RAKE.

Transactions of the Bombay Natural History Society—three numbers

PRESENTED BY THE AUTHORS.

Thirty-two Papers on Ornithology by W. Brewster.
On a collection of Mammals from the Island of Trinidad, with descriptions of New species by J. A. Allen and Frank M. Chapman.
On the Birds of the Island of Trinidad by Frank M. Chapman.
On two new Neotropical Mammals by Oldfield Thomas.
Miscellaneous pamphlets on Economic Entomology by F. M. Webster.
Trinidad Botanic Gardens—Bulletin of Misc. Information No. 18-22.
The Land & Freshwater Mollusca of T'dad by R. J. Lechmere Guppy.
Report of the Supt. of the Botanic Gardens, Trinidad, for 1893.

IN EXCHANGE.

Bulletin of the Botanical Department, Jamaica by W. Fawcett.
Notes from the Museum of the Institute of Jamaica.
Journal of the Institute of Jamaica.
Bulletin of the American Museum, (New York) Vol. V. 1893.
Canadian Entomologist.
Annual Report of the Smithsonian Institution for 1891 (Washington).
American Museum (New York), Annual Report for 1893.
Belfast Naturalists' Field Club Annual Report & Procs. 1892-3.
Proceedings of the Rochester Academy of Science Vol II Parts 1 and 2.
Manchester Microscopical Society Annual Report 1892.
Trans. of the Hertfordshire Nat. His. Society Vol VII Parts 8 and 9.
The United States of Venezuela in 1893 by Toledo & Ernst.
Field Club.
Bull. des Séances et Bull. bibliographique de la Société Ent. de France.

PRESENTED BY THE U. S. DEPT. OF AGRI., DIV. OF ENT.

Insect Life Vol V No 5 Vol VI Nos. 1 2 3.

PURCHASED.

Science Gossip.
Flower—Osteology of the Mammalia.
Parker and Bettany—Morphology of the skull.

The Club collection has progressed very favourably during the year, although it is far from complete. All the funds and energy are now centred in the Mammal Collection. The cost of this amounts to $205.61 which is rather high as compared with $19.69 devoted to other departments, but I think the Club is to be congratulated on having formed a collection, which now constitutes the chief (with the exception of the Leotaud Collection

of Birds and the Keate Collection of Shells) and only attraction
to the Victoria Institute and the one to which the Institute
owes its numerous visitors. There is yet much work to be done
—the Reptile and Insect Collections especially are not complete
but " Rome was not built in one day," and the Members of the
Club who contribute are all amateurs, who are always busily
engaged in professional or mercantile duties, with few leisure
moments for Natural History. Amongst the contributors to the
collections special mention is to be made of Mr. H. Caracciolo,
(President) Mr. A. B. Carr, and Mr. W. E. Broadway. The
following is a list of the specimens in the collection.

MAMMALIA.

2   Red Howlers, *Mycetes seniculus*, L.
1   Ocelot, *Felis pardalis*, L.
1   Wood-dog, *Galictis barbara*, L.
2   Mangrove dogs, *Procyon cancrivorus*, Cuv.
2   Squirrels, *Sciurus aestuaus hoffmani*, Peters.
1   Spiny rat, *Loncheres guianæ*, Thos.
2   Pilorees, *Echimys trinitatis*, All. Chpm.
2   Agoutis, *Dasyprocta aguti*, L.
1   Albino Aguti.
1   Peccary, *Dicotyles tajacu*, L.
2   Tamanduas, *Tamandua tetradactyla*, Linn.
1   Little Ant-eater, *Cyclothurus didactylus*.
1   Armadillo, *Tatusia novemcincta*, L.
1   Manicou, *Didelphis marsupialis* L.
1   Manicou gros yeux, *Philander trinitatis*, Thos

AVES.

1   Hawk.
1   Soldado bird (from Venezuela).

REPTILIA.

2   *Caiman sclerops*, Schneid.
     Polychrus marmoratus, Linn.
     Ameiva surinamensis, Laur.
     Amphisbœna fuliginosa Linn.
     Amphisbœna alba, Linn.
     Mabuia agilis, Raddi.
     Epicrates cenchris, L.
     Corallus cookii, var *ruschenbergi*, Gray.
     Eunectes murinus, L. South America.
     Geophis lineatus, D. and B.
     Liophis cobella, Linn.
     Liophis reginæ, L.
     Liophis melanotus, Shaw.
     Coluber corais, Cuv.
     Coluber variabilis. Wied.
     Coluber boddærti, Seetz.
     Herpetodryas carinatus, L.
     Leptophis liocercus, Wied.
     Homalocranium melanocephalum, L.
     Dipsas cenchoa, L.
     Oxybelis acuminatus, Wied.

Scytale coronatum, Schneid.
Elaps lemniscatus, Linn.
Elaps riisei, Jan.
Leptognathus nebulatus, L.
Crotalus horridus L. (South America).
Bothrops lanceolatus, L. (St Lucia).

PISCES.

Pristis antiquorum Lath, young specimen, stuffed.
Carcharias terræ-novæ ? Richards stuffed.

INSECTA.

12 Cases of different orders.

During the past year no official Club rambles have taken place, although one futile effort was made to inaugurate one at Easter. What the reason is, I cannot say but no doubt, the difficulties in travelling and the limited time of many of the members form two of the causes of our laches in this respect. Nevertheless individual members are constantly making excursions which always result in a rich harvest of specimens, all of which are duly recorded in the pages of the *Journal*. The following is a list of some of the excursions undertaken by members and the number of days spent in each.

. Mayaro—East coast 14 days. High woods of Caparo—3 weeks and 1 week. Caura—Northern range of hills, 3 days. Maraval, Blue Basin, Carenage, Santa Cruz, Maracas, St. Anns,—1 day excursions being valleys and places near town.

The following is a list of donations and the names of their donors for the year 1893-4. The thanks of the Club are due to those who have thus contributed to the collection.

MAMMALIA.

1 Wood-dog skin *Galictis barbara*, Linn, presented by Mr. A. B. Carr.
1 Otter skin *Lutra insularis*, F. Cuv. presented by Mr. C. J. Thavenot.
1 Albino agouti (stuffed) *Dasyprocta aguti*, L. presented by Mr. P. Gonzales.
A Cavy from Venezuela (stuffed) presented by Mr. A. M. Sucre.
A Peccary *Dicotyles tajacu*, Linn. presented by Mr. J. Shine Wilson.
A Peccary skin presented by Mr. A. B. Carr.
A Little Ant-eater *Cyclothurus didactylus*, L. presented by Mr. M. Harding-Finlayson.
A Little Ant-eater with a young one presented by Mr. J. Russell Murray.
A Manicou gros yeux *Philander trinitatis* presented by Mr. J Guilbert.

AVES.

Nest of a Corn bird *Ostinops decumanus* presented by Mr. J. Guilbert.
A horned Owl *Megascops brasilianus* presented by Mr. J. Guilbert.
A small Owl *Glaucidium phalænoides* presented by Master J. O. Urich.
An American barn owl *Strix pratinicola* presented by Mr. E. W. Lack
An American barn owl presented by Rev. Father Allgeyer.

REPTILIA.

2   Alligators *Caiman sclerops* presented by Mr. N. A. St. Hilaire.
    A two-headed snake *Amphisbæna fuliginosa* presented by Mr. G.
    Fitzwilliam, Princes Town.
3   Young *Epicrates cenchris* presented by Mr. J. H. Hart.
    A *Corallus cooku* var *ruschenbergi* presented by Mr. E. Lange, junior.
    A Mangrove snake *Liophis cobella* presented by Mr. A. Devenish.
    A Lora *Leptophis liocercus* presented by Rev. Father Clunes.
2   Coral Snakes from Venezuela *Elaps corallinus* presented by Mr.
    H. J. Baldamus
    A clouded snake *Leptognathus nebulatus* presented by Mr. E.
    Lange, junior.

PISCES.

A young saw fish *Pristis antiquorum* presented by Mr. H. Caracciolo
A sweet water Eel, presented by Mr. J. Graham Taylor.

INSECTA.

A fine collection of butterflies numbering 17 species, named, pre-
sented by Mr. W. E. Broadway.
A collection of beetles presented by Mr. A. Lamy.
Caterpillars were presented by Revd. C. H. Stoker, Tunapuna.
Revd. E. F. Tree, Couva   Lance Corporal Wilkinson, Couva.
Moths were presented by Mr. M. J. Leotaud, Mr. E. Lazare, Revd.
Father Clunes, Princes Town.
Beetles were presented by Mr. Edward Chase, Tacarigua.   Revd.
E F. Tree, Couva.   Mr. Luizon.   Mr. L. Libert, Couva.   Mr.
J. Guilbert, Couva.   Mr. C. A. Meltz.   Revd C. H. Stoker.
A Wasp's nest from Venezuela presented by Mr. H. Caracciolo
Male & female Parasol Ants *Atta cephalotes* presented by Mr. J. J.
Hobson.
A giant Locust presented by Master Dick Kernahan.

MISCELLANEOUS.

2   Giant centipedes *Scolopendra* sp. presented by Mr. J. de la Sauva-
    gere.
    A Mygale spider presented by the Hon. G T. Fenwick.
    Stem of a *Gorgonia* presented by Mr. W. Scott Kernahan.
    Eggs of *Bulimus oblongus*, Garden snail, presented by Mr. W.
    J. L. Kernahan.
    Head ornaments made of the elytra of Buprestide beetles from
    Venezuela presented by Mr. E. St. Vraz

In conclusion I may congratulate the Members upon the
continued success and the marked progress of the Trinidad Field
Naturalists' Club.

F. W. URICH,
Hon. Sec. and Treasurer.

Trinidad,
3rd August 1894.

TRINIDAD FIELD NATURALISTS' CLUB.

Officers and Committees for the year 1894-95.

Patron :

His Excellency Sir F. Napier Broome, K.C.M.G

President :

Dr. B. N. Rake.

Vice-Presidents:

H. Caracciolo ; Sir John T. Goldney, Kt.

Hon. Secretary & Treasurer :

F. W. Urich.

FINANCE COMMITTEE.

Dr. Rake, P.	P. L. Guppy, jr.
H. Caracciolo, V.P.	F. W. Urich.
T. I. Potter,	

BUSINESS COMMITTEE.

Dr. Rake, P.	Hon. Lieut Col. Wilson, C.M.G.
H. Caracciolo, V.P.	W. S. Tucker,
Syl. Devenish,	T. I. Potter,
R. R. Mole,	F. W. Urich.

PUBLICATION COMMITTEE.

Dr. Rake, P.	Prof. Carmody,
H. Caracciolo, V.P.	R. R. Mole,
Syl. Devenish,	F. W. Urich.

MEMBERS OF THE TRINIDAD FIELD NATURALISTS' CLUB.

Honorary Members.

	Date of Election		Date of Election
Boettger, Professor Dr. O.	6 10 93	Riley, Prof. C.V., Ph. D.	2 3 94
Brewster, William	4 5 94	Smith, G. Whitfield	13 11 91
Butler, A G , F.L S., F.E.S.	1 7 92	St. Vraz. E.	8 8 91
Crowfoot, W.M., M.B., F.R C.S.	3 3 93	Stone, Sir J B., F.L.S., F R.G.S etc.	7 7 93
Gatty, S.H.	19 8 92	Thomas, Oldfield, F Z S.	6 3 92
Gunther, Dr.A., F.R.S. etc	1 7 92	Vogt, Professor C ,M.D.	3 3 93
Hamilton, Hon C.B.	5 2 92	Warming, E , Professor	4 12 91
Howard, L O.	6 7 94	Wilson, H.F.	1 7 92
Kirby, W.F., F.L S., F E.S.	2 9 92	Wells, S.	2 9 92
Mitchell, P. Chalmers	2 12 92	Whitehead, C.E,	7 10 92
Morris, D.,M.A., C.M.G.	6 10 93		

Corresponding Members.

Bock, E.	2 6 93	Jeffreys, W.J.	3 3 93
Chapman, F.M.	5 5 93	Lassen	8 8 92
Cockerell, T.D.A., F.E.S., F.Z.S.	5 5 93	Leomsen	8 8 92
		Rodriguez, Dr. en Med	2 6 93
Deyrolle, E.	7 10 92	Terry, John, F.R.G S. etc.	5 1 94
Elliot, H.V , Lieut R.N.	3 3 93	Wright, Rev. E. Douglas	6 4 94
Ganteaume, Harris	8 8 91		

Town Members.

	Date of Election		Date of Election
Broome, H.E, Sir F. N., K.C.M G.	6 5 92	Horsford, Hon. D. B.	2 2 93
Agostini, Edgar	4 11 92	Koch, W. V. M., M.B., C.M.	2 6 93
Agostini, J. L.	2 3 94	Knox, C. F., M.R.C.S.	5 1 94
Alcazar, Hon. H. A.	2 2 94	Lota, A., M.D.	7 6 92
Anduze, Jules	2 3 94	Llaños, J. R.	2 3 94
Archer, Julian H.	7 7 93	Libert, C.	7 10 92
Arnott, T. D.	7 7 93	Lovell, Hon. Dr. F.H., CMG	2 2 94
Bain, F. M.	7 7 93	Maingot, F.J., F.G.S.	7 6 92
Baldamus, H. J.	5 1 94	Miller, J.	6 4 94
Borberg, A.	7 7 93	Mole, R. R.	10 7 91
Broadway, W. E.	10 7 91	Murray, J. R., F.B.S.S.	2 10 91
Broome, L. E.	7 4 93	O'Connor, P. A.	2 2 94
Caracciolo, H , F E.S.	10 7 91	Potter, T. I.	10 7 91
Carmody, P., F I.C., F.C.S.	8 1 92	Prada, E., M R C.S.	2 3 94
Carr, T. W.	3 11 93	Rake, B. N., M.D. (Lond)	19 8 92
Creagh-Creagh, G.	5 1 94	Reed, G. J.	7 6 92
Cumberland, S. A.	1 9 93	Rousseau, J. T., B.A.	2 3 94
Collens, J. H.	5 2 92	Saurmann, C.	2 3 94
Devenish, Syl , M.A.	8 1 92	Scott, C. W.	10 7 91
de Montbrun, J. L.	6 10 93	Seccombe, G S , L.R.C.P.	2 3 94
Dickson, J. R., M.B., C.M.	1 6 94	Smith, R. H. S.	5 1 94
Dumoret, R.	7 6 92	Sorzano, Til	4 11 92
Eagle, F.	8 1 92	Spooner, F. H. V.	6 10 93
Eagle, S.	2 9 92	Taitt, Alfred	10 7 91
Ewen, E. D.	2 9 92	Tate, Henry, B.A., F.I.C.	2 9 92
Garcia, Hon. G. L.	2 3 94	Taylor, J. G.	2 2 94
Gaywood, A.	6 4 94	Thavenot, C. J.	6 5 92
Gerold, E.	4 3 92	Thomson, J.	6 7 94
Goldney, Sir J. T.	6 1 93	Tripp, E.	6 10 93
Gordon, Hon. W. G.	7 7 93	Tucker, W. S.	5 1 94
Guppy, P. L.	10 7 91	Urich, F. W., F.E.S.	10 7 91
Guppy, R. J. L., C.M.Z.S.	6 1 93	Walker, C. W.	8 8 91
Harding-Finlayson, M.	5 1 94	Warner, A., B.A.	5 1 94
Hernandez, F.	4 8 93	Wehekind, V. L.	2 3 94
Hoadley, John	1 9 93	Wilson, Hon. Lieut.-Col. D., C.M.G.	7 10 92
Holt, Rev. E. J.	6 4 94	Wilson, Edwin C.	2 3 94

Country Members.

Carr, A. B.	19 8 92	Knaggs, R. H. E., M.R.C.S.	5 5 93
de Verteuil, L.	7 6 92	Laing, Alex. S.	5 1 94
de Verteuil, F.A., M.R.C.S.	6 10 93	Mahoney, F. J.	7 10 92
Eakin, J. W., M.B.	5 1 94	Meaden, C. W.	7 6 92
Goode, J.	6 1 93	Mitchell, J. W. L.	2 12 92
Guilbert, J.	19 8 92	Smyth, Hon. J. Bell	4 11 92
Greenidge, J. S.	2 9 92	Wight, A. W., M.R.C.S.	2 6 93
Guppy, Hon. R.	2 9 92	Wilson, J. S.	7 10 92
Hobson, J. J.	4 12 91	Woodlock, A., L.R.C.S.	7 10 92
Kernahan, G. J.	6 10 93		

TRINIDAD FIELD NATURALISTS' CLUB.

TREASURER'S STATEMENT FOR THE YEAR 1893-4, ENDING 31ST JULY, 1894.

Dr.	$	c.
To Balance from last year	103	54
" Members subscriptions and entrance fees ...	340	23
" Receipts from *Journal*	195	68
" Nett proceeds of Promenade Concert .	51	21
	$690	**66**

Cr.	$	c.
By expenses for Mammal collection, mounting specimens and cases	205	61
" Maintenance of Reptile & Insect collections	19	69
" Printing *Journal*	235	76
" Printing and Stationery	57	67
" Postage	21	89
" Expenses incurred at meetings; (lighting including expenses of last Annual Meeting)	51	47
" Subscription of 5 Members to Victoria Institute to 24th May, 1895	25	00
" Miscellaneous (Porterage, Attendants, collecting subscriptions, etc.,)	46	97
" Balance in hand	26	60
	$ 690	**66**

Examined and found correct,
On behalf of the Finance Committee.

H. CARACCIOLO,
President.

F. W. URICH,
Hon. Sec. & Treasurer.

TRINIDAD,
1st August, 1894.

Mr. Devenish proposed and Colonel Wilson seconded the adoption of the Report.—Carried. The Secretary presented the Publication Committee's report which is appended.

PUBLICATION COMMITTEE'S REPORT.

IN accordance with Rule 29 the Publication Committee have the honour to present their Report for the year ending 31st July, 1894.—

At the last Annual meeting Dr. Rake was appointed a member of the Committee in the place of Mr. Urich, who became ex-officio a member when he was elected Hon. Secretary of the Club. Otherwise there has been no change in the constitution of the Commitee.

The Committee have during the year issued six numbers of the *Journal*, containing 178 pages of reports of proceedings and papers—an increase of 51 pages upon the number of last year. The reports are still up to date, but the Committee regret to have to announce a still greater arrears of Club papers, of which no less than 11 are awaiting publication. During the past twelve months fifteen original papers have been printed in addition to the report of the Annual Meeting which occupied no less than 22 pages, and several articles published elsewhere have been also accorded a place in the *Journal.* The Committee do not at present see their way to bring the publication of all the papers now on their hands up to date for some time to come. Probably when the Mammal Collection has been completed the Club may be able to make an effort to clear up the arrears which are now slowly but surely accumulating.

The Committee congratulates the Club upon the completion of the first Volume of the *Journal* last February. The new series, the Committee believes, to be an improvement upon the first one and it trusts that the second volume, which will probably be concluded in February 1896, will redound with as much credit to the Club as did Volume I.

The Committee have again to thank the Editors of the Press for the liberality with which they threw open their columns to the reports of Club Meetings.

The Committee appends for the information of the Club a list of the papers awaiting publication, their authors and the dates on which they were read or laid on the table.

(Signed on behalf of the Committee).

H. CARACCIOLO,
President.

LIST OF PAPERS AWAITING PUBLICATION.

An Ascent of Tucuché, T. I. Potter, 6th October, '93·
Our persecuted friends and cherished enemies, F. W, Urich, 6th Dec., '93
A visit to the High Woods of Caparo, R. R. Mole, 2nd February, '94
A Preliminary list of Trinidad Insects, Dr. Rake and Messrs. Caracciolo
 and Broadway, 2nd March, '94·
Notes on some Trinidad bats, H. Caracciolo, 4th May, 1894.
List of some Trinidad Lepidoptera, not previously recorded, H. F. Wilson.
Notes on some Trinidad Orchids, T. I. Potter,
Notes on the Cocoa beetle, A. B. Carr, 1st June, 1894.
Notes on the Poor me One.   ,,   ,,
Note on the breeding of the small Night-Jar, B. N. Rake, 1st June 1894.
Notes on some Trinidad Butterflies, P. L. Guppy, 6th July, 1894

The President proposed and Colonel Wilson seconded its adoption.—Carried.

The Club then proceeded to the election of its officials for the year 1894-5 for result of which see details preceding the Secretary's List of Members.

Mr. Alfred Taitt proposed a hearty vote of thanks to Mr. Urich for the admirable way in which he had discharged the duties of his office during the past year and the capital report he had presented that evening.—Mr. Potter seconded.—Carried.

The Secretary proposed the transference of Mr. W. E. Broadway's name from the list of Town Members to the Honorary Roll. He said that Mr. Broadway was one of the founders of the Club and had always been one of its most energetic members. He had rendered a great deal of valuable service by the readiness and accuracy with which he had identified botanical specimens for other members and he had also been most successful in entomology, particularly butterflies. His presence was greatly missed that evening—it being the first meeting of the club which had taken place at which Mr. Broadway was not present. —Mr. R. R. Mole seconded.—Carried with applause.—On the proposition of Mr. C. W. Scott, seconded by Mr. J. Graham Taylor, Mr. James Wilson (ter) was elected a Town member.

Colonel Wilson on the request of Mr. Devenish read an address by the Business Committee (on behalf of the Club) to Mr. Caracciolo the retiring President.—Mr. Caracciolo gracefully replied in suitable terms and invited the new President to take the chair.

Dr. Rake thanked the Club for the honour conferred upon him and said that while he could not hope to emulate the sterling work Mr. Caracciolo had done he would do his best to promote the welfare of the Club.—Mr. Caracciolo then vacated the chair and Dr. Rake assumed the presidency. The Committees were then elected, for details of which see the lists preceeding the List of Members. The Club adjourned at 9.30.

6TH SEPTEMBER 1894.

PRESENT : Mr. H. Caracciolo, Vice President, (in the chair) Messrs Syl Devenish M.A., J. T. Rousseau B.A., Charles Libert, G. J. Reed, F. Hernandez, C. J. Thavenot, J. R. Llaños, T. W. Carr, W. S. Tucker, John Hoadley, S. A. Cumberland, R. R. Mole, H. J. Baldamus, John Thomson, T. I. Potter and F. W. Urich, Hon. Secretary.

The minutes were read and confirmed.

Mr. Mole moved (1) that the Hon. Secretary be directed to write letters of condolence to Mr. Gatty (who recently lost his wife by fever in Singapore); to the parents of the late Mr. Arthur Gaywood and Mr. Henry Tate, B.A., F.I.C., Acting Government Analyst, to the widow of the late President Dr. Rake, expressing the profound sympathy which the members feel with them in their sudden bereavements. (2) That as a mark of respect to the late Dr. Rake, the Presidency remain unfilled until next Annual Meeting. (3) That as a further mark of respect to the memory of their departed friends the meeting forthwith adjourn until October.—The Vice-President seconded and the meeting adjourned.

CLUB PAPERS.

A LIST OF TRINIDAD BUTTERFLIES AND MOTHS.

By H. F. WILSON.

THIS List is compiled from specimens collected during a stay in Trinidad from April to June 1892. Many of the butterflies were taken during a single walk with Sir Frederick Pollock from St. Juan to the Maraval Valley over " The Saddle." The moths were mostly caught at light in Woodford House, while others were taken at St. Ann's by Lady Broome, who was good enough to interest herself in my entomological pursuits.

RHOPALOCERA—OR BUTTERFLIES.

Twenty-three or twenty-four, of the species collected by myself (including one given me by Dr. Rake, and two by Mr. Gatty) have already been recorded from Trinidad by Dr. W. M.

Crowfoot (Journal No. 7 page 173) but the following are not in that list :

1. *Catopsilia sennæ*, Linn.
2. *Danaus leucogyne*, Butl.
3. *Caligo oberon*, Butl.
4. *Euptychia hermes*, Fabr.

To which may be added the following, sent to me by Lady Broome in the autumn of 1893.

5. *Taygetis inornata*, Feld.
6. *Euptychia arnæa*, Fabr.
7. *Mechanitis plagigera*, Butl.

HETEROCERA—OR MOTHS.

SPHINGIDÆ.

8. *Chærocampa (Theretra) tersa*, Linn.
9. *Chærocampa tyndarus*, Boisd.

Green-hawk moth new to the Island, taken at light in Woodford House.

10. *Pseudosphinx tetrio* Linn.

Bred from Caterpillars feeding on frangipani.

ZYGŒNIDÆ.

11. *Macrocneme*, (sp.)

Taken by Lady Broome.

12. *Saurita cassandra*, Linn.

Brown moth, body spotted with red and green,

SATURNIDÆ.

13. *Hyperchiria* sp.

Taken at light in Woodford House ; yellow fore-wings, reddish under-wings with large eye.

CASTNIDÆ.

14. *Castnia palatinus*, Cram.

Taken by Lady Broome.

ARCTIIDÆ.

15. *Ecpantheria* (or *Isia*) *eridanus, var.* Cram.

Taken in the drawing-room at Government House. White moth orange belts on abdomen.

16. *Urania leilus*, Linn.

Given me by Mr. Taitt.

17. *Sematura empedoclaria*, Hubn.
18. *Euchætes insulata*, Walk.
19. *Erebus odora*, Linn.
20. *Mapeta xanthomelas*, Walk.
21. *Melipotis ochracea*, Guér.
22. *Euthisanotia timais*, Cram.
23. *Protoparce rustica*, Fabr.
24. *Protoparce cingulata*, Fabr.
25. *Syrnia iphianassa*, Cram.
26. *Letis hercyna*, Drum.
27. *Thysania zenobia*, Cram.
28. *Bronchelia matronaria*, var. Guér.
29. *Histæa meldolæ*, Butl.
30. *Phaloe larzae*, Boisd.
31. *Pessina leontia*, Stoll.
32. *Thymele eurycles*, Latr.
33. *Hymenia recurvalis*, Hubn.
34. *Phædropsis*, sp.
35. *Byssodis*, sp.
36. *Macaria*, sp.

Taken by Lady Broome and sent in autumn 1893.

Besides the above, several other butterflies and moths taken by me in 1892 were selected by Mr. W. F. Kirby for the British Museum, but he is unfortunately unable now to give me their names.

22nd March, 1894.

AN ASCENT OF MOUNT TUCUCHE.

By T. I. POTTER.

MOUNT Tucuche, El Tucuche, or Las Cuevas, the highest mountain in the Colony, is as we know situated at the Northern end of the lovely Maracas Valley, and from the South Western slope of the main peak the well known Maracas Waterfall, the haunt of many pleasure parties and one of the attractions of visitors, springs into the Valley below. The mountain is 3,012 feet above the level of the sea, according to the determination of Messrs. Wall & Sawkins, and comprises three distinct peaks. Having read Mr. Crueger's most interesting account of his ascent of Tucuche many years ago, given in Dr. de Verteuil's useful History of Trinidad, and Messrs. Wall and Sawkins' and other writers' accounts of the beautiful scenery and delightfully refreshing atmosphere which exist at the summit of this mountain, Messrs. Urich, Broadway and myself made a solemn compact that we should on a certain day attempt the ascent of Tucuche. Accordingly, having made arrangements for spending the night

previous to the ascent in the Maracas Valley, and as near to the
foot of the mountain as possible, we fixed upon one of the last
days of February for our excursion It was about the middle of
the dry season, and the weather promised to keep fine and to
favour our attempt when we left Port-of-Spain, by the 1.30 p.m.
train for St. Joseph.

Our party consisted of three members of the Club (whose
names I have mentioned above) my brother, and two boys who
carried the small amount of impedimenta necessary.   Arriving at
St. Joseph we climbed the hill, on which the former capital of the
Island is situated, after the fashion of the old Spanish built towns,
and branched off on to the Maracas Road, the first part of which,
for a thoroughfare, is very steep.   From the highest point of this
road we got a fine view of Tucuche away to the north, the three
peaks standing out clearly against the northern sky.   We were still
in the old town, and southward from this point we could also
see the Caroni, where in "days of old" Sir Walter Raleigh
had sailed up and thus had surprised Don Antonio de
Berrio the founder of St. Joseph. Not far behind us was
the Barrack Square, where in 1837 the mutiny of the black
troops occurred, and where the ringleaders were executed.
But we did not think of these things, as we trudged
forward, for Tucuche was the goal, and soon we had passed
through St. Joseph and its suburb "Buena Vista," descended
the other side of the hill, and were marching along the Maracas
country road.   Messrs. Urich and Broadway with their boy
intended to pass the night with Mr. Speyer, but my brother
and myself with our boy, had been kindly accommodated with a
night's lodging by Mr. V. Gomez of La Florida, a few miles higher
up the Valley.   After crossing the Maracas River five or six times
we reached Mr. Speyer's residence about 3.30 p.m.   Here we made
a short rest, and having arranged that we should meet at 6.30 on the
following morning, we left the rear guard of our party at their
sleeping quarters and proceeded to ours.   At 4.45 p.m., and after
having again crossed the meandering Maracas River twice, we
arrived at La Florida's Manager's quarters, and my brother and
myself were very kindly attended to by Mr. Gomez.   From this
spot the three peaks of Tucuche could be easily distinguished.   The
eastern and western ones are conical, while the centre or main
peak, apparently the highest, rises gradually to its summit.
From the gallery of the house at "La Florida" I could plainly
see the top of the famous Maracas waterfall, distinctly visible at
that hour of the afternoon, by the bright reflection of western
sunlight from the volume of water as it leapt into the unseen
and unknown depths below.   The night was pleasantly cool, as
the dry season's nights generally are, and when we arose
in the morning the thermometer registered 69° F, at 6 o'clock.

Day broke gloriously without a cloud or blemish and everything seemed favourable to our ascent. We were soon ready, and within a few minutes of 6.30 we heard the signal shout of our friends. They had procured the assistance of three guides, so our party now consisted of nine persons After the usual saluta- tions and after seeing that nothing had been forgotten, we started from the La Florida gap at 6.40 a.m. for the top of Tucuche. Striding along the Las Cuevas Road at a steady pace, the three guides leading, we passed the Road to the Waterfall on our right and many cocoa plantations. Suddenly, after a thirty minutes walk, we turned into a track on the right, leading through Mr. Zepero's cocoa estate, and soon halted on the bank of a clear brook. Here each man filled his water bottle, and took his first drink of water, for on the further side the ascent began. Advancing quickly we climbed through cocoa and provision grounds until we found ourselves in what we were told was Mr. Eligon's cocoa estate. The track took us to the overseer's quarters, where we met two men from whom we obtained permission to proceed and did so without further tarrying. This was the last house or building on the road and soon we found the ascent getting steeper. We pushed our way with little difficulty through tufts of *Heliconias* and stretches of *rastrajo* in which the order *Mela- stomaceæ* was well represented. Once more we emerged into a cocoa piece, apparently a squatting occupancy from its isolated situation, and in the young cultivation on the upper part of this piece we made our first halt at 7.55 a.m. This was the last bit of cultivation met with in the ascent. The air was cool, and its temperature, taken by a small thermometer which I carried, was 70° F. We rested here for about ten minutes, and admired the lovely landscape to the South and West of us. Before starting we refreshed ourselves from a deliciously cool stream (one of the sources of the Maracas River) which trickled through this garden. As the rest of our road lay through the virgin forest, each one of the guides, and those who carried hunting knives, unsheathed his weapon and prepared to cut his way forward. Resuming our upward march, we crossed a patch of high weeds and plunged into a thicket of huge *Heliconias*, under the shade of which grew in great quantities a pretty white *Begonia*. It was here that we met with the steepest bit of the ascent, and soon each man was using hands as well as feet in clambering up the steep and yielding side of the ridge. Overcoming all difficulties, and taking advantage of everything in our favour, we soon gained the crest of the main ridge which leads to the summit of the mountain. Without waiting to take breath after the stiff climb, we pushed forward, finding ourselves in a cooler and purer atmosphere than that in which we had been hitherto. The vegetation also showed a change at this place. Patches of a silvery green moss *(Selaginella* sp.) covered parts of the ground, and mosses and lichens decorated many of the tall tree trunks around us. Trees of imposing grandeur surrounded us, and on all sides the tapering columns of their trunks seemed like the pillars of Nature's temple the green roof of which was very many feet above. Not a sound of animal life was heard however, which fact seemed strange as we had expected to find game of some kind. Our way now lay along the top of the ridge, and as there was little or no underwood we progressed rapidly. On our left a precipice several hundred feet in depth yawned, and ran parallel with the track for a considerable distance, while on the right the ridge rolled away at an incline of about 45° to the valley on the North East. We next halted at a spot on this ridge, not far from the summit of the mountain, and where it was convenient to leave some of our baggage, chiefly plants, etc., collected on the way up. Taking advantage of this halt, Messrs. Urich and Broadway and myself attempted to explore our immediate vicinity, and we succeeded in making some interesting captures, mine being a curious *Epidendrum* and a beautifully marbled or mottled green and brown *Tellandsia.* Having carefully stowed away our treasures until our return, we moved foward once more, and now entered an entirely new zone. The thick silvery green moss now carpeted the damp and peaty soil, while other species covered the stems and branches around, and hung about in festoons, like masses of black hair or seaweed. The tall trees gave place to smaller ones, and these again to a small bamboo and many beautiful *Manicarias* and *Geonomas,* members of the palm family. The soil was yielding and very rich being composed chiefly of peat and leafmould and the cold air was heavily charged with moisture, which rendered everything in it dripping and damp. Now and then we traversed small forests of *Geonomas* (Anari palm) and in the last of these I found growing to the stem of one of them a *Maxillaria* with pure white fragrant flowers, and on the moss which covered the ground some large and beautiful blossoms of an *Utricularia (U. montana.)* Damper and cooler grew the atmosphere as we proceeded, and on creeping low under an arch, formed by the trunk of a fallen tree and a mass of climbing plants, we emerged on to a small patch of open ground near the base of a rather weather beaten tree (for winds seem to be very boisterous on Tucuche,) in what appeared to be cloud-land. Here our guides stopped, and informed us that we were on the top of Tucuche, and that further we could not go. We seemed to be standing on an island in the midst of a mighty ocean, for whichever way we looked we saw nothing beyond our immediate surroundings but cloud. Tellandsias with curious flower stalks, creepers with scarlet shoots, mosses and lichens grew around us in wild profusion and covered the almost leafless stem and trunk of the stunted tree under which we stood. In spite of what we were told by the guides we made efforts to proceed if possible, but as every way led downwards from this spot, which seemed above the world, as it were, we concluded that we stood on the highest land in the colony and congratulated ourselves accordingly. We arrived at this spot at 9.45 a.m., having taken three hours to make the ascent from La Florida Gap. As it was near-ing ten o'clock, and as we had had our appetites sharpened by the keen mountain air and the exercise, breakfast was served as soon as we had rested a little ; this being soon over, we began looking out for specimens. While doing so, we saw two *Papilio argesilaus*, one or two beetles and flies, but the great mass of clouds in which we were, prevented, I daresay, our seeing many insects on this occasion. At about noon I took the temperature, when the sun seemed inclined to break through our cloudy envelope, and the mercury stood at 76° F. in the open air. While my brother assisted Mr. Broadway with his botanical specimens, of which he made a rich harvest, Mr. Urich and myself pro-ceeded to investigate the cause of certain mysterious sounds which are to be heard proceeding from the ground as soon as the traveller enters the cooler region of Tucuche. These sounds resemble that produced by the rapid suction of air into a narrow open cylinder or tube in which a partial vacuum is created, and is caused by the sudden descent of a worm or similar creature in the ground, from the top to the bottom of its hole. At each tread in the cooler region of Tucuche we had heard these sounds around us, and marking a spot whence this sound came we dug for about two feet into the wet, soft clayey soil, and unearthed two curious red leeches which apparently had made this noise. A most remarkable feature of our trip was the marked scarcity on Tucuche of animal life. We saw no representative of any of the mammals, and we did not even see a snake or a lizard. After procuring what we could of the few living things to be found on this elevated spot, we made pre-parations to descend, and just before starting we duly recorded our visit, sealing up the document in a bottle, which we stuck in a hole in the old tree. At this moment the clouds showed signs of breaking, and in the rifts we got glimpses, like dissolving views, of the glorious panorama which lay at our feet on the northern side and away to the south. The bays of Maracas and Las Cuevas with their intervening headlands made very pretty pic-tures and on a clear day the view alone from Tucuche would amply repay the toil of a climb to the summit. Now the clouds close over us again, and as we have to catch the evening train to

Port-of-Spain, we begin the descent at 12.30 p.m. Stopping for a moment at our temporary depôt to assemble our treasures, and pack things up properly for the descent, we take almost the same route back, and soon find ourselves in the lower region of cocoa estates and provision gardens, and at 2.45 we are once more on the Las Cuevas Road. At 3 p.m., my brother and myself reached our station at La Florida, and we all meet again at the St. Joseph's Station in time to catch the last train for Port-of-Spain, after having thoroughly enjoyed ourselves in this successful ascent of Tucuche, and being quite ready to repeat the trip on a future occasion.

Appended is a list of some of the plants found on Tucuche by Mr. W. E. Broadway and determined by Mr. J. H. Hart of the Botanic Gardens. They are as follows :—

*Margravia umbellata, L.
Esenbeckia castanocarpa, Gr.
Leguminosa, sp.
Melastomaceæ, sp.
†Panox attenuatum, Sw Didymopanax.
Rubiaceæ, several species
Compositæ
Neurolæana lobata, R. Br.
Apteria setacea, Nutt.
Æchmea, sp.
Solanum jamaicense. Sw?
Scleria latifolia, Sw.
Cyperaceæ, sp.
Ichnanthus pallens, Doell.
Begonia, sp.
Pleuroththallis, sp.
Arthrostylium excelsum, Gr.
Philodendron.
Acrostichum alienum Sw.

Selaginella sp.
Polygala paniculata L.
Alsophylla aspera R. Br.
Polypodium serrulatum Mett.
   Id   trichomanoides Sw.
Gleichenia pubescens H.B.K.
Hemetelia grandiflora. Spreng.
Adiantum tetraphyllum, Willd.
Hymenophyllum ciliatum, Sw.
Polypodium taxifolium, L.
Musci possibly Hypnum.
Hepaticæ.
Alsophila sagettæfolia, Hook
Polypodium decussatum, L.
Spiranthes elata, Rich.
Cranichis muscosa, Sw.
Maranta Touchat, Aublt.
Jungermannia, sp.
Sphagmocea, sp.

* This plant was collected by Crueger on the same mountain in 1847.
† Nearest approach to English Ivy in the Tropics.

6th October, 1893.

NOTE ON THE BREEDING OF THE NIGHT JAR.

Nyctidromus albicollis Gm.—*Caprimulgus albicollis* Léotaud.

By the late Beaven Rake, M.D. (Lond).

ONE afternoon early in May when Mr. Tate and I were out with our butterfly nets, we put up a night jar from some recently cleared land. I remarked to Mr. Tate that if the bird followed the custom of its European relative the eggs ought to be on the bare ground, and that their colour would probably resemble that of the surrounding earth and stone.

We failed to find the eggs that day, but about a fortnight later Mr. Tate happened to be walking over the same ground and found a single egg. We went together the next day, and put up the bird again. There was no difficulty in identifying it as *Caprimulgus albicollis*. There was still only one egg, laid in a slight depression in the ground, close to a stump and under the shade of a tuft of leaves.

The egg differs from that of *Caprimulgus europæus* in having far more red about it, both in the ground tint and in the mottling. Indeed most of the colouring is in shades of salmon colour and reddish brown. There are a few gray marks towards the larger end. The egg is smaller than that of the European night jar. By reason of its reddish colour it is easy to distinguish the egg as it lies on the ground. This is not the case with the European species.

Dr. Leotaud in his book on the birds of Trinidad describes the eggs of this species as white. It may be that he wrote his description from a pair of eggs in which no colouring matter had been deposited. This accident, it is well known, may happen to any egg from disease or defect in the mother bird.

I have heard of other eggs of *Caprimulgus albicollis* at Maraval, and they have been described as being of the colour of the egg now shown. The usual number of eggs is two as in other members of the Goatsucker family.

1st June, 1894.

NOTES ON THE CACAO BEETLE
(Steirostoma depressum)

By A. B. CARR.

THIS beetle, which is known to the Spanish peons as *Congoroche*, belongs to the Longicornes. In its younger stages it lives in trees, especially in cacao trees, and as of late years its numbers seem to have increased, I think a few notes on its life history and the means of combating it would be interesting to the Members of the Club, and useful to Cacao Planters.

THE LARVAL STAGE : The egg is generally laid about the month of February or March. The places selected for this purpose being either some little crevices in the bark of the tree or in its fork ; or in a wound made in trimming. The larva when first hatched is small (about ¼" long) growing rapidly until it reaches a length of 1½ to 2 inches, the latter length being but seldom seen, as at that stage it burrows a hole into the heart of the tree, when and where the most serious trouble begins. The heart is followed for about four inches to six inches when after some weeks the insect having undergone its metamorphosis issues forth as the perfect beetle.

THE PERFECT INSECT : The beetle is particularly fond of soft and thick-barked trees, such as the wild Chataigne (*Pachira aquatica*, Aubl., this is it's favorite resort) Forest Mahoe (*Stercalia cariboea*, R. Br.) Cannon ball tree (*Couroupito guianensis* Aubl.) and our cultivated cacao. They do not seem to destroy any of the forest trees named, probably from the fact of their being attacked when fairly well grown, but unfortunately for the planter, when the female beetle resorts to the cacao groves, she selects the young, soft and juicy trees (say those from two to six years old), with the result that they are killed if not attended to. While the grub carries on its work of distruction *under* the bark of the tree, the perfect insect eats round the bark, say at a depth of 1/16" depriving the plant of a good deal of its vitality. She first of all eats systematically along and around each branch, generally beginning at the fork and eating upwards to the tops of the branches. She spends most of her time on one tree, and if not caught, will move on to the next nearest, and so on *ad infinitum* if left to her own sweet pleasure. The dry season is particularly favourable to the beetle, as during that time it lays its eggs. It is curious how fond the beetle is of returning to any plant that, through some chance, escapes destruction after its first attack. It is known to select such trees in preference to any other. This I suppose is due to the sweet gum which flows from recent wounds and which, as is well known possesses some peculiar attraction to the beetle. The months in which this pest is at its worst are February to May, both inclusive. The flight of the beetle is very short about 20 yards and slow, so that it can be easily caught whenever seen. Following trimmings of the Bucaré Immortel, the ravages made by it are well marked, there being at such a period at least twice the usual number of trees affected. It is therefore necessary for the planter to have his trees frequently examined when the trimming of the Immortels is over.

TREATMENT : The beetle seems to have but few natural enemies : I have noticed only the Cacao-pickers, very wrongly called Mangeur de Cacao, the Tick-bird or Merle Corbeau and the big-billed Qu'est-ce-qu'il-dit. Ants seem to kill the young grubs. Heavy rains are an agent in destroying the larvæ before they get under the bark. In the dry season hand-picking is very useful. The presence of the grub can always be known by the frass (which in entomology, means the debris of the stems and bark ejected by the beetle) protruding from the stem.   The grub can then be cut out, and a dressing of coal tar put on the wound so made. In fact a dressing of coal tar should always be put on cacao and shade trees after they have been pruned, for recent pruning always attracts the beetle.   If the beetle is too deep in the stem of the tree to be cut out conveniently, a solution of cyanide of potassium and water (1 to 20 parts) is very effective when injected into the burrows.   Failing cyanide, kerosine oil and common soap (1 to 5 parts) answers well.

1st June, 1894.

THE TRINIDAD SPECIES OF PERIPATUS.

TRINIDAD Peripati have frequently been referred to at Club Meetings but as there appears to be considerable confusion amongst the members with regard to their nomenclature it was thought that the following description by Professor Adam Sedgwick, M.A., F.R.S., would be interesting as well as useful :—

" Dr. J. v. Kennel found two distinct species of Peripatus in Trinidad ; one of these he calls *P. Edwardsii* and the other *P. torquatus*.   His description of both is unfortunately extremely meagre."

The species which he calls *Edwardsii* possesses twenty-eight pairs of legs.   The generative opening is between the legs of the penultimate pair, and the generative organs present the characters of the Neotropical species.   Dr. v. Kennel was kind enough to send me two of this species in spirit, and I am able to supplement his description.   One of these specimens had 31 pairs of legs and the other 30, from which it appears that Kennel, like so many other zoologists who have examined *Peripatus*, has not been very careful in counting the legs.   The dorsal surface was of a chocolate colour, the ventral surface being a light brown. The papillæ and ridges of the skin presented the features characteristic of the Neotropical species.   The bases of the primary papillæ are conical as in *demeraranus*.   The jaws also presented no points of difference from those of the species from Caracas, excepting that possibly the number of minor teeth was rather larger : in one I found as many as eleven.

I think there can be no doubt that this is a distinct species and I propose to call it and define it as follows :—

Peripatus trinidadensis (Edwardsii, v. Kennel).   Peripatus with 28 to 31 pairs of ambulatory legs, and a large number of minor teeth on the inner blade of the jaw. The basal portions of the primary papillæ are conical.

Peripatus torquatus (von Kennel). Peripatus of large size, with 41 to 42 pairs of ambulatory legs. The head is marked off from the body by a bright yellow band on the dorsal surface.

The larger species is named *P. torquatus* and v. Kennel gives the following description of it. " The females reach the length of 15 c.m., with a diameter of 8 m.m ; while the males have a length of about 10 c.m. The colour of the dorsal surface is red brown, the middle line of the back being somewhat darker and paling off towards the sides. The head with the tentacles is black and is marked off from the body on the dorsal side by a bright yellow band, which often shows a small interruption in the middle line. There are 41 or 42 pairs of legs.

OBITUARY NOTICES.

THE Publication Committee regret to have to record in the present number the death, on August 24th, of Dr. B. N. Rake, President of the Trinidad Field Naturalists' Club. A notice of Dr. Rake's career will appear in the December *Journal*. The Committee feel that they are not in a position at the present moment to do more than chronicle the sad event, but they hope to publish a notice which will show—it is feared most inadequately —their high appreciation of the late Dr. Rake, and the sorrow which they feel at the sudden, sad and early termination of his brilliant and useful life.

THE Committee are also much grieved to have to record the death of Mr. Arthur Gaywood. Mr. Gaywood joined the Club on the 6th April this year. He came to several of the meetings and would probably have been a regular attendant. Mr. Gaywood was present at the August meeting. He died of the prevailing malignant fever on the 13th August. He was 25 years old.

IT is also the sad duty of the Committee to have to add to the list of our departed friends the name of Mr. Henry Tate, B.A., F.I.C., Acting Government Analyst, who died on the 18th August, with all the evidences of the most malignant type of remittent fever.

Mr. Tate was a son of Alexander Tate, Civil Engineer, of Longwood, Belfast. His cousin, Captain Stewart, now commands

H.M.S. *Mohawk* on the West Indian and North American Station. Mr. Tate was educated at Queen's College, Belfast, and took the degree of B.A. in the Royal University of Ireland with honours of Chemistry. In 1892, soon after taking his degree, he was appointed Assistant Government Analyst in Trinidad. He came to the Colony in May, 1892, and in the winter of 1893-94 went home for four month's further study at King's College, London, after which he passed an examination and obtained the diploma of Fellow of the Institute of Chemistry. At the time of his death he was for the second time acting as Government Analyst, and Professor of Chemistry in the two Colleges, during the absence of Professor Carmody on vacation leave. He had also acted for shorter periods.

Mr. Tate was so quiet and unassuming, that he was well known to comparatively few, but those who knew him recognised beneath the surface a warm Irish heart full of refinement and modesty, and a devotion to conscientious work far in excess of his physical capacity. There is little doubt that his constitution, always delicate, was weakened by working over time in the Government Laboratory, and at the Annual College Examinations in Chemistry. He was thus unable to resist the onset of hemorrhagic fever.

Mr. Tate joined the Club on the 2nd of September, 1892 ; he was an enthusiastic entomologist, and several times exhibited rare butterflies at the meetings.

Communications and Exchanges intended for the Club should be addressed to the Honorary Secretary, Port-of-Spain, Trinidad, B.W.I.

Price 6d.　　　　Annual Subscription, 3/.

Vol. 2. DECEMBER, 1894. No. 5.

J'engage donc tous à éviter dans leurs écrits toute personnalité, toute allusion dépassant les limites de la discussion la plus sincère et la plus courtoise.—LABOÛLBÉNE

Trinidad·Field·Naturalists'·Club.

NATURA MAXIME MIRANDA IN MINIMIS.

Publication Committee:

President.

H. CARACCIOLO, F.E.S., *V.P.*:　P. CARMODY, F.I.C., F.C.S.,

SYL. DEVENISH, M.A.,　　R. R. MOLE,

F. W. URICH, F.E.S., *Hon. Secretary.*

CONTENTS :—

Dr. Rake	116
Report of October Meeting	127
Friends and Enemies	129

TRY THE

WEST INDIAN CIGARETTE Co's

GOLDEN DREAM

CIGARETTES

SOLD EVERYWHERE. 12 IN A PACKAGE

FOR

SIX CENTS.

These Cigarettes are made from best quality and selected BRIGHT VIRGINIA LEAF TOBACCO and are guarranteed to be absolutely pure and unadulterated. These Cigarettes are also put up in Tins of 100 each, thus assuring the consumer as to their being in perfect order.

If you perfer to roll your own Cigarettes try a $\frac{1}{4}$ lb Tin of

GOLDEN FLEECE CIGARETTE TOBACCO

Cheapest and best in the Market.

TELEPHONE 232.

For Sale at WHOLESALE AT FACTORY ST. ANNS ROAD

OR

HUNTER, SKEETE & CO., LOCAL AGENTS.

BEAVEN NEAVE RAKE, M. D. (Lond.), M. R.

Late President Trinidad Field Naturalists'

☞ All Communications and Exchanges intended for the Club should be addressed to the Honorary Secretary, Port-of-Spain, Trinidad, B.W.I.

JOURNAL

OF THE

Field Naturalists' Club.

VOL. II. DECEMBER 1894. No. 5.

DR. RAKE.

> Death loves a shining mark, a signal blow.
> —*Young, Night Thoughts.*

DR. RAKE was the son of the late Mr. Thomas Beaven Rake, Surgeon of Fordingbridge, Hampshire. He was born on April 28th, 1858, and in 1872 he went to Oliver's Mount School, Scarborough, and at once distinguished himself in his school career never returning home for his holidays without prizes. In 1876 he matriculated at London University and after a year at home, as an apprentice to his father, he went to Guy's Hospital as a preliminary scientific student in October 1877. There he soon gave evidences of traits which were pre-eminently his characteristics throughout his career—an extraordinary capability for perseverance and thoroughness, and at the same time the power of attracting the good-will and respect of his tutors and fellow students. He gained a prize given by the Physical Society for an essay on the Localisation of the Functions of the Brain, and amongst other triumphs took the Joseph Horace Prize, 1877, and the Gurney Horace Prize, 1879. He passed the First M.B. with honours, and the M.B. degree with first class honours in medicine, and honours in obstetric and forensic medicine. In 1879 he took the M.R.C.S., and the L.R.C.P., the following year. In 1882 he became M.D. London with marks qualifying for the gold medal. He was house surgeon at Guy's in 1882, and house physician in 1883, and was for sometime resident obstetrical physician. When he finally left Guy's he spent twelve months on the Continent visiting the most celebrated hospitals, and passing through courses of clinics under the most talented medical men of Berlin, Vienna, and Paris. On his return to England he was appointed house physician to the Victoria Hospital for Children, Chelsea. So well had he upheld the honoured name of Rake at Guy's (where his father studied 30 years before) that his younger brothers who followed him needed no other introduction than that they bore his name. He was offered by the Colonial Office the post of superintendent of the Leper Asylum in this Island in 1883. Soon after his arrival he married—his wife coming out to Trinidad as soon as he had made a home for her. In June of the following year he experienced a severe attack of malarial fever from which, contrary to the expectations of the physicians who attended him, he escaped, but with impaired health, and he never recovered his former strength. Dr. Rake had not been long in the Island before his talent and persevering industry gained for him the highest opinions of the Government and the medical fraternity. He set himself to work with his accustomed energy to become acquainted with the little understood and loathsome disease with which those entrusted to his care were infected, and ere· long sent his first communication on the subject " Tuberculous Leprosy of the Tongue and Larynx" to the London Pathological Society. This was in 1885. During the following year he made seven communications to the Society, and subsequently forwarded many instructive specimens and valuable observations and experiments on bacilli and inoculation. In this way and by his Annual Reports on the work of the Asylum he quickly established a world-wide reputation as an authority on Leprosy, his researches into which occupied much of his leisure time. Not content with the ample material by which he was surrounded in Trinidad, his biennial vacations were spent in visiting the centres of leprosy in other countries in Norway, Spain and Tangier. The Leprosy Commission, appointed subsequent to the interest which was roused by the death of the saintly and heroic Father Damien, led to the late Sir Andrew Clarke, on behalf of the Royal College Physicians (which was requested to nominate one of the Commissioners) nominating Dr. Rake who accepted it with the permission of the Colonial Office. Still further honour awaited him, for when the five Commissioners met they elected Dr. Rake their president. The Commission visited India and travelled all over that vast country; and the report of the Leprosy Commission has become a standard work on the disease. The Indian climate further impaired his already undermined constitution, and at one time it was doubtful whether he would ever see England again, He however recovered sufficiently to go home, but his stay was only a short one, for he wanted to get back again to his work · amongst the lepers in Trinidad.

It may not be out of place here to refer to the Annual Reports which Dr. Rake issued during his administration of the Leper

Asylum. They are eight in number, and every one of them furnishes the strongest internal evidence of downright, hard, earnest, intelligently-directed work. Not only did Dr. Rake make experiments himself, but he gave a fair trial to the experiments of others in the treatment of his patients. All through his reports one can see that the sole object of all he did was to arrive at the truth, and in order to arrive at that goal a readiness to give up any and every theory or opinion he had previously formed directly he himself, or what is perhaps harder in such cases, any one else proved that it was untenable. As one reads these reports one cannot help being struck with his indomitable zeal in the search after knowledge, in the face of the greatest difficulties and with a total forgetfulness of self, a singleness of purpose, a great, noble, ardent desire to arrive at some method, to devise some treatment by which the fearful disease of leprosy might be extirpated, and failing that, methods by which, if absolute cure was impossible, its ravages might be retarded, and the pain and anguish of the unfortunate sufferers alleviated. To the lay mind the reports are interesting but not pleasant reading, for they open a new vista of human suffering, of heroic well-directed efforts to aid and to succour, backed up by the life-long martyrdom of the saintly ladies who carry out so unhesitatingly the treatment prescribed.

Turning to the first report, (February 1885) we find that Dr. Rake, after describing the results of the use of Gurjun and Chaulmoogra Oils, says : " The general health and nutrition of "lepers improves so decidedly in most cases under careful diet "and nursing, that it is difficult to form an opinion as to the value of any medicine." He then describes the effects of arsenic in two cases, in one of which the results were a great improvement in health, while the other experienced benefit from the treatment. A number of amputations and operations are then described with a detail which show the earnestness with which the new Superintendent had commenced his life-work. His next report, which, with tabular statements, covers 35 pages, deals with the distribution of bacilli, and he draws attention to the fact that seventeen observations on material taken from vaccine vesicles or pustules in lepers failed to show any bacilli in any of them, and he adds : " A point of interest with reference to the alleged " communication of leprosy by vaccination, though, as will be shown " later, I am not yet prepared to accept the leprosy bacillus as " pathogenic." Then, after describing the method adopted to find the bacillus—a modification of Ehrlich's—magenta without a contrast stain—he says errors of experiments are so easily made that one series of observation is not enough to decide the point. He then goes on "If bacilli are as common in the

"internal viscera as some writers seem to think, they are at any
" rate, much harder to find by ordinary methods." Whereas they
were frequently found in tubercles. He then describes several
experiments at inoculation of animals with leprous bacilli. A
cat inoculated three times, showed, four months later, no abnormal symptoms or bacillus in heart, muscle, or blood from the
heart. A fowl similarly inoculated, and actually fed on leprous
material, surrounded the introduced tubercle with a false membrane beyond which no bacilli were found eleven months later.
After narrating these experiments and others he quotes Koch's
four conditions which, if fulfilled, prove that a certain disease is due
to a particular micro-organism, and then he says " It is thus
" evident that we are still a long way from proving conclusively
" that leprosy is a bacillary disease." Meanwhile, Dr. Rake had
formed an opinion that a good deal might be done in the way
of local palliative treatment by excision of tubercles, and then
he modestly describes the results in two cases which led to improved personal appearance and a lessening of the tendency to
ulceration, and photographs before and after the operations
illustrate the result. In his return for the following year, dated
March 1887, he suggests that " it would be interesting to com-
" pare the percentage of lepers with the progress of cultivation
" and drainage, decrease in the price of fresh food and improved
" sanitation generally." Further on he quotes a case bearing on
the question of the hereditary nature of the disease, but he does
not appear to have considered it one of sufficient value to form
any scientific basis upon. In the report of March 1888, he
states that the bacillus had been found in 32 internal viscera
and in 41 other parts, the most frequent situation being the
femoral gland. No bacilli were found in the viscera in any
case before the disease had lasted two years. He then goes
into the question of contagion. " This question has been very
" much to the fore during the year not only in the medical journals,
" but also in the public papers. The College of Physicians
" however has issued a report in which it sustains with very
" slight modifications the position which it took up in 1867.
" My own views founded on personal observation and
" experiment have already been published (British Medical
" Journal August 20th and Sept. 17th, 1887) I have, however, a
" large number of experiments still going on or in prospect and
" shall of course have no hesitation in changing my opinion if I
" find scientific ground for so doing. One point which I think
" is of considerable importance was well brought out in a leading
" article in the Times of Nov. 8th, 1887. Commenting on a letter of a
" somewhat alarmist nature which appeared the same day, it was
" suggested with great reason that the apparent increase of leprosy
" in many places during the last few years is in great measure due

" to its increased recognition now by medical men as compared
" with former years. The possibility of introducing leprosy by
" vaccination, which has also been a good deal discussed during the
" year, is subsidiary to the question of contagion ; that is the
" communicability of leprosy by any means from one person to
" another. This again is but a part of a far wider subject. The
" point to be examined is whether the bacillus is the cause of
" leprosy. If further research should show that the pure cultivation
" of the bacillus can produce leprosy in animals we shall only
" have advanced one step. The next question would be the
" conditions under which the bacillus gains admission to the body.
" Tubercle bacilli as Cohnheim has proved are daily entering the
" lungs of every town-dweller but there must be general or special
" influences depressing the vitality or damaging the structure of
" lung tissue before they can multiply and set up irritation and
" proliferation of endothelium, blocking of alveoli and growth of
" lymphoid material. So with leprosy : It is possible that the
" leprosy bacillus is constantly entering the body though whether
" in air, water or food we do not yet know seeing that its presence
" or multiplication outside the body has not yet been satisfactorily
" demonstrated. There must, however, be other factors at work
" to determine what individuals shall become lepers. Here the
" influence of bad food, want of ventilation, filth and wasting
" diseases such as phthisis and syphilis would seem to come in.
" The question of contagion then resolves itself into that of the
" possibility of deriving the bacillus from another individual rather
" than from air, water, or other sources. Here the question
" must halt until further discoveries throw more light on it."
This report also deals with the operation of nerve stretching,
which he performed 100 times on sixty patients, and in cases of
ulceration it was, as a rule, attended with more or less good effect
—the ulcers often healing in a few days. In several cases the
operation caused a relief from pain, but its results in cases of
anæsthesia were not so encouraging and no effect could be traced
in tuberculation. Of the 100 cases, roughly speaking, good
resulted in half. The report which Dr. Rake wrote in February
1889 gives a short account of his visit to several asylums in
Southern Europe. He then writes : " During the past year the
" contagionists have had great food for self-congratulation in the
" announcement in the *British Medical Journal* of November
" 24th of the development of tubercular leprosy in the Hawaiian
" convict Keanu inoculated with leprous material by Arning on
" September 30, 1885. When, however, we come to examine
" this question dispassionately, what do we find ? That a man
" living on an island infested with leprosy was inoculated three
" years ago with the disease and now has developed it. But in
" that time he may have acquired leprosy in a dozen different

"ways, in air, food, water, &c., or it may have been in his family.
"True, the man was ascertained so far as possible to come of a
"clean family, and he has been isolated in gaol since the inocu-
"lation.   Still any one who has attempted to take the statement
"of lepers will appreciate the value of family history, and in a
"country where leprosy is rampant are we sure that it can be
"shut out by four walls? I repeat what I said in my last report
"that an experiment of this kind to be scientifically perfect
"must be performed in a country free from leprosy, and in an
"individual who has never left that country and whose imme-
"diate ancestors have always stayed at home.   Even if we admit
"the worst that the disease is directly inoculable beneath the
"skin in some cases, it is no argument that it is contagious in
"the ordinary sense of the term like typhus fever or small-pox."
During this year the percentage of deaths 7·47 was the lowest
ever noted in the Leper Asylum, and Dr. Rake added probably
many other Leper Asylums.   He goes on : "There were no
"deaths from dysentery during the year and the cases were
"trifling in number and severity. When we remember that dysen-
"tery was raging in the district where the Asylum stands we
"have, I think, an argument in favour of the late epidemic hav-
"ing been at least aggravated by neglect of proper food and
"cleanliness.   It must not be forgotten that lepers are very
"liable to ulceration of the large intestine.   The poison of dys-
"entery ought therefore to find with them a peculiarly favour-
"able nidus."   During this year many experiments were made
in cultivation of *bacillus leprœ* and he arrives at the following
conclusions : "An inquiry of this kind is practically endless,
"so varied are the conditions of temperature, time, nutriment,
"medium, living animal tissue, or putrescent substance and so
"many are the observations necessary to avoid or lessen the risk
"of errors of experiment, such as they are, however, my con-
"clusions are the result of four years' work, and I here sum-
"marise them :—(1) At a tropical temperature and on the
"ordinary nutrient media I have failed to grow *bacillus leprœ.*
"(2) In all animals yet examined I have failed to find any local
"growth or general dissemination of the bacillus after inocula-
"tion, whether beneath the skin, in the abdominal cavity, or in
"the anterior chamber.   Feeding with leprous tissues has also
"given negative results.   (3) I have found no growth of the
"*bacillus leprœ* when placed in putrid fluids or buried in the
"earth."   In the report of February 1890, he returns to the
"burning question of the day," the question of contagion, to
which an impetus had been given by the death of the hero-
martyr Father Damien, the discovery of a leper selling meat in
Whitechapel, Dr. Abraham's paper before the Epidemiological
Society, and the formation of the Leprosy Committee and National

Leprosy Fund. He regarded Father Damien's case in the same light as that in which he looked upon Keanu's. "The question," he says, "is narrowed down to this : Can a healthy " person more readily derive the bacillus from an infected human " being, or from food, air, water or some host which contains the " bacillus or its spores? In any case, whether or not we accept " Father Damien's case as proving contagion, it becomes, I think, " an argument for segregation, for every leper may very possibly " become a centre for the dissemination of bacilli or spores in " his immediate surroundings and by reducing these centres the " spread of the disease may possibly be checked. We do not yet " know what is the life history of the bacillus outside the body, " but it is quite possible that there is an intermediate spore " stage which has not yet been recognized. This theory of an " intermediate stage or host might explain the difficulty or im- " possibility of direct communication of the disease from one " subject to another, as in the analogous case of tape worms and " other cestoda, or again in the case of the *Filaria sanguinis* " *hominis* where the mosquito is the intermediary host. Against " this suggestion it will of course be argued that cestoda and " næmatoda are animals whereas bacilli are plants. This, how- " ever, will not upset the possibility of a spore stage and what " nidus the spores may find outside the body is only of secondary " importance." He then quoted Mr. Hutchinson who had no doubt that lupus and cancer might be transplanted from one person to another, and then cites Hanau's successful implantation of a cancerous growth on a rat, which, if true, fulfilled Hutchinson's prophecy. Referring again to Keanu's inoculation he cites cases where 31 healthy individuals were inoculated without success but adds : "Of course one positive case is worth any number of " negative ones, but the latter ought certainly to be quoted side " by side with the former." In his last paragraph on this subject Dr. Rake was not prepared to admit leprosy is con- tagious in the ordinary sense, but he added that segregation might do good by lessening the number of infected foci. His views on an intermediate spore stage led him to make experi- ments with earth from lepers' graves As a result he concluded that the rods found had no connection with leprosy as they were identical with similar rods found in soil a mile away. Salt pork and fish, pigeon peas, bad fish, were also examined with negative results. Dr. Rake then tried protective and antagonistic inocula- tion but without any definite results. About this time he expressed a wish that some condemned criminals might be offered the alter- native of inoculation by which some important additions might be added to present knowledge of pathology and treatment. He also gives an account of a simpler treatment than nerve stretching, viz., the passing of a bistoury through an ulcer on the sole of the foot to the dorsum and cutting straight forward through all tissues bringing the bistoury out between the toes ; afterwards stuffing the incision with lint to prevent immediate closure of the wound which he then left to granulate up from the bottom. These operations were attended with good results. The report which Dr. Rake wrote in February, 1893, when he had returned from the Indian Leprosy Commission (an absence from the colony of one year and eight months) also dealt with the communicability of the disease, and he mentions a case which seemed to suggest a possible communication, but he considered its value fractional, for no evidence was to his mind positive in a country where leprosy is endemic. " Scientific proof," he adds, " will not " be obtained until criminals are inoculated with leprous material "in a country free from endemic leprosy." He then passes on to the question whether there has been an increase in leprosy. The conclusion he arrives at is that " in Trinidad there has " probably not been much change one way or the other, but if " anything there has been a slight decrease in the proportion of "lepers to the general population ; that in India there has been " a marked absolute decrease ; in Norway there was a steady " decrease long before compulsory segregation was thought of and " that in the North Western States of North America where " leprosy was not endemic, the immigration of Norwegian lepers " has not been attended with any prejudicial results to the rest " of the community, nearly all the imported cases having died "out and no new cases having occurred." The remainder of the report is made up of a summary of the six years' work and a number of valuable tables. The last report was written in February of last year. In this paper he narrates his experience in the use of Thyroid Extract in the treatment of leprosy and summing up the result says that the most powerful preparation had been used without producing any appreciable effect on the disease. He had also been trying blood serum therapy but the results had been negative, but he expresses himself as not being without hope that blood serum from lepers might prove of use in early cases of the disease.

Throughout the reports there are numerous suggestions for the improvement of the comfort and the mental well-being of his charges and Dr. Rake did not forget to try and interest the general public in his patients. He made appeals to his friends for pictures for the walls and he started a library of instructive and entertaining books with the help of a grant by the Government, and he made a further requisition on the public which resulted in a large number of illustrated papers and books being presented to the Institution, with games and musical instruments which were much valued by the boys, In one report he refers to the interest of the coolies in pictures of

scenes in their own country, and the pleasure some tennis
rackets and balls the Governor had sent to the Asylum had
afforded. On the occasion of Her Majesty's Jubilee, pigs were
roasted whole, and fireworks were let off to the great amusement
of the inmates. He advocated many improvements in the buildings, and instituted a system of gardening for the inmates, the
produce of which was bought by the Government for con-
sumption in the Asylum, and in one report he mentions that
though light work is beneficial bodily and mentally, there was
a tendency to extremes, several of the lepers in their love of
gain, had overworked themselves with the result that they got
sores upon their hands and feet. All through the reports there
is evidence of the absorbing interest which he had in the victims
of this loathsome disease, and a constant endeavour to alleviate
their suffering and benefit them not only physically but morally.

His admiration of the noble work of the Dominican Sisters
who have charge of the nursing of the Institution was unbounded,
and each year he refers to their devoted assistance in dressing
evil-smelling ulcers, their care in operation cases, their co-
operation in the preparation of statistics, and their ever ready
untiring help in experimental and other researches, and admin-
istrative work. In the report for 1890 he writes thus: "It
" would be a work of supererogation to repeat the praises of the
" self abnegation of the Dominican Sisters. From time to time
" they are attacked by those who should know better, and the
" charges made are always found on enquiry to be utterly
" groundless. The sickening work which the Sisters are con-
" stantly doing, and the unmerited insults which they so often
" endure, without complaints, from some of the most degraded
" inmates would long ago have caused women of lower calibro
" to turn aside from the work. Much has been written and
" spoken of Father Damien, and rightly so, but these devoted
" women had begun their mission five years before he set out
" from Belgium, so that they are comparative veterans in the
" campaign of succour to

    " This poor rib grated dungeon of the holy human ghost,
    " This house with all its hateful needs, no cleaner than the beast,
    " This coarse diseaseful creature which in Eden was divine,
    " This Satan haunted ruin, this little city of sewers . . . . ."

" Narrow bigotry assails them from time to time, but their's is
" a work which rises superior to the petty distinctions of creed
" or nationality ; it is in the truest sense humanitarian." Again
in 1893 he writes : "It is impossible to describe in words the
" devotion of the Dominican Nuns, to whom falls all the more
" arduous work in connection with the treatment of the
" patients. Next month they celebrate the twenty-fifth anni-
" versary of their arrival in Trinidad. Of the original band of

" Sisters only one survives—the Dispenser, Sœur Marie Augustine.
" During a period of twenty-three years she has only spent eight
" days outside the Asylum walls—a record probably never
" approached in the history of leprosy. That she may long
" continue to adorn the post she so worthily occupies must be
" the wish of all who have the privilege of her acquaintance.
" When we remember the chaos which reigned at the Leper
" Asylum five and twenty years ago, we may well say of her,
" as of all the Sisters of Cocorite *si monumentum quæris*
" *circumspice.*" Truly the noble tribute of a noble heart to noble
hearts. Dr. Rake not only manifested his appreciation in
words but constantly pleaded with the Government for better
accommodation of these Dominican ladies, whose rooms he described as little better than cupboards.

So far an attempt has been made to give a sketch of Dr. Rake's
attainments and life work, but it was not alone in the character
of a great authority on leprosy that he was known to the Field
Naturalists' Club. He had been an ardent student of Nature
from his boyhood. At Fordingbridge he was well known as a
naturalist to whom neither bird, beast, reptile, insect or plant
came amiss, and during the year 1876-7 he searched the neighbourhood of his home in quest of Natural History objects of
every description. From Trinidad he was constantly sending home
to his father curios which he obtained in his expeditions. He
was a true sportsman being passionately fond of rod and gun.
It was not to be wondered at therefore, that he soon identified
himself with the Trinidad Field Naturalists' Club of which he
became a member in August, 1892. He was a regular attendant
of the meetings at which his observations and remarks were
always interesting and to the point, and at almost the first
meeting he had something valuable to contribute with reference
to the much discussed question of the " Mosquito Worm"
(*Dermatobia noxialis*) which he had sent to the British Museum
for identification long before present controversialists had troubled
their heads about the matter. He was, as has been before stated,
interested in all branches of Natural History and went to
considerable trouble to make the monthly club gatherings
attractive ; sometimes he would exhibit trap door spiders and
their nests, at others, bees (*Trigona*), or insects akin to the ant-
lions or *Myrmeleontidæ*. Now it would be the simple announcement of a hasty glimpse of a rare butterfly (*Papilio argesilaus*)
flashing by as he drove to the Asylum. At other times a valuable
book or the transactions of a kindred society, or a porcupine so
young that its quills had not grown, or the eggs of a rare bird would
form his contribution to the programme of the evening. His first
paper read before the Club was a series of most interesting Natural
History Notes made during his visit to India, which was

published *in extenso* in the *Journal*. A perusal of the fourteen pages it occupies gives one but a faint idea of how wonderfully keen an observer Dr. Rake was. Everything interested him, from the ant-lions which the native boys jerked out of their pits with pieces of string, the queer little families of frogs at Mahabalipuram up to an old brass gun, and the gigantic tomb of Gol Gombaz in Bijapur. Dr. Rake, however, principally occupied himself with butterflies, and together with his friend Mr. Gatty, then Attorney-General, made a large collection which was named by Dr. Crowfoot. These appeared as a preliminary list in the pages of the *Journal* of April 1892. In March last year he contributed a highly instructive paper on "The Schizomycetes," which must be fresh in the memories of most of our members and subscribers. His last communication was a "Note on the Breeding of the Night Jar" read at the meeting last June and published in the October number. Dr. Rake always regretted his inability to take part in the various excursions which members of the Club are frequently undertaking, but his duties at the Asylum were too arduous, he was, however, always pleased to hear about them, and on one occasion joined several of the members in a butterfly hunting expedition. Soon after the new rules were adopted by the Club in September 1893 a strong feeling was manifested among the members that Dr. Rake should be the President for the year 1894-5, and it was therefore not a matter of surprise to any one at the memorable meeting of August, 1894, that Dr. Rake was elected by a very large majority to the Presidental Chair, and few who were present at that gathering will forget the modest demeanour with which Dr. Rake accepted the office, promising that he would do his best to promote the welfare of the Society, or the applause amidst which he assumed his seat, for he had ingratiated himself with all the members by his quiet, unostentatious, demeanour, and the respect which he accorded to the opinions and views of others. This was the last meeting he attended. The island was at that moment passing through an epidemic of what the faculty were pleased to call "malignant," but which laymen spoke of as "yellow" fever and Dr. Rake whose family had returned to England a week or two before, as the most distinguished medical man in the Colony, was called into consultation in numerous cases, and long will be remembered the assiduous care he exercised over his charges, and the tender sympathy which he extended to many an afflicted household. From the very first he, with his accustomed thoroughness, began the study of the fell disease which swept away many so quickly so silently, and mysteriously—in the cases of his own patients he attended seven or eight times a day, and several times during the night, and all this in addition to his regular work at the

Asylum, the preparation of an article for a new standard book of reference, and the writing of his address for the approaching annual meeting of the Club. In the midst of this period of mental strain and anxiety the news came that his father had died. This had a dreadfully depressing effect upon his sensitive temperament. Then Mr. Henry Tate, the Assistant Government Analyst, who lived with him (also a member of the Club) was stricken down by the pestilence, and Dr. Rake undertook his nursing, sitting up at night with him and administering his medicines with his own hand. Only those who have passed through this awful ordeal of nursing in a yellow fever case know what such a self imposed task means. Ultimately Mr. Tate succumbed, as so many more Europeans had done. Dr. Rake although the weather was damp, and he was in want of rest and sleep made all the final arrangements, and attended the funeral which took place during a heavy rain-fall. He returned to his lonely home and then it was discovered that he too had an attack of fever. During his illness he was visited by almost all the members of the profession. He was accorded all the attention, the nursing, and the skill which the Island could command. Anxious that his promise should be kept he directed the Hon. Secretary of the Club to forward his last photograph to a friend, and in the two conversations he had with him, referred to affairs in connection with the Society. The greatest sympathy was manifested on all sides. The Sisters at the Asylum, who looked upon him as a personal friend, begged to be allowed to nurse him, but other though not less experienced hands had been previously entrusted with the work. From the first there had been very little hope and on the 24th of August, the day on which the Annual Meeting was to have taken place, the Naturalists' Club met to follow the remains of their President to their last resting place. Such was the sadly abrupt end of a life pregnant with the possibilities of glorious achievements and of permanent benefit to humanity. Cut down so suddenly in his early prime, so soon after he had entered the path to an exceptionally brilliant professional career, he had even then been crowned with the laurel wreath of Fame. He was pre-eminently fitted for the peculiar line of work he had chosen. He loved it intensely, but his learning and attainments did not inspire him with contempt for the work and experience of others, and provided a plausible reason could be shewn for a particular method of treatment he was willing to give it the fullest and fairest trial—in that lay an important factor in his prospects of success. As it is, short as his career was, Dr. Rake's life-work will never be forgotten in the annals of medicine, and his memory will ever remain green with those who had the honour of knowing him, while his name will always be revered by the Trinidad Field Naturalists' Club.

REPORT OF CLUB MEETINGS.

5TH OCTOBER, 1894.

PRESENT : Sir John Goldney, V.P., in the Chair. Messrs. J. R. Llaños, H. Caracciolo, T. W. Carr, John Hoadley, Thos. I. Potter, J. T. Rousseau, Syl. Devenish, John Thomson and F. W. Urich, Hon. Secretary.

Mr. F. de Labastide was elected a Town Member of the Club.—Mr. Urich read a paper on the fungus-growing ants of Trinidad. The species referred to were *Atta sexdens* L, *Atta. cephalotes* L, *Atta (Acromyrmex) octospinosa*, Reich, *Atta (Trachymyrmex) urichii*, Forel, *Apterostigma urichii*, Forel, *Apterostigma mayri*, Forel and *Sericomyrmex opacus*, Mayr. The habits of these ants were discussed and the construction of their nests described. Reference was also made to the means employed for destroying them and their natural enemies were enumerated. Artificial nests, microscopical slides of the fungus and the ants themselves were exhibited in connection with this paper.—Mr. Russell Murray made some remarks about his method of dealing with these pests and Mr. Potter mentioned how he destroyed nests of *A. octospinosa* in masonry. Mr. Russell Murray also mentioned how nests of wood lice *(Termes sp.)* could be exterminated with just a pinch of arsenic put into a small hole made at the top. According to Mr. Murray wood lice are very ghoulish in their habits. The members of the nest who devoured the arsenic, would be eaten by the rest of the inhabitants and in this way the arsenic would pass through the whole nest and its entire population be poisoned in succession.— Mr. Caracciolo exhibited a beetle *Antichira chalcea* of which he said several were found feeding on Cashew fruit, but the curious fact about them was that as soon as he approached them they let themselves drop to the earth and disappeared in such a mysterious fashion that he was almost led to believe that they had dropped purposely into holes in the ground situated immediately underneath the fruit they were eating. He also exhibited a fine specimen of the large bat *Vampyrus spectrum* and several other specimens of Chiroptera, amongst which figured *Hemiderma brevicaudum*, Wied, *Artibeus planirostris*, Spix, *Glossophaga soricina*, Pale, and *Phylostoma hastatum*.—A cray fish (*Scylla* sp.) from the Gulf was placed on the table.—Mr. Potter showed 3 specimens of *Peripatus* from Venezuela which were found in peat sent with some orchids from that country.—A curious insect popularly called a Lantern Fly, *Phrictus serratus*, Fahr., having a large protuberance on its head from which it is said a strong light is emitted at night, was placed on view by the same member. The specimen was taken at St. Ann's. Mr. Caracciolo said that he had kept a live specimen of this insect in confinement and it did not emit any light and he doubted very much whether the insect was capable of producing light, and probably the light attributed to these insects really came from the " candle flies " *(Elateridæ).* A coral snake, *Elaps riisei* presented by Mr. Maingot was exhibited. The Secretary on behalf of Mr. C. J. Thavenot exhibited a large skin of a Trinidad otter *Lutra insularis* which he killed at Princes Town. Mr. Thavenot was in a tree watching for agouti when the otter made a rush at one which had just been shot and immediately fell a victim to Mr. Thavenot's second barrel. After deciding that the monthly meetings of the Club should take place on the second Friday of the month, instead of the first, as hitherto, the meeting adjourned at 9.30.

CLUB PAPERS.

FRIENDS AND ENEMIES.

BY F. W. URICH, F.E.S.,

IN the great struggle for existence which is perpetually going on between animals of different species, various orders of plants, and animals and plants, there are many interesting facts which are not commonly known to the great majority of the human race, but it is generally admitted that we owe a great proportion of the immunity we enjoy from the annoyances of certain creatures to the presence of others who prey upon those who constitute themselves the unrelenting persecutors of man. Were it not for these humble friends the human race would eke out but a miserable existence and in some cases would perish miserably altogether. As a rule our benefactors are recognized and protected, but in this category there are included many impostors who have only found favour through being possessed of outward beauty or grace, while in reality they are constantly engaged in one form or another in doing harm to mankind. Others again we class among our enemies, because of their rude, uncouth forms, uncanny appearance, or because their habits are but imperfectly understood and their actions, which are really beneficial in their results, are regarded as destructive and mischievous. It is chiefly the cause of our persecuted friends

which I intend to plead to-night and in doing so I hope you
will neither find me too long or too tedious. As I have just
hinted, those animals which are neither (in our estimation) pretty
or graceful and which lead secluded or nocturnal lives are regarded
with suspicion. Everything which creeps, crawls, or flies in
obscurity, never mind how innocent it is, no matter what good
qualities it may possess, is condemned and slain without trial
or enquiry of any kind and in many cases the slayer considers
he has done a valiant action and rendered himself worthy of the
homage of the human race. Undoubtedly many animals,
by their appearance, on a first examination rouse our
repugnance, but why is Man learnedly styled, *Homo sapiens*,
if he is not to use his reason and to enquire into the things of
life, and amongst them, the habits and characteristics of his
humble friends of the Lower Creation? Amongst mammals,
Bats are perhaps the most abhorred and despised. In all
countries and in all times have they been regarded with dislike
and suspicion. The Jews thought them unclean and accursed
creatures. The Greeks borrowed from them the wings of their
harpies and the Christians adorn their pictures of the
Evil One with the pinions of a bat. The poets too, have not been
behind hand in calumniating these unfortunate night-flyers.
Shakespeare makes his witches add "wool of bat" to the in-
gredients in the boiling cauldron ; the worst evil Caliban could
wish his master was that bats should light on him, and numbers
of other sweet singers of our own and other countries have evil
things to say when "the weak-eyed bat with short shrill shriek
flits by on leathern wing." But what is the truth about bats?
At the first glance they are not very taking in appearance, but
a closer examination will show that they are, many of them,
beautifully marked and that they are covered with extremely
fine fur ; that they are wonderfully and beautifully adapted for
their peculiar mode of life, and that they deserve a better fate than
to be knocked down and killed when they happen to lose their
way and suddenly find themselves in our dining rooms. For
their good offices they are contemned and killed. When we
light our lamps at night, the light attracts numbers of mosquitoes and other disagreeable insects, who love to pursue their
mystic ærial dances in the bright rays, and the bats, knowing this,
for they are intelligent animals, swoop in amongst them to
catch their evening meal. Our President will no doubt be able
to tell you how many species of bats, there are in Trinidad which
are positive enemies of mankind, in that they are addicted to the
habit of bleeding our animals and robbing our fruit trees, but he
will also tell you that these are more frequently found in the country
districts and that the large majority of the vast army of small
bats about town are insectivorous and do no damage to either

man or his industry in any way and should be classed amongst our friends and protected accordingly. Another night flyer is even more decidedly still our friend—the Owl,—but how is he treated ? What does the Bard of Avon say of him ? He classes him with goblins and elvish sprites ; he is the omnivorous and fearful owl of death, and owl's shrieks herald the death of a notorious felon. And yet the owl is a creature whose presence can do nothing but good and whose absence, in some cases, means ruin, disaster and starvation. But mankind shoots and destroys this constant friend with an assiduity which is unremitting. Man has few worse enemies than rats, squirrels and mice and owls live almost exclusively upon these destructive little rodents. It is generally believed that Trinidad has no field rats, yet an American who was here the other day, in a few weeks captured no less than nine different species of these animals in a very small area. Undoubtedly much of the damage done on Cocoa estates which is put down to squirrels is the work of rats. Rats are amongst the most prolific of mammals and if their natural enemies are killed they increase to an enormous extent in a very short time. To show you of what importance owls are, and these remarks also apply to hawks, I will mention this : Up to very recently owls and hawks were shot indiscriminately by the farmers of the United States and things went on so until the Owls were almost as extinct as the Dodo and in a few years would have been so altogether. With the decrease in the number of Owls and Hawks followed an enormous increase of field rats and a proportionate decrease in crops. The whole country suffered and the Government had to take the matter in hand. Under the advice of eminent naturalists, at vast expense, a complete collection of the Owls and Hawks of the United States was made. Their life histories were studied carefully, good drawings were made of them, full descriptions of their habits and modes of life were distributed free over the United States and at last the people began to see that the owls and hawks were their friends and not their enemies, although they did, sometimes, steal an occasional stray chicken. Only last year there was a great outcry in the North of England at the vast increase in the number of voles and it was soon ascertained that the cause was the persecution of the owls and hawks which were killed because they occasionally varied their dinners with young pheasants or partridges. What shall we say of the Owls of Trinidad ? They are called Jumby Birds. Should one hoot out his glee at having caught a particularly fine rat or mouse our simple country folk immediately say he is foretelling the death of some one in the house. Certainly the note is not a very cheerful one and because it is heard at night its harsh disturbance of the general stillness is resented by the superstitious.

Seen by day the owl's great round yellow staring eyes and curious threatening antics when molested, add to the general impression of weirdness with which he is surrounded. For these reasons he is shot without pity, when of all living things his life should be protected by Ordinance and every means in our power. They are the most useful of the birds of prey, because their habits coincide with the rats and mice in being nocturnal. From birds we will descend still further. Of all created things and of all misunderstood things, reptiles are the most abhorred, the least understood. Their obscure life, their cold touch, their slow, stealthy, or sudden and impulsive movements, their creeping, hopping, crawling, their odd attitudes, their glassy stare, their apparent sudden changes of appearance have raised for them a host of enemies, whereas amongst the vast army of animals there are few which enact such an important part or are more untiring in their exertions for the benefit of man. Here is a notable example; of what earthly use is yonder hideous alligator* with his basilisk eyes and grinning rows of jagged teeth? Yet in some of the Southern United States, it has been found necessary to inflict a fine of $25 upon any one killing an alligator for the next two years and to establish a close season in order that the sugar cultivation may be saved. The alligators had been shot for their hides, and their chief item of consumption, the rats on the estates bordering the lagoons and rivers, had increased to such an alarming extent that sugar cultivation could hardly be carried on and legislation had to step in to save the "horrid" alligators in order that the cane cultivation might not be abandoned altogether. Now I don't for one moment mean to say that the alligators of Trinidad are so useful as those of the United States, but no doubt they do good in their way and perhaps, if their numbers were appreciably decreased we might wish to see them back again. But let us turn our attention to a very common little lizard. Most lizards according to popular belief are poisonous and the common gecko† is a very Borgia amongst the saurians, if vulgar report is to be believed. They are called "wood-slaves," "Vingt-quatre heures" or "Twenty-four hours," yet they are perfectly harmless and at the same time, like many other harmless and useful things, remarkably plain and common-place to the unthinking observer. They have clumsy forms, fat sprawling feet, large staring, hideous eyes which flash like black diamonds in the light of the candle and their tails are covered with minute, thorny looking excrescence. Like the bat and owl, the poor, little, ugly, useful gecko has endured centuries of calumny and yet goes on his way,

---

* *Caiman sclerops,* Schneid

† *Hemidactylus mabuia,* Mor de Joun.

warring on insects obnoxious to man, with a zeal which deserves a better recognition than a broken back, or life crushed out by an unsympathetic heel. The Greeks called the Gecko, "Stellio" and Aristotle tells us that the Stellio lives in vaults and graves, creeps on walls and ceilings, falls into the food, creeps up the nostrils of donkeys, causes them to refuse their food and finally kills them by its bite. Pliny assures us that a most dangerous compound is obtained by drowning a gecko in wine or stifling it in pomatum. The wine or pomatum so treated causes freckles and he says "Many give this gecko treated wine or pomatum to pretty girls with the wicked intention of spoiling their beauty." He also gives the remedy to be applied in the case of the unlucky fair one, who is thus the object of jealousy, and the antidote is the yellow of an egg mixed with honey and saltpetre ! Geckos are credited with producing skin eruption if they run over one with their " soft flabby viscous little toes." They are supposed to poison food by their touch and the country people in Trinidad tell you that when they see you approaching they first " take your colour" and then jump upon you and stick to you until death releases you twenty-four hours after this extraordinary performance. Hence the name. Now, the next time you see a gecko, you may take my word for it that he will not jump on you and if you can master your emotions sufficiently, try to quietly watch him or catch him in a pocket handkerchief and put him in a wide-mouthed bottle and you will then be able to examine the wonderful construction of those same " flabby viscous little toes " which enable him to climb so cleverly up a pane of glass and on smooth walls and ceilings. The toes are divided into a number of lobes or fleshy pads. Each lobe has a row of small bristles on its ridge which cause it to resemble the sucking disk of a leech. When the gecko presses his foot on a smooth pane of glass or on a wall the air between the foot and the glass or wall is driven out and when the presure is removed the inner surface of the foot is raised by the elasticity and rigidity of the bristles, special muscles also come into play and thus a vacuum is created and atmospherical pressure secures the foot-hold. It is absurd, however, to suppose that the gecko can stick to an object with such force as to prevent its removal without the aid of a hot iron, which is the correct thing to use on such occasions —according to popular belief. Geckos feed largely on cockroaches—the smaller kinds—and other insects. These are more plentifully met in old houses and Master Gecko, who has an admirable appetite, naturally finds himself in these localities too. Now, Master Gecko as he runs hither and thither about his work, like many other creatures, in the fulness of his heart sings. But then his range of voice is not a large one being

limited to one note and it is something like *tack-e-tack-e-tack*
Consequently houses where Geckos are common are pretty certain
to be known as haunted. And in the middle of the night when
people hear Master Gecko's *tack-e-tack* they cover up their
heads with a sheet and lie trembling waiting for the ghost to take
himself off, when after all the cause of alarm is nothing more
than a harmless little lizard. Another lizard* found in our
garden is also a useful little creature living almost entirely upon
the little white cottony insects† which creep about the crotons and
other plants. This lizard is not subject to much persecution, but
another and larger one, commonly known as the "Chamæleon"‡
is slaughtered without mercy. With the Gecko, they also rejoice
in the appellation of "Twenty-four hours" and they are supposed
to be fearfully poisonous. This lizard possesses the power,
together with the gecko and anole, of changing its colour and it is
to this faculty that its misfortunes are to be attributed. There
is nothing strange about this changing of colour, many animals
such as fish, frogs etc. are able to do it. It is generally believed
that the lizards change colour to suit their surroundings;
recent investigations, however, have shown that this is not always
the case. The change of colours is effected by many agents and
is dependent on the nerves. The agents which would come into
play are hunger, thirst, anger, fear. The change of colour is also
affected by heat, light, dampness etc. With regard to how these
animals are able to change colour I will tell you what Professor Semper says: "One example—the skin of the frog will
"suffice for all cases. The skin consists of two distinct portions,
"the epidermis and the cutis. The former is entirely composed
"of cells and the innermost layer contains cylindrical cells; the
"cutis is chiefly fibrous and encloses nerves, large cavities for
"glands and cell elements. These last are commonly filled with
"pigment, and the remarkable changes of colour in the frog's
"skin depend entirely on the distribution of these highly ramified
"pigment cells and their power of shrinking under certain kinds
"of irritation. The pigment in these contractile cells,—known
"as the chromatophores—is different in different individuals and
"in different parts of the body, yellow, brown, black, sometimes
"even red or green. Besides, the colour of the chromatophores
"varies with the state that they happen to be in, and differs during
"contraction and expansion. Heineke, for instance, has shown
"that in *Gobius ruthensparri* the chromatophores that are yellow or
"greenish-yellow, when destended become orange-coloured when
"contracted while the orange or red ones when shrunk become

---

* *Anolis alligator*, Dum. Bibr.

† *Orthesia insignis*.

‡ *Polychrus marmoratus*, Linn.

"brown or even black. These chromatophores are distributed
"in the skin with a certain regularity; in this particular, rep-
" tiles, fishes and amphibians show hardly any or no difference.
"The true chromatophores lie in different layers in the cutis;
" close to the epidermis, light coloured yellow cells occur, be-
" neath them the red or brown, and, in the deepest layer the
" black. In some spots the pigment cells of one kind or the
" other may be wholly wanting; sometimes the black ones form
" a close mass in one spot, while in others the red or yellow pre-
" dominate, but very few spots are devoid of pigment altogether.
" It is on this distribution and stratification of the chromato-
" phores and their alternate expansion and contraction that the
" pattern (so to speak) depends, which the frog's skin displays at
" any given moment. If all the chromatophores are relaxed,
" brown or black will predominate and in spots where light-
" coloured chromatophores lie in patches their hue will be dulled;
"if they contract, while the light ones are still extended, these
" latter will be more conspicious." So you see these changes
of colour can be satisfactorily explained and there is nothing
supernatural about them. The Chamæleons sit quietly on the
trees or move about slowly and the nature of their food, viz:
butterflies make them particularly useful about gardens. Yet
the ladies go into ecstasies over the lovely butterly whose larvæ
devour the flowers while they shriek with horror at the hideous
lizard which is of infinite service to them in killing the beautiful
enemies of their gardens. We now come to the snakes, "the
tribe accurst" according to the Bible. How persecuted and
yet how useful these poor reptiles are. There are some that are
poisonous but I can assure you that it is but rarely one meets
a venomous snake. Only harmless useful serpents, except perhaps the coral snake,* are met about town and the neighbourhood.
Snakes have always played a great part in the fables and beliefs
of nations. Not only the fables of the Jews and Christians,
but those of other nations contain something about snakes;
by some they are dispised, by others feared, others love them and
others again worship them, but by few are they understood.   Of
the numerous snakes in Trinidad which are useful I will only
mention a few. The Cascabel Dormillon† or Tree Boa, is one of the
most useful in that he kills rats and squirrels on cocoa estates. Unfortunately he is also killed in return for his good services, wherever
seen.   I think Mr. O'Reilly's description of this snake would
perhaps better interest you than any words of mine. He says:
'·The Cascabel Dormillon merits our attention for many reasons.
': He is the victim of the worst calumny. His very name is a lie.
"Cascabel Dormillon, means sleeping rattlesnake. Now he is

---

* *Elaps riisei* Jan. and *E. lemniscatus*, Lnn.
† *Corallus cookii.* Gray.

" not a Rattlesnake at all, but a Boa, and as for sleeping, 'tis true he
" takes a nap by day, like other gentlemen, who have nothing to do,
" but after dark he is wide awake enough, like an owl, or an
" Insurance agent on the night of a fire. He is an ugly-looking
" brute but so are many good-hearted fellows we meet every
" day. As for venom, he is not half so bad as a mother-in-law
" or a traitor friend. He takes a starry night for a ramble,
" catches his rats by the sweet silver light of the moon and meets
" his sweetheart in the shady dell. Every one in Trinidad says
" he is frightfully venomous. The Venezuelans say the same.
" They regard him as one of the worst characters in the country.
" They called him " Mapanire." Now what is the truth? His
" true name in serpent lore is *Corallus*. He lives altogether in
" the trees, rarely coming to the ground. He is one of the best
" climbers in the world." There are other useful snakes, such as
the Cribo,* the *Scytale coronatum* or ratanero. The cribo it is
true will take chickens, but he lives principally on rats and I
think we should excuse him if he sometimes takes a chick by
mistake. After the snakes come the toads and frogs. Very
little good has ever been written of toads and frogs. Milton
represents the evil one in the guise of an enormous toad
whispering temptations into the ear of Eve as she lay asleep :

" . . . . . . him there they found.
" Squat like a toad, close at the ear of Eve,
" Assaying by his devilish art to reach
" The organs of her fancy, and with them forge
" Illusions, as he list, phantoms and dreams;
" Or if, inspiring venom, he might taint
" The animal spirits, that from pure blood arise
" Like gentle breaths from rivers pure."

Shakespeare was particularly ingenious in attributing bad
qualities to this useful amphibian. " Adversity is like a toad
ugly and venomous " ; he sings of venom toads, and poison
never hung on a fouler toad ; a poisonous hunch-backed toad ;
slaves toads and rogues are linked together ; while no one
could be more loathsome than a toad and so on. On the
authority of Shakespeare and numerous other writers and
tradition the toad then is unanimously declared to be
poisonous. Now the toad has a certain acrid secretion in his
skin which he takes care to discharge when any other animal
interferes with him, but he is not at all venomous and as he
can neither bite nor scratch he is a ver helpless creature although
such wonderful powers and spiteful ytemper are attributed to
him. That he is useful there is no doubt, for he devours all the
enemies of the kitchen gardener, snails, beetles, flies, grubs of all

---

* *Coluber corais*, Cuvier.

kinds and once I saw one gorging a big centipede* which he stuffed
into his huge mouth with his front feet in a very clever though
clumsy fashion.   No doubt scorpions often share the same
fate.   Here I would like to mention two popular fallacies
with regard to these two arthropods.   It is usually said that
the centipede stings and the scorpion bites.   But the facts are
exactly the reverse.   The centipedes bite with their mandibles
and the scorpion wounds with the sting at the end of the long
abdomen.   Jack Spaniards† are also said to bite, when they
only sting.   Tree frogs and in fact all frogs are accused of
being venomous just as the toads are, on equally poor grounds.
Unfortunately the amphibians have one bad habit and that
is a practice of holding nightly concerts, but this is more than
made up by the good they do us.   Passing over the great army
of useful insects, without which, according to Mr. F. M. Webster,
" puny humanity would scarce be able to live out a miserable
existence" we come to earthworms.   There are few things more
despised than they, and they are accused of eating the roots
of plants.   But Darwin tells us that in their burrowings they
plough the land in an excellent manner for seedlings and that
their value to agriculture cannot be estimated, so great is it.
    I have run on so long that I have not time to go into
details as to some animals which, while they are protected,
are really enemies of man and his labour, and I shall only be able
to mention a few of them.   First and foremost is the deer.‡   The poor
deer, the noble deer, the sobbing deer, with its grace-
ful delicate form, bounding footstep and large languishing eyes,
which dumbly implore pity.   The deer is protected in most
countries and our deer are no exception to the rule.   Why, it is
difficult to say, unless it be for the reasons above quoted.   We
slay our rat-eating reptiles and birds, creatures which never
harmed a cocoa pod or sugar cane, we shoot our wood peckers,
because they hunt out the insects which are burrowing into the
bark of our fruit trees.   We kill and persecute our geckos, lizards
and toads, because of their strange forms, regardless of the
incessant warfare they wage on insect pests.   But the deer which
comes into our cocoa plantations eats up the buds, breaks the
young trees, rubs off the bark, is protected by Ordinance and
allowed a free run of our plantations and the man who shoots it
in the close season is liable to a fine for having done the country
a service in freeing it of one destructive animal.   One other
instance is that of the butterfly, which adds so much to the
beauty of the landscape, but those who plead so vigorously

---

* *Scolopendra* sp.

† *Vespsis* sp.

‡ *Cariacus nemorivagus.*

against the impa'ement of butterflies and complain of the many enemies they have, forget that these things of beauty were once the hungry caterpillars which devoured our best trees, ferns and plants and played havoc with our flower gardens, and always knock down the Chamaeleons who do their best to capture them as they go to lay their eggs on their favourite food plants. But my time has gone. I intended to have enumerated the Dragon fly or addersman, as he is called in some parts of England, whose sole mission is the destruction of gnats, mosquitoes and butterflies, doves and pigeons which devastate our grain cultivations and a dozen other things, but as I before said, the time is gone and I can only conclude, by hoping that something I have said may be useful to you and that it will serve to induce you to disregard popular prejudice, which demands the destruction of many of our best, but hitherto persecuted friends.

December 6th, 1893.

THE "POOR ME ONE."

(Nyctibius Jamaicensis Gm.)

By A. B. CARR.

ON a moonlight evening in the month of February 1893, between 7 and 8 o'clock, I was attracted by the beautiful, though sad, call of the "Poor me One" (the supposed small ant eater, of which more anon), and on taking up my gun and proceeding in the direction of the sound, I observed, perched on a dry stump, somewhat taller than the young Immortel trees in its immediate vicinity, a big bird sitting, after the manner of goatsuckers, upright with its tail pressed firmly on the stump. I had not long to wait, ere the bird uttered its call, which consisted of six notes pitched in a minor key, thus :—

This was repeated again and again, at short intervals. When satisfied that my soloist was the author of this peculiarly sweet and sad song, with which for so long a time the small ant-eater, *Cyclothurus didactylus,* had been unduly accredited, I shot and secured my bird, which on examination turned out to be our largest Geat-sucker, *Nyctibius jamaicensis*, Gm. On questioning some of the older members of our peasantry I found only two cases in which it was known that the call came from some bird, but what bird they could not say.

The " Poor me one " calls only from February to June, both months inclusive. It is strictly a nocturnal bird, feeding on night beetles, the large fire-fly being its chief victim. The bird answers readily to a poor imitation of its call and can be made to follow one at will. It is very unsuspecting, and will not move until one is within a few yards of it. Its prey is caught on the wing, and after each "catch" the bird returns to its perch. The dimensions are : length $13\frac{3}{4}''$, stretch of wing $28''$, length of tail $7''$, stretch of mouth from tips of mandibles $4''$. This tremendous gape serves as a net with which to catch its prey. The " Poor me one " never calls on dark nights nor is it seen in the day. It is a graceful flyer, flapping its wings but seldom. The sound of the call travels over a distance of at least 800 yards, but when at that distance, or even at considerably less, say from 4-500 yards, . not more than three to four notes can be distinctly heard. On getting within 2 to 300 yards the last two of the six notes are somewhat similar to the faint cooing of a domesticated cock-pigeon, wooing his spouse. Both sexes call, and are alike in plumage. They seem to be more partial to the low than to the high lands.

Caparo,
    May 27th, 1894.

REMEMBER

That when purchasing the necessaries or luxuries of life, it is to your interest to secure the best possible value for your money. *The man who squanders his means* by paying above the market rate for articles in daily use is inviting ruin, which will not be slow in answering the invitation. For this reason sensible people who desire to die solvent, patronize when making their purchases

THE

Popular Establishment owned and managed by Mr. P. A. Ramsey, situated at the south-eastern corner of Upper Prince Street and Brunswick Square. The Proprietor of this well-known place of business has made it his duty to see that his customers are supplied with *the best and freshest Drugs, Chemicals and Medicines,* the stock of which is replenished by every steamer arriving from Europe and the United States, thus affording the

CREOLE

population of the country an opportunity of procuring the same quality of drugs as would be supplied to them in London, Paris or New York, and in order to secure their being accurately dispensed, the Proprietor has secured the services of *a Staff of Skilled Druggists* who perform their duties under his personal supervision, hence affording a double guarantee to those who favour him with their prescriptions. And in order to further popularize this favourite

PHARMACY

the Proprietor keeps on hand a large and varied stock of the best known *Patent Medicines, Toilet Requisites, Perfumery,* and other articles usually kept in stock in a Pharmacy. In order further to encourage Cash Customers, a quantity of *Pictures, Fans, Toys,* and other

ACCEPTABLE SOUVENIRS

Are daily *given away gratis* to ready money purchasers.

Civility and Attention to all ! ! !

A NIGHT BELL

Is attached to the premises and all calls are promptly attended to.

Communications and Exchanges intended for the Club should be addressed to the Honorary Secretary, Port-of-Spain, Trinidad, B.W.I.

Price 6d.   NEW YO Annual Subscription, 3/.

Vol. 2.   FEBRUARY, 1895.   No. 6.

J'engage donc tous à éviter dans leurs écrits toute personnalité, toute allusion dépassant les limites de la discussion la plus sincère et la plus courtoise.—LABOULBÈNE

Trinidad·Field·Naturalists'·Club.

NATURA MAXIME MIRANDA IN MINIMIS.

Publication Committee:

President.

H. CARACCIOLO, F.E.S., *V.P.*:   P. CARMODY, F.I.C., F.C.S.,
SYL. DEVENISH, M.A.,   R. R. MOLE,
F. W. URICH, F.E.S., *Hon. Secretary.*

CONTENTS :—

Report of Club Meetings :
November, 1894	...	...	...	139
December	...	...	...	...	142
January	...	...	...	...	142
The Race Abhorred	...	...	...	...	143
In Memoriam—Hon. R. Guppy	...	...	...	144
Tobago Herpetological Fauna	...	...	...	145
Banana Disease	...	...	...	...	146
Visit to High Woods of Caparo	...	...	...	147
Lappé Hunting	...	...	...	...	151
Agouti ,,	...	...	...	...	154
Quenk ..	...	...	...	...	158

☞ All Communications and Exchanges intended for the Club should be addressed to the Honorary Secretary, Port-of-Spain, Trinidad, B.W.I.

JOURNAL

OF THE

Field Naturalists' Club.

| VOL. II. | FEBRUARY, 1895. | NO. 6. |

REPORT OF CLUB MEETINGS.

9TH NOVEMBER, 1894.

PRESENT: Sir John T. Goldney, V.P., in the Chair, Professor Carmody, Messrs. Syl. Devenish, M.A., Lechmere Guppy, jnr., R. R. Mole, C. J. Thavenot, T. I. Potter, S. A. Cumberland, H. J. Baldamus, F. V. Spooner, T. W. Carr, F. Eagle, and F. W. Urich, Hon. Secretary. Mr. William Lunt was present as a visitor. The following gentlemen were elected members of the club: Mr. G. Brown Goode (Washington), Honorary Member, and Mr. D. Munro, Town Member.—Letters were read from the Director of the British Museum and the Secretary of the Institute of Jamaica, thanking the club for copies of the *Journal*. A letter was read from Mr. Brown Goode, Secretary of the U.S. National Museum, offering the services of the scientific staff of the Museum, and announcing that the Club had been placed on the list to receive all the publications of the Smithsonian Institution. The Chairman directed the Secretary to write a letter of thanks to Mr. Goode and also to state that the Club would be glad to exchange specimens and publications with the Smithsonian Institution. A question asked by Mr. Goode with reference to "Snakes swallowing their young" was referred by Sir John to Mr. Mole who was asked to communicate with Mr. Goode. A letter from Mr. G. W. Smith, Honorary Member (Grenada), was read, in which he promised a paper on an excursion he had recently made to

"Feddon's Camp." A communication was read from Mr. Caracciolo, V.P., expressing his regret and inability to attend the meeting.—Mr. Mole read the following note: "Most of the members present must have heard Mr. Devenish relate how he on one occasion saw a Cribo eating a young Grouper. It is probable that many snakes are habitual fish-eaters. The common ringed snake of Europe is credited with being one of them; while the Viperine snake (*Tropidonotus viperinus*) of Southern Europe, of which we have a specimen here, prefers fish to anything else. It was while feeding this snake that I discovered that one of our Trinidad species, called by the Creoles Mangrove Mapepire—albeit harmless—is also a fish-eater and from my observations on that occasion I anticipate that the other *Liophidæ* of Trinidad, viz., the Beh-belle-Chemin and High Woods Coral are also piscivorous. At any rate I have brought specimens here so that we may ascertain the fact for ourselves. *Elaps riisei* of which the one in the bottle, is a very fine example, caught by Colonel Man in his quarters, is also pasionately fond of water and perhaps we may find out that he too is more or less a fish connoisseur. Another fish-eating snake of Trinidad is one we have not yet obtained—*Helicops angulata*—of which Mr. O'Reilly had two specimens. The sea snakes of the Indian Ocean are exclusively fish eaters." The following snakes were then placed in a glass sided box, which covered a large glass jar in which there were about 150 small fish: *Tropidonotus viperinus* (from Spain), *Liophis cobella*, *Liophis reginæ*, *Liophis melanotus* and an unidentified specimen. The snakes were soon actively engaged in catching the fish which they frequently brought out of the water to devour and then returned to the jar for more. In a short time very few fish remained while the increased size of the snakes shewed how they had benefited by their meal. The experiment was interesting inasmuch as only the Spanish snake had fed on fish previously and it was demonstrated for the first time that *L. cobella* and *L. reginæ* are most expert fishers.—Sir John Goldney related how, when in Singapore, cobras were frequent visitors in his dining room and how quickly the room was vacated until means could be found to expel the unwelcome intruders. On one occasion a favourite retriever in attempting to worry a cobra was bitten right in the pupil of the eye, where the poison had very little effect, and the dog recovered. On another, when driving, he surprised a snake in the road and his groom threw a stone at it and decapitated it. The snake was taken to the curator of the museum at Singapore, who could not identify it. It was afterwards sent to Calcutta, from thence it travelled to the British Museum and the British Museum people wrote out to Singapore and asked "but where is its head?" a very difficult question to answer, after an interval of so many weeks. That snake was a puzzle up to this day and was labelled with his name in the Calcutta Museum.—Mr. Urich mentioned that amongst some ticks recently sent for determination to Dr. Marx, was *Ambyloma adspersum*, a genus which seems to prefer living on reptiles, and which infested the ordinary toads found in Port-of-Spain. According to Dr. Marx the same species has been recorded from Barbados, Nicaragua and the Phillipine Islands—a rather wide and remarkable distribution of the species. A discussion on the tick pest of Jamaica ensued in which Sir John took a lively part, mentioning amongst other things, that the zebu cattle did not suffer at all from these pests. It was curious, he said, to see how some cows were attacked and others not; Tobago cattle were comparatively free from attack.—The Secretary read the result of an examination of some Trinidad Borers by Mr. W. F. Blandford; they were *Xyleborus confusus*, Eich, found at Caparo by Mr. A. B. Carr, doing damage to cocoa trees; *Thione championi* from Caparo, not a destructive insect, but rather a parasite on some other Borer. A new species allied to *Xyleborus capucinus* found by Mr. Potter at St. Ann's where it had destroyed a young cocoa tree and a *Platypus* not identified for want of more specimens.—Mr. Mole exhibited a small galap (or mud tortoise) with a jointed plastron by which it shut itself in when it had withdrawn its head and feet beneath the shelter of its carapace; also a kinkajou (*Cercoleptes caudivolvulus*) from Venezuela which was much admired for its soft fur, gentle manners and curious gestures.—The Secretary placed on the table some starfish, dredged up from the second Bocas by the Hon. W. Gordon-Gordon, which seemed to be *Asteroporpa annulata*.—The Secretary exhibited two young mapepires *(Lachesis muta)* which were taken from their mother by Mr. H. Hutton at Poole; there were a great many more, probably 40.—Mr. Mole said it was a curious fact nearly every snake in Trinidad was "mapepire something or other," but on this occasion, when the labourers discovered a real "mapepire," they called it a tigre—a snake which is considered tolerably harmless, and it was only Mr. Hutton, who knew what a terrible reptile they had encountered and warned the men off so that he could shoot it.—Mr. Cumberland called attention to a young wood dog, *Galictis barbara*, from Venezuela which by its playfulness and quaint antics caused much amusement .—Mr. Lechmere Guppy showed a large case of fine butterflies recently taken, amongst which figured some specimens of a very delicate and beautiful structure which had made their appearance in Port-of-Spain lately and which many old collectors had never previously seen; he also laid on the table some well prepared birds skins by Mr. Gray and an egg of the mud tortoise *Nicoria punctularia*; also a curious growth from the interior of a bamboo joint. Mr. Mole placed on the table a fine specimen of *Elaps riisei* caught by Colonel Man. Two specimens of the burrowing snakes *Glauconia albifrons* were also on view.

The meeting adjourned at 9.30 p.m.

14TH DECEMBER, 1894.

PRESENT: Sir John T. Goldney, V.P., (in the Chair,) Professor Carmody, Messrs. H. Caracciolo, V.P., Syl. Devenish, M.A., C. W. Meaden. T. W. Carr, E. D. Ewen, W. S. Tucker, J. Graham Taylor, John Hoadley, Thos. I. Potter, R. R. Mole, F. Eagle, J. H. Collens, J. Russell-Murray, Til. Sorzano, and F. W. Urich, Hon. Secretary.

Sir John Goldney addressed the meeting with regard to a proposal of amalgamation with the Victoria Institute and after a long discussion it was decided that on receiving a formal and written invitation from the President of the Victoria Institute (Sir J. Goldney) that four Members of the Club should meet four Members of the Victoria Institute for an informal conversation about the best way of bringing about a union of the two societies.—A vote of condolence was passed to Mr. R. J. L. Guppy on the loss of his father, the Hon. Robert Guppy, and the Hon. Sec retary was directed to write a letter expressing the Club's sympathy.—Some regulations, for the lending of books from the Library were adopted.—Mr. Mole exhibited a Mocassin snake *Tropidonotus fasciatus* which was fed with frogs. Mr. Potter showed a *Mygale* spider caught at St. Anns in the act of eating a centipede. Mr. Meaden called attention to a Noctuid moth, caterpillars, pupæ and imagos (all of which he produced) damaging pigeon pease shrubs by boring into the bark. Mr. Urich laid on the table a Moth, (*Empretia* sp.?) caterpillars of which had done damage to some young palms in his garden.—There were several notes of interest to be read but owing to the length of the discussion on the union of the Victoria Institute and the Club they were postponed to a future meeting and the Club adjourned at 9.30 p.m.

11TH JANUARY, 1895.

PRESENT: Mr. H. Caracciolo, V.P., (in the Chair,) Prof. Carmody, Messrs. E. D. Ewen, R. J. L. Guppy, C.M.Z.S., J. R. Llanos, T. W. Carr, T. D. Arnott, T. I. Potter, E. Tripp, W. S. Tucker, John Hoadley, C. J. Thavenot, C. W. Meaden, J. T. Rousseau, M.A., J. H. Collens, F. V. H. Spooner, R. R. Mole, F. de Labastide, S. A. Cumberland and F. W. Urich, Hon. Secretary.

The following gentlemen were elected Town Members of the Club: Dr. F. A. Rodriguez, M.B., C M., and Mr. John Barclay.

All scientific business was put aside in order that the following proposals by Sir John Goldney, President of the Victoria Institute, with reference to union of the Club and the Victoria Institute should be fully discussed:

I propose that if the F.N.C. are willing to unite with the Victoria Institute:

1. That the name should be The Victoria and Field Naturalists' Institute.
2. That the F.N.C. should have a majority in the Committee.
3. That the Committee elect its Secretary and Treasurer, and the heads of the different Sections.
4. That the *Journal* of the F.N.C shall be continued under its present management at the expense of the United Institute.
5. Subscriptions to be paid to the funds of the United Institute, and to be apportioned by the Committee.
6. The property in the collection of the F.N.C. not to be affected by the Union.

After discussion it was finally "resolved that Sir John Goldney be thanked through the Hon. Secretary for his efforts to promote the welfare of the two institutions and that he be informed the Club thoroughly appreciates his motives, but the Club is of opinion that the time is not yet ripe for such a union." Mr. Guppy made a proposal that the two societies should hold their meetings together.—A similar reply to that given to Sir John Goldney's proposition was ordered to be made to Mr. Guppy.

THE RACE ABHORRED.

THE "Race Abhorred," in other words "Snakes," was the title of a lecture delivered by Mr. R. R. Mole under the auspices of the Club in the Victoria Institute on the 26th October. Sir John Goldney presided and there were present the Acting Governor (the Hon. C. C. Knollys C.M.G.) and Mrs. Knollys, Lady Goldney, Mr. and Mrs. Clarence Bourne, Mr. C. F. Monier-Williams, the Hon. Dr. Lovell, Professor Carmody, the Rev. Canon and Mrs. Doorly, and about seventy ladies and gentlemen. The Chairman in his opening remarks spoke of the excellent and successful efforts which the Club was making to form a good collection of the Trinidad fauna without Government aid, and the good work the Society was doing in popularizing the study of Natural History.

The lecture was then delivered. It was illustrated by a collection of living snakes, prepared fangs, skins, bottled specimens and blackboard diagrams. At its conclusion the usual votes of thanks were cordially passed. The proceeds were devoted to the Mammal Collection.

IN MEMORIAM.

THE late Hon. Robert Guppy, M.A., Member of the Legislative Council, died at his residence, Piedmont, San Fernando, on the 11th November, 1894. He was third son of Samuel Guppy, Esq., of Bristol, and Arno's Court, Somersetshire. Mr. Robert Guppy was born in Bristol on 22nd January, 1808. He was educated in Leicester and London and the College of Henry IV., Paris. He entered Pembroke College, Oxford, and graduated B.A. in 1829 and M.A. in 1830; kept terms at the Middle Temple; called to the Bar November, 1832, and went the Oxford Circuit. He travelled in France, Switzerland and Italy. He came to Trinidad in 1839 to retrieve the properties of Messrs. Protheroe and Son, and was for a time connected with the sugar interest. Sir Henry McLeod in 1840 appointed Mr. Guppy a J.P. and Road Commissioner. In 1844 he was sent by instructions of the Colonial Minister to Sierra Leone to enquire into the prospects of obtaining a sufficient supply of labourers from that settlement. His report and advice had a great share in promoting the introduction of coolies from India. Upon his return from Africa Mr. Guppy acted as Stipendiary Justice for the Naparimas and Savana Grande, to which the Couva district was afterwards added. This acting appointment continued for fifteen months, at the expiration of which Mr. Guppy was appointed first Town Clerk of San Fernando under the Ordinance, just then passed, constituting a Municipal Corporation in that town. Mr. Guppy was subsequently appointed Warden of the North Naparima Ward Union; and, having had ample experience of the necessity of better means of communication, with the cordial assistance of Mr. William Eccles, Mr. Richard Darling, Dr. Philip and other local proprietors, he promoted and carried out the Cipero Tramroad (a railway) by which Princestown has become the most prosperous internal town in the island. Mr. Guppy was Secretary, General Superintendent and Manager of the concern until 1867. In 1862 Mr. Guppy resigned his post as Warden, and resumed his practice as a Barrister. Mr. Guppy introduced the electric telegraph into the island in 1868. He was for many years Mayor of San Fernando. Despite his many and valuable services to the Island however, he was not called to the Legislature until March, 1893, when he had almost reached the end of his long and useful career.

Always in sympathy with anything which had for its object the propagation of knowledge amongst the inhabitants, he identified himself with the Naturalists' Club in September, 1892, and attended several of its meetings. Mr. Guppy was celebrated throughout the Island for his hospitality and many of the members will recollect with pleasure the admirably arranged pic-nic to which he invited the Club three years ago.

Mr. Guppy's name was universally honoured and respected and his death severed a link which bound the present with the past, and which covered more than half the century during which Trinidad has been under English rule.

CLUB PAPERS.

A CONTRIBUTION TO THE HERPETOLOGICAL FAUNA OF THE ISLAND OF TOBAGO. ·

By Professor Dr. O. Boettger, Frankfurt-on-the-Main.

A few days ago Mr. Albrecht Seitz brought me a small collection of Reptiles and Batrachians from Tobago, which are in many ways deserving of attention. In the first place only a few of the species given were previously known from the Island, which faunistically belongs to one of the least known West Indian Islands, secondly many of the species are conspicuous by corresponding exactly with the Trinidad species, as we know from Mole and Urich's important contributions to the Herpetology of Trinidad.

The species collected are :

REPTILIA.

1. *Boa constrictor*, (L). Brought over alive.

2. *Drymobius boddaerti*, (Sentz,) typ.

Mole and Urich, Journ., Trinidad F. N. Club, vol. ii, page 85. (*Coluber*).

22 Maxillary teeth; in one specimen 3 on each side, in the other 2 Postoculars

Squ. 17 G 2/2 V 188 A 1/1 Sc 120/120 & 1.
,, 17 ,, 2/2 ,, 188 ,, 1/1 ,, 128/128 & 1.

3. *Spilotes pullatus*, (L) var B.

Boulenger, Cat., Snakes Brit. Mus., vol. ii, page 23.

Mole and Urich, l.c., page 84, (*Coluber variabilis*).

Entirely black, only the ventrals of the anterior portion of the two-thirds of the body (about from 140—160th ventral) partly yellow.

Squ. 16 G 2/2 V 224 A 1 Sc 108/108 & 1.
„   16 „ 2/2 „ 224 „ 1   „ 108/109 & 1.

4. *Oxybelis acuminatus*, (Weid).

Squ. 17 G 3/4 V 188 A 1/1 Sc 178/178 & 1

BATRACHIA.

5. *Bufo marinus*, L.

From this small collection we can already draw the conclusion with some certainty, that Tobago, which is very likely, was in former times joined to Trinidad or that at any rate it obtained its Reptiles and Batrachians by active or passive migration from this Island.

December, 1894.

BANANA DISEASE.

A DISEASE, probably fungoid, has been attacking plants of the *Musa* tribe throughout this Island and probably in other Colonies. The roots are first affected, the plant begins to look sickly, then the centre shoot becomes brown and rotten and the plant soon after dies. Various attempts have been made at different times to combat this disease which at one time worked great havoc in banana plantations in Demerara. The simplest and most effective of these attempts is that discovered by Mr. J. W. L. Mitchell, of Tacarigua, whose own account sent to the Club and, by the kindness of the Editor, first published in the *Port-of-Spain Gazette*, we append :—

Tacarigua, Sept. 1, 1894.

COMMON SALT, A CURE FOR FIG DISEASE.

DEAR SIR.—I have been trying experiments for some months past and find the application of 1½ to 2 oz. salt a perfect cure. Suppose a tree to be diseased and half dead, with only 3 green leaves and a centre, apply the salt sprinkled at the junction of the leaves and stem, and some round the roots. Then water well, if there is no rain; the rapid decline of the tree is at once stopped, and it begins to put out new leaves.

Yours truly,

(Sgd.) J. W. L. MITCHELL.

A VISIT TO THE HIGH WOODS OF CAPARO.

By R. R. Mole.

SOME months ago I read a paper before this Club in which I attempted to describe " A Day's Insect Hunting at Caparo. "* The events recorded on that occasion were only a part of the incidents of an excursion which lasted over several days. This paper is the result of the extension of the numerous notes which I made at the time and which I hope, even if they are not so novel to many of the members as they were to me, will nevertheless be not altogether uninteresting as the experiences of one who is not used to hunting anything more formidable than such small deer as snakes and reptiles generally, and was, until that time, unacquainted with the joy of the hunter when " from its covert starts the fearful prey. "

The morning was a somewhat dark and lowering one as shortly after eight o'clock, Mr. Urich and myself, under the care of Messrs. Albert and Arthur Carr, with a Spaniard rejoicing in the somewhat suggestive and not inappropriate name of Boney, set out down the Caparo Road accompanied by a stalwart cocoa contractor named Sammy, who carried the only gun with the party, and an old bayonet on the end of a long stout staff. The point of this curious spear was thrust into a corn husk as a preventative of accidents. We were followed by a pack of dogs of no particular or uniform breed, size, or colour, and chiefly remarkable for their mangy coats and long legs. These animals answered to the names of " Cook," " Melon," " Gertrude," " Blackie," " Bull " and " Spotty." One ill-favoured cur was dignified by the high sounding appellation of " Juan Carlo," while the eighth and last was " Jubilee." " Cook" a sandy coloured dog was the largest of the lot and was likewise in the best condition. He was the most valuable—his special quarry being the peccary, or quenk,† as the wild hog of the woods of Trinidad is popularly named. After following the main road for sometime we bore off to the left, behind Mr. Metford's plant-ation, on a track which runs through the woods in the direction of Freeport and soon afterwards, first ascending a steep hill, entered the High Woods. I must confess that I was disappointed at the first blush. After reading the wondrous descriptions of the late Canon Kingsley and others, I expected much more than the gloomy forest, destitute of life, by which I found myself surrounded. Dead leaves, bare trunks, here and there a creeper. Above a dense mass

---

*No. 2, Vol ii., T.F.N.C. *Journal.*
†*Dicotyles tajacu,* L.

of foliage shutting out the sky from our view. I looked in vain for the marvellous effects of light and shade caused by the golden sunshine above; the gorgeously winged butterflies and moths were nowhere to be seen; the impression of being in a glorious minster aisle with feathered choristers pouring out their hearts in thrilling song upon the roof was not my experience. Of these things I saw nothing. But, as we penetrated deeper and deeper into what were to me pathless wilds I found much to interest, instruct, amuse. Our companions had seen it all again and again but they were generous enough to point out each little fact which they thought would be entertaining to a stranger used only to the asphalted streets and ways of town. So well did they succeed that I soon forgot to note the clever manner in which I was being piloted, and only listened to the treasures of woodsman's lore which they so freely imparted. There was the tree, they said, which yielded the copaiba* balsam of our drug stores, and then they showed me how it was collected—by cutting a huge notch in the bole, and hollowing out the lower part of the incision so as to form a basin. A collector had been here before us and had fastened strips of bark and leaves over the notch to shield it from the air. We peeped inside and there saw about half a pint of liquid ready for bottling. Another tree of which I could only see the trunk was the Bois Canari† the ashes of which are used in making the clay for the manufacture of the canaries‡ of our creole kitchens. An enormous trunk $4\frac{1}{2}$ feet in diameter was that of the Balata§ and I recollected with pleasure the delicious fruit I have only once or twice tasted. Frequently these noble trees are hewn down simply for their fruit—the growth of generations, the number of which are perhaps more than I should like to hazard mentioning, for fear of being thought guilty of exaggeration, are destroyed for the enjoyment of an hour. Many of the trees in the forest are so buttressed round that one wonders how any one could have the patience to hew them down, but the difficulty is soon surmounted by the ruthless woodsman who builds a scaffolding 12 or 14 feet high round the tree and then cuts the trunk where it is more slender than it is at its base. Although we were apparently so far away from anywhere in particular, we had not got away from evidences of man's ingenuity, for we stumbled across a forester's bucket for carrying water, constructed of the gigantic leaves of the Palmiste Mahoe,‖ and made so cleverly that it would hold two gallons of water without a leak. To find a ravine close by was only natural after

---

*Copaifera officinalis.   †Hirtella silicea.
‡Canari—earthenware pots used in lieu of saucepans in the West Indies pronounced Can-er-ee.
. §Mimusops globosa, Gaertn.   ‖Leaf spathe of Oreodoxa regia.

picking up such an utensil. In the ravine some one had been catching fish. A dam of palm leaves and pieces of wood had been constructed, stopping the flow of the water, except by one narrow aperture in which a fish pot was still lying. When the forester wishes to vary his dinner of game with fish he simply constructs one of these dams and places his fish pot. Then going some way above he splashes and beats the water, so driving the fish before him until they enter the fish pot from which escape is well nigh impossible. Further on we found a sort of lean-to hut made of palm leaves, and here we discovered the business of these gentlemen of the forest who have within its precincts all the material and implements necessary for their accommodation and subsistence. The mass of fibre lying on the ground close to the hut had been obtained from the terite* and is used in making baskets, and particularly those large covered baskets† in which country people carry their clothes. The basket maker had invented an ingenious contrivance of three sticks to sit upon, and had still further made his task a comfortable one by placing upon his curious stool a bundle of terite fibre as a cushion. They were evidently not without literary tastes, these basket makers of the High Woods, and they are loyal withal, for we found an old London newspaper adorned with a portrait of Her Majesty, which the basket maker had evidently been reading and admiring just before he set out on the errand which had deprived us of the pleasure of meeting him. The woods even furnish their inhabitants with writing materials, as Mr. Carr practically demonstrated by plucking a wild lily leaf‡ and with a pointed stick or thorn writing on its under surface a few words which would be decipherable for days afterwards. While we were inspecting some fine specimens of the cannon ball tree§ the dogs which had been busily roaming about started away with sundry shrill yelps and were immediately lost to sight. Away they went, until their voices became almost inaudible in the distance, then as they returned— for the agouti (which they were in chase of) invariably returns to the vicinity of the spot where he was first started—their noisy music became plainer and more distinct, until we could hear the animals rushing through the undergrowth near where we were standing, and then the noise got fainter and fainter as the pursued and the pursuers tore their mad course in the opposite direction. Eventually the agouti managed to elude the dogs and we resumed our march, noting on the way a couple of large

*Ischnosiphon parkerii ?
†Carib baskets.
‡Pancratium caribbæum.
§Couroupita guianensis.

ameiva lizards.* My companion, to whom the woods are as an open book, seemed to note everything, and pointed to the marks, made only an hour or so before, by the peccaries when they were lazily rubbing their sides against the trees. In trying to obtain further traces of the peccaries we were obliged to wade through a large patch of long, broad-bladed plants which we found to our cost were well described by the soubriquet of "razor grass."† Near here, I was told, a coolie had a singular adventure with a mapepire. The East Indian with some companions was out at night on an armadillo hunting expedition and having a lighted lantern in his hand, was standing talking to a companion when he became suddenly conscious of the unpleasant fact that the head and forepart of the body of an enormous mapepire‡ was slowly rising out of the under-growth, the glittering eyes and rapidly vibrating tongue was stealthily approaching the light and paralyzed the Coolie with fear. The reptile seemed to be trying to discover what gigantic fire-fly had entered his domain. The Coolies were so terrified at the sight of the blunt, rough scaled head, the flickering tongue and crotaline pits that they could not act, but a Creole, who had accompanied them, was more ready witted, and raising his trusty cutlass he delivered a swishing blow which cut off the inquisitive one's head. Next morning when they returned to the place they found the severed head firmly grasping in its enormously fanged jaws the stem of a young tree. The razor grass crossed we entered the terital§; underneath the large leaves a fine Bombyx silk moth was flitting which Mr. Urich carefully secured. Here animal life was more in evidence, for we caught a glimpse of a yellow crested green parrot|| flying noisily away, and heard the clang of the toucans¶ and cooing of the doves or ramiers.** A tree close at hand bore long scratches upon its bark which were unhesitatingly declared to be those of a tiger cat†† but pussy herself was not at home. Soon afterwards we found ourselves among some wild cacao trees and then we ascended an incline which gradually became steeper, and at length emerged upon a beautiful plateau extending over an area of several acres. Such a spot our friends thought an admirable one for a cocoa estate, it being a situation not liable to flooding in the wet season, a great desideratum, but one likely to be overlooked by the inexperienced, who are liable to select

---

*Ameiva surinamensis.
†Scleria, sp.
‡Lachesis muta, Linn.
§The terite grows in large patches which are locally called " terital."
||Amazona amazonica, L.
¶Ramphastos vitellinus, Licht.
**Columba speciosa, Gm.
††Probably Felis pardalis.

low-lying ground which, though dry enough in the first part of the year, would be absolutely untenable when the rains set in. Here we sat down on a log to rest and began talking over our return, all hope of finding peccaries being gone. As we talked the dogs, with the exception of the veteran Cook, started after another agouti and were off in pursuit in a moment, but they soon lost it; after their return another agouti appeared but was soon run down and killed. Gertrude, Jubilee, Spotty and Bull soon found another and were speedily out of hearing. Being tired we did not join in the chase and called to the dogs to return. A long while elapsed and then Spotty came limping through the bushes with his face covered with blood. Examination proved that he had had his eyelid ripped up in a fight with some animal, and opinions were varied as to whether it was the act of a lappe or an ant-eater. Sammy was sent to ascertain what had become of the other dogs and to find, if possible, their quarry. Half-an-hour later we heard our messenger calling to us to follow as the dogs had killed something. It is needless to say we started off as fast as possible. As I was suffering from a lame foot I was soon out-distanced by my companions, who ran like deer, but I still managed to follow, sometimes rolling down steep hills and always much impeded by certain curious, and extremely annoying creeping palms, armed with long thorns and which are capitally described by their local name of " Wait a minute "*—many were the occasions on which I waited several minutes. After awhile I neared the verge of the wood and rushed into a thicket which seemed to skirt the forest. I could not find an opening but as I saw Boney cutting a path for himself I followed him and soon emerged on a clearing surrounded by forest. Great trunks of felled trees were lying in all directions and effectually prevented any sprinting on my part. On the right dense volumes of smoke were rising, for fire had been set to the underwood and this, with the blazing sun combined to make the atmosphere decidedly sultry. Despite his years and emaciated appearance Boney tripped along in front as lightly as a goat. But I was not prepared to follow him when he suddenly ran along a slippery sapling which bridged a sort of hollow, and disappeared on the other side. Thought I to myself "when I set out this morning I was prepared to do " many things I had never done before but I never expected " rope dancing was one of them." The situation was far from pleasant, the hollow was deep, and at the bottom was soft mud and dirty water. This was not a bad thing to fall into if one must fall. But the mud was covered over with a thick mass of thorny brambles which were simply terrifying. On the right hand the fire was crackling and blazing within a couple of yards

*Desmoncus major.

of the narrow bridge. It was the case of a fiery Scylla and a thorny Charybdis, combined with the sword-blade bridge by which good Mahommedans reach Paradise. I started on my journey, at first slowly, step by step, but presently finding the heat on my right almost insupportable, I attempted to move faster, a frantic struggle to regain my balance ensued but without avail, and I fell with a heavy crash into the thorny bed on my left. Not being minded to become a nineteenth century St. Laurence I struggled vigorously and extricated myself at the cost of sundry scratches to my hands and rents to my garments, and eventually, considerably the worse for wear, I rejoined my companions who were standing round a dead animal which looked like an exaggerated guinea pig, of a brown colour with white spots, which I was told was a lappe.* The beast weighed about 30 lbs. and was in fine condition. From the appearance of the mammæ (which bye-the-bye I noticed with surprise were situated immediately between the fore legs, like those of the elephant) she had evidently young ones at home. It appeared that the dogs first started an agouti, which had run into the lappe's hole. This latter animal cannot stand canine barking and yelping and invariably bolts from its retreat on such occasions. This lappe had done this and was taking refuge in another burrow when it was shot by a half naked individual who was prowling about and who hearing the noise made by the dogs had come up in time to see the chase. The lappe was only severely wounded and again ran away but was soon afterwards pulled down by the dogs who would not allow the man to approach it. It was the lappe which had wounded Spotty's eye and ripped up Gertrude's ear. While Boney prepared the game for convenient transit to the forest we sat down to the huntsman's snack—biscuits and sugar! On the return journey much of the ground we had passed over was retraced —my bridge excepted. In the forest a Mahoe† sapling was cut down and its bark stripped off in long bands. With these Boney cleverly tied up the lappe's feet and so arranged it that two long bands formed a pair of braces, through which Sammy passed his arms so that the animal rested comfortably upon his back like a knapsack, leaving his hands free. Several birds were observed, amongst them being a wood-pecker, called by the country folk *Le Charpentier.*‡ It is said that this bird adopts a rather cunning method to obtain food. He is supposed to make a hole in a tree and then leave it for some time. After a week or two he returns, knowing that insects of various kinds are sure to have taken possession of it. Arrived

---

Cœlogenys paca, L.
†*Sterculia caribbœa.*
‡*Dendrobates kiikii*, Mahl.

at his trap he puts his beak in the hole and makes a loud rattling noise which so frightens the insects that they all rush out when they are promptly swallowed by the wood-pecker. This story may be true but it seems to be at variance with the general habit of the family which is supposed only to make holes in trees which have already been bored by insects. Tired out we made our way home, as fast as possible, arriving about four o'clock, after a tramp of at least twelve miles in the High Woods.

That evening as we sat in the comfortable rockers or swung in hammocks we chatted over the events of the day, now and again jumping up to catch the fine moths and beetles which flew into the house attracted by the light and once or twice we captured inquisitive tree frogs* which would mysteriously appear on the table and turn their great heads and large brilliant eyes towards the many insects dancing round the lamp.

Early next morning, as we rose, after a most comfortable night, and threw open the windows we heard the roar of the Howling Monkeys, now high and distinct and then dying away to a mere whisper. The animals must have been fully three miles off but we could hear them quite clearly and we were much surprised at the noise these comparatively small creatures are capable of. Mr. Carr, who has often shot them, told us that, when close to, their roar surpasses in volume that of the lions in Central Park, New York. On one occasion a wounded male sat on a stump and roared so loudly it unnerved him for the moment and it was some little while before he could control himself sufficiently to give the poor brute an effectual quietus. Besides being an ardent sportsman, Mr. Carr is a keen observer of Nature and does not always shoot the animals he encounters. Once, he and a friend managed to get near a tree in which a troop of Howlers were disporting themselves. In a huge forked branch gravely sat a fine male† in the prime of life. His chin was decorated by an enormous beard. Round him jumping, swinging, climbing, were a number of females and young ones. It was a family party and perhaps Mr. Howler had only just returned from a long journey. The ladies were particularly attentive to their lord and as each of them passed him she invariably caressed him about the face, beard and shoulders, darting away immediately, only to return the next moment and repeat her antics. The scene was quite a revelation as to Monkey family life and was watched for a long time with the greatest interest and amusement by the sportsmen who were good enough to refrain from putting a tragic end to so much innocent mirth and so many merry gambols.

---

*Hyla crepitans Weid ; Phyllomedusa burmeisteri, Blgr.
†Mycetes seniculus.

After coffee we sallied forth with the purpose of capturing agouti.　We crossed the Mamural River and ascended a small hill which is celebrated in the neighbourhood for the number of mapepire which have been killed on it, but it was not our good fortune to see one although we examined the ground very carefully in the hope of finding this terrible viper, so aptly named Bushmaster by our fellow colonists in Demerara.　A little further on and our friends pointed out the foot marks or rather the formidable claw marks of the Trinidad ant-eater, or matapel as he is locally called.　You are all acquainted with Dr. de Verteuil's story illustrating the strength of this animal and Mr. Broadway very recently saw a tough fight between a young one* and his dogs. We not only saw the marks of the claws in the soft earth, but also the work they were able to do in tearing up an old tree trunk for the ant-eater had been feasting on woodlice.　We were also shown fresh traces of some other animals.　The armadilloes† had been grubbing vigorously during the night, but had retreated to their holes with the dawn of day and we did not meet a single belated reveller wending his way to his burrow.　The dogs, who had been roving about, now started off in full cry after an agouti. Away they went until we could hardly hear their voices.　Then closer until the woods rang again with their noisy clamour and we could hear them rushing through the bushes close to us. Then farther off—and suddenly their note changed.　The agouti was declared to be in a hole.　We ran towards the spot and soon arrived at a Bois Mulatre‡ or wild Tamarind tree which had been half blown down by the wind.　The tree was hollow right up the centre and the agouti had entered at the root and got high up into the trunk.　Having carefully stopped up the entrance —not without difficulty, for the dogs were franticly trying to squeeze themselves into it—Mr Arthur Carr went up the tree and passing a long slender stick through a hole where a branch had once been, ascertained the position of the agouti—some five feet higher up.　The ready cutlass in skilful hands soon made an aperture while the agouti could be heard chattering his teeth at a prodigious rate.　This note of fear, however, soon changed to loud and piercing shrieks when the terrified rodent felt his hind feet seized, and these were redoubled when he was slowly, gently, but firmly, drawn out of his hiding place. *P-e-e-e-e-r*! P-e-e-e-r!! Pierre! Pierre!! screamed the agouti. "Listen; he is calling upon his patron saint," said Mr. Carr and strange as it may seem St. Peter is said to be the especial guardian of the agouti from the close resemblance of his

---

Tamandua tetradactyla.
†*Tatusia novemcincta,* L.
‡*Pentaclethra filamentosa,*

cry of terror to the word "Pierre." The little animal was closely tied and deposited in a strong haversack, when, finding he was not being ill-treated, he soon subsided into silence. Poor fellow, he did not long survive his adventure. He was incarcerated in a wire cage, from whence a month or six weeks later he escaped, but he had been taken too far from his native haunts and after wandering about a little while, he voluntarily ran back to the prison he had learned to know as his home. There the dogs caught him and tore his poor little body to pieces. The agouti secured we resumed our wanderings and were shown many objects of interest. Amongst them the Mamuse liane* which the thrifty planter splits into three and then weaves into cacao baskets. A fragrant scent of vanilla led us to spend some time looking for the plant, but without success, although the smell pervaded a considerable area. We noted a wild guava tree† the fruit of which is said to be delicious and is much prized by woodsmen detained in the forest after they have consumed their provisions. We also saw several Taparons or Cannon ball trees. Three or four more agouti were raised by the dogs but only one was secured—shot by Mr. Carr. Some fine cedars‡ were pointed out to us; also a Guatecaire,§ from the cones of which the Indians make their pipe-bowls—rather small and inconvenient ones to the European smoker's way of thinking. As we walked along Arthur Carr called out that he could "see a Mat in a Jigger Wood tree."‖ The jigger wood is a tree which apparently grows up and down at the same time. Of course, most trees send down their roots into the earth and grow upwards simultaneously. But the jigger wood gives ocular demonstration of how this is done—the roots springing by hundreds some little way above the earth's surface and shooting down directly into it, raising the heavy trunk as they do so a considerable distance, much in the fashion of the mangroves. The tree is sometimes called Bois negresse and looks as if it was a vegetable stilt walker or a gigantic broom. But there was a mat in the jigger wood. A mat I may remark is a large species of ground lizard, sometimes reaching three feet in length and stout and strong in proportion; he is a handsome fellow wearing a gay coat of black and yellow coloured mail, and is the possessor of a good strong set of teeth. He has been dubbed by scientists *Tupinambis nigropunctatus.* His flesh is said to be very good eating and not unlike chicken in colour and flavour. Our mat had run away

---

*Carludovica plumierii
†Eugenia sp.
‡Cedrela odorata.
§Lecythis idatimon.
‖Bravaisia floribunda.

from the dogs and had clambered up some fifteen feet and hung clinging with his long crooked claws to some knotty protuberances in the trunk. From this point of temporary safety he was anxiously watching the dogs and ourselves. He was a perfect specimen, some two feet in length, and the morning sun shining on his brilliant coat showed him off to the greatest advantage. He was beyond our reach, even when we had climbed up part of the way, and the only method of capturing him appeared to be to knock him down. The fall would not damage him but then the dogs would soon finish him if they were allowed to, so we cast about for a better plan to secure him alive and unhurt. Several were suggested but rejected as being too hazardous. At length Mr. Urich, who is fond of experiments, made a startling proposition. A long stick was cut and split at the end. In this was fastened a big piece of cotton wool which Mr. Urich steeped in chloroform. This was presented to the mat's nose and we watched the result with curiosity. After a second or two the head drooped, the bright eyes partially closed and the feet were relaxing their hold. We waited underneath with an open bag ready to catch him as he fell. But the lizard's drowsiness was only of a moment's duration, he raised his head and tried to scramble higher but lost his hold, fell, and just as we were pouncing upon him dived into the midst of the numerous roots and was lost to sight in an instant. Not deterred, however, we went to work with our cutlasses lopping off the roots and trying to find the half stupefied saurian. Soon one desisted, then another, and then a third, until only two remained at work. Suddenly there was a rush, and away went the mat helter skelter, tail in air followed by the dogs and ourselves as quickly as we could travel. The run was only a short one, for the lizard, after covering fifteen or twenty yards, threw himself into a ravine some eight feet deep and sought refuge in a small stream at the bottom, where he remained under the surface perfectly motionless. The water being clear we poked him until he began to swim with horizontal, undulatory motions, the limbs close to the sides, up and down, still keeping beneath the surface. At length he swam into a shallow place and holding him down with the flat blade of a cutlass he was soon secured and consigned to a bag, but not before he had inflicted sundry long scratches upon his captor's hands and wrists. Any of our members wishing to see this identical lizard may do so when next in London by visiting the Reptile House in the Zoological Gardens, Regent's Park, where he arrived last October in splendid condition. He is an exceptionally fine reptile, being perfect in claws and tail which is not the rule with lizards which are much addicted to fighting. As we returned home we passed near the place where the Mamural stream joins the Caparo river. Here there was a somewhat open spot with an exquisite bit of tropical scenery. A huge Mountain Rose, or Cooperhoop* covered with parasites and lianes overhung the Mamural river. The varied mass of greens, browns and yellows of the Enaré,† Terite, Manae and other palms, the large leafed Seguines,‡ the numerous mosses and dried leaves and grass, seemed to present every imaginable shade of colour under the golden rays of brilliant sunshine. The gentle murmur of the streams in which hundreds of little silvery fish were merrily swimming and splashing, added to the charm of what I considered my first glimpse of the beauties of tropical forest scenery. Amongst the insects we found on this occasion in the vicinity of this lovely spot were two species of *Eciton?* or hunting ants, some young *Acridœ* living in the rolled up leaf of a terite, and a wasp like butterfly. We were shown, amongst the many plants which are valued by the frequenters of the forest, two kinds of Manac, one of which grows singly, the other in stools, but to the inexperienced eye there is no other distinguishing point.§ The latter is said to be a capital antidote to snake poison. There is a curious superstition connected with the portion which is said to have virtue ; only those roots which grow towards the rising sun are said to have any efficacy—the other parts being perfectly useless.

The next day was the last of our visit to the High Woods of Caparo and our kind hosts determined that one more effort should be made to show us how the wild hog of the Trinidad woods is hunted. We started off at the usual hour, 8 o'clock, by way of a ridge of hills dividing the Grand Ravine Valley from the Mamural Valley, which we had visited the day before. Our party consisted of the Messrs. Carr, and Urich, myself, Boney and Sammy. We were followed by eight dogs. Shortly after we started an agouti was disturbed, but the dogs were called off and not allowed to waste their strength, and our time, in its pursuit. The forest scenery was much of the same description as that which I had seen on the occasion of our first day's excursion. Upon one of the trees—I forget now what it was—we found vanilla‖ in flower and a few moments were spent in listening to Mr. Urich's explanations as to how fertilization is accomplished by the aid of birds and insects. Then we began climbing some exceedingly steep but moderately high hills covered with keen edged razor grass. From the summit of one of them we overlooked what had been a tremendous landslip in which hundreds

---

Brownea rosa.
†*Geonoma énaré.*
‡*Philodendron* sp.
§*Euterpe oleracea* and *E. surculosus.*
‖ *Vanilla planifolia*

of acres had slid down at least 200 feet, making a huge gap
through which we could see the forest extending for miles and
miles, right up to the Montserrat settlements—a fine view of a
billowy, leafy sea of green.   Tracks of the armadillo were in
plenty, and at one spot our companions told us how they had, a
month or two before, blocked up an ant-eater in a hole, intending
to take him on their return, but the wily brute had, by the
exercise of his great strength, removed all obstacles and effected
his escape.   A noise here attracted us which resembled the
whirr of a wheel, lasting for some seconds and then terminating
in a sharp shrill whistle—the cry of the *merle laque jaune*⁰ or
yellow corn bird.   But our thoughts were soon diverted from
this remarkable note, for we had suddenly come upon unmistakeable traces of our game.   There were the quenk signs ; there they
had been drinking ; there were their hoof marks ; there they had
been scratching their muddy backs against the trees.

The dogs quiet down and begin running hither and
thither with rapidly waving tails.   Every one is growing excited,
and now the dogs concentrate their attention on one particular spot.   They begin to take a definite course.   Melon
and Cook, with noses lowered, now trot along more confidently.
The scent is warm.   The other dogs begin to follow.   Presently
a sharp yelp from Melon, then Cook replies.   We have been
imperceptibly increasing our pace and now we are rushing
through the woods at headlong speed.   The dogs whimpering and
yelping are far ahead.   My comrades seem to have disappeared—
only Albert is near.   My wounded foot distresses me, but I say
to myself " hold on, I have never chased peccary before—perhaps
never shall again."   I am going at my top speed ; now tearing
through the razor grass with hands held high to shield my face ;
now leaping down a steep bank, next stumbling against a half
buried root.   On we rush, Carr ever in front, calling
first to me and then to the dogs—" Come on, come, come !"
" Good dog ! Good dog ! houp ! houp ! good dog ! " " Come on
Mole ! !"   The dogs are going their best and so am I.   Crash—I
find myself in a thicket of " wait a minute " thorns. I tear myself
free and immediately slip on a moss-covered root, measuring my
length on the ground.   I scramble up, considerably demoralized,
and run down a steep hill at the peril of my neck ; but the chase
is at the bottom ; when I get there it is at the top of the next
one.   Breath comes thick and short and I am dripping with
perspiration but on, and on I stagger; now tripping, now held
back by lianes into which in my headlong flight I have
incautiously run.   Ever on ahead is Carr, calling on
" come " but I am so exhausted I can't reply.   I set my teeth,

Ostinops decumanus Pall,

the ascent is trying, but still on and on I labour—what matters my foot; what matters short breath, what matters it I can hardly see; what matters anything in the wide world, so long as *I* am in at the death. The hill is surmounted and again we are rushing helter-skelter into another valley. I can scarcely see; I can hardly hear for the beating of my own heart. I only know that we are in full chase and that if it lasts much longer, I cannot. If I can only climb that next hill I may perhaps be in at the end. Never was there so steep a hill, never were the lianes thicker, never were there so many importunate thorns, never were there such a quantity, each one lodging in my clothing with the tenacity of a minute grappling iron ; never were my legs, used only to the unimpeded paths of town, so weary ; never was my strength so feeble. At last I reach the hill-top, ready to sink with fatigue, lungs almost bursting with laboured exertion. I must give up; but Carr shouts " He has stood up—listen to the music ! l " Such a grand canine chorus in the valley beneath ; it rends the air, clear and distinct ; it puts new life into me and mustering up my little remaining strength for one supreme effort, I fly down the precipitous descent, catching at the trees as I pass for momentary support. What a sight ! At the bottom of the ravine the hill ascends sharply on three sides of us ; to the right there is a gentle descent. In a hollow, over which grows a Bois Mulatre, with his hind quarters against its roots, with fiercely bristling grey and black hair, flashing black eyes, and gleaming white teeth stands our game—at bay ! His undaunted front is opposed by five or six fierce dogs, yelping and jumping, now rushing forward, now retreating, as the peccary with wondrous activity turns this way and that champing, snapping, gnashing his formidable tusks with rage. First one way and then another turns the furious beast but escape is impossible and, finding himself surrounded the brave animal prepares to sell his life as dearly as he can. The dogs are not afraid, they know their business well, they have simply to keep the quarry at bay ; but the quenk is an old boar, resolute withal, and means mischief to his canine persecutors. He rushes forward now and again and as he does so, the dogs threaten his flanks. Bull tries to seize his throat and the quenk turns to repel the attack. Immediately Gertrude jumps in from the other side—Alas ! poor Gertrude, your last battle !—the pig turns as quick as thought and drives his tusk right into her breast ripping it up with one deep keen stroke and permanently dislocating her shoulder ; another rip in the neck follows as quickly, and with a piteous howl poor Gertrude drags herself away with blood flowing from four or five wounds. Melon and Bull dash simultaneously at the peccary's throat and are each received with two lightning-like right and left rips of the terrible tusks. Jubilee tries his luck but the peccary drives him back. Cook, cunning old veteran, though no less brave, couples his courage with prudence and coolly scrambles up to the bank above ; then, when the others are making a demonstration in front and the quenk's attention is fully fixed, he takes the opportunity and leaping boldly down on top of the pig seizes him by the back of the neck where he holds on with bull-dog tenacity.   The others close in and now there is a grand rough-and tumble in which hair and fur, dead leaves and sticks fly, and blood flows, and the air is filled with squeals and yelps.   First pig then dogs are getting the mastery, but the latter are at a disadvantage as their teeth do not inflict any serious wounds, while the boar's tusks, so deftly used, draw blood at every stroke.   The dogs snarl, growl, yelp and bite and the furious gleaming tusks rip with merciless effect and chatter with frightful rapidity.   It cannot last much longer and it is a question after all whether the pig will not be able to free himself of his antagonists and leave the field the victor, for Gertrude is disabled and Bull is down under his feet *hors de combat*.   It is a brave fight but the odds in favour of the dogs are suddenly increased ; piggy is defeated almost in the moment of victory.   Something leaps from the bank right into the mass of maddened animals and in a second the fight is over, the quenk is down, for Arthur Carr has his foot on his neck.   Bull extricates and drags himself away.   Sammy rushes forward and drives the lance sure and deep into the brave boar's heart and the plucky pig dies without a groan.   The dogs are beaten off but not without difficulty.   They are mad with excitement and wounds and are anxious to tear the dead foe to pieces.   Boney comes up with a cool business-like air and disembowels the game throwing the entrails to the dogs who are thereby somewhat pacified.   I am too tired to move, but my companions are as fresh as possible ; like true huntsmen their first care is for their canine friends.   Poor Gertrude is lying on the bank with glazing eyes and the blood flowing in a steady stream from her chest. This is stopped by the application of a poultice of the half digested food found in the quenk's stomach.   She tries to rise but falls down again.   Besides this serious wound she has her lip cut open, a wound on the top of the neck and another one in her side.*   Melon   has   three   small   rips   under   the   front   legs and a scratch over the eye while a portion of his jaw-bone is laid bare.   Bull has a gash on the ribs on his right side, another under the right fore leg and a third on his crown.   Jubilee has one neck wound and two under the fore legs.   Blackie has a cut behind the ear, one in the neck and a deep one in the throat. Cook the wary one, escaped with two small scratches—but then

---

*Gertrude lingered on for a few months and eventually died in spite of the assiduous care of her masters, with whom she was a great favourite.

he seized the foe from behind. The dogs all wounded hunting was over for the day. For my part I was heartily glad for I could not have run another yard be the excitement ever so great. After a long rest, during which Boney strapped up the quenk with thongs of bark, we started homewards, Arthur Carr making an impromptu hammock, in which he placed the wounded Gertrude, slung it round his neck and gave her additional support with his arms, carrying her, oh, so tenderly—a mother with a sick baby could not have been more careful than he, as he avoided rough and broken ground, and the many obstacles which one meets in the forest, for at every jolt—and there were some unavoidable ones—the wounded creature moaned piteously. Then Sammy put his arms through the bark braces and shouldered the quenk and away we marched for home, capturing a manicou* and six young ones on the road, arriving at *Itcarrdonum* about one o'clock, after a journey of about six miles in the woods.

2nd February, 1894.

Didelphys marsupialis, L.

NOTES ON TICKS.

THE presence of Ticks upon Cattle has been attracting a great deal of attention during the last few years. Although these pests have been known in the West Indies for a long time, it was pointed out a very few months ago that the ticks were among the least known of the West Indian fauna. Active steps have, however, been taken in various directions to remedy this state of things. The Curator of the Institute of Jamaica in particular has been collecting details and specimens from various sources, and has published a series of interesting leaflets which show the gradual acquirement of knowledge on the ticks infesting the cattle in that island. In our own as well as a great many other tropical countries the cattle suffer a good deal from these irritating and weakening pests, and any discussion upon them which will lead to their diminution should be of great service.

In the Supplement to the Leeward Islands' *Gazette* of April, 1892, a short article appeared on this subject. In that paper I gave a general description of the animal and its congeners : their position in the animal kingdom was defined, and the main distinctions were drawn between the class Arachnida, to which they belong, and the much more numerous class of Insects. The life-history and habits of the individual tick were detailed, and the article concluded with a determination of the large Antigua Gold Tick, received from Mr. A. D. Michael. The species was then provisionally named *Hyalomma dissimile* (a Mexican form), but more material was required, as the specimens at Mr. Michael's disposal consisted of two imperfect females which had been forwarded to me before reaching the Leeward Islands.

I have since, at Mr. Michael's request, made full collections of males and females, old and young, of both the Gold Tick and the smaller, common Blue Tick. The resulting determinations were interesting. There appears to be a well-founded belief that the Gold Tick was introduced into Antigua from the West Coast of Africa. About thirty years ago some Senegal cattle are said to have been brought over to the Leeward Islands and first disembarked at St. Kitts. Their destination was however Antigua, and upon their arrival here they were found to be infested with a large tick. This, I suppose is the reason for the Gold Tick being also called the St. Kitts Tick, although it is said not to be found in that Island. I mentioned these facts to Mr. Michael, and suggested a particular examination of West African species, having regard to the local tradition.

Mr. Michael wrote : "*St. Kitts Tick.* I informed you before that this belonged to the genus *Hyalomma*, but that, not having the male, and not knowing the colouring, I could not be certain about the species. Having now the advantage of the male specimens and of your discription, and also the young females, I can, I think, identify it with certainty. I was right about the genus, but the species which I suggested it might be, turns out not to be the correct one. I think there is no doubt whatever that your St. Kitts Tick is the *Amblyomma venustum* of Koch, and would now be called *Hyalomma venustum.* The interesting part of this is that Koch described and named the species in 1847 from a single male specimen *which he received from Senegal.* I do not know of any subsequent record of the creature."

With regard to the common Blue Tick, Mr. Michael writes : "*Creole Tick.* I am not quite so certain as to the species of this tick. The specimens have not come in quite such good condition as those of the other sort, and it is not such a well-marked species. There seems to be only one male, and I think I can identify it as another of Koch's species. It would now be considered to belong to the genus *Rhipicephalus*, and I think there is little doubt that the specimens belong to *Rhipicephalus concinna* of Koch. Koch had two males and two females, but did not know where they came from."—Leeward Islands' *Agricultural Journal*, July, 1894.

(To be Continued.)

Communications and Exchanges intended for the Club should be addressed to the Honorary Secretary, Port-of-Spain, Trinidad, B.W.I.

Price 6d.   Annual Subscription, 3/.

Vol. 2.   APRIL, 1895.   No. 7

J'engage donc tous à éviter dans leurs écrits toute personnalité, toute allusion dépassant les limites de la discussion la plus sincère et la plus courtoise.—LABOULBÈNE

Trinidad·Field·Naturalists'·Club.

NATURA MAXIME MIRANDA IN MINIMIS.

Publication Committee:

President.

H. CARACCIOLO, F.E.S., *V.P.*:   P. CARMODY, F.I.C., F.C.S.,
SYL. DEVENISH, M.A.,   R. R. MOLE,
F. W. URICH, F.E.S., *Hon. Secretary.*

CONTENTS :—

Report of Club Meetings :

February	163
Bats	164
Trinidad Butterflies	170
Fungus Growing Ants	175
Notes on Ticks	183

" MIRROR " Office, Port-of-Spain.

☞ All Communications and Exchanges intended for the Club should be addressed to the Honorary Secretary, Port-of-Spain, Trinidad, B.W.I.

JOURNAL

OF THE

Field Naturalists' Club.

VOL. II.   APRIL, 1895.   No. 7.

REPORT OF CLUB MEETINGS.

15TH FEBRUARY, 1895.

PRESENT : Sir John Goldney, V.P. (in the Chair), Messrs. H. Caracciolo, V.P., F. de Labastide, R. R. Mole, A. T. Reocb, John Hoadley, T. I. Potter and F. W. Urich, Hon. Secretary. Mr. John Ponsonby (London) and Mr. H. C. Bourne were the visitors present. Mr. R. S. Reid was elected a Town Member. A letter was read from Dr. G. Brown Goode (Washington) thanking the Club for his election to Honorary Membership. Mr. Caracciolo described an excursion to Macqueripe Bay. Mr. Mole read some additional notes on "The Caudal Appendage in *Crotalus horridus*." The Secretary exhibited an abnormal egg from Guanapo presented by Mr. Albert Lucien. The egg, laid by an ordinary hen, was unusually large, it contained another perfect egg inside. Size of egg :—$3\frac{1}{2}$ ins. long, $2\frac{1}{2}$ ins. broad. Inner egg ordinary size.

15TH MARCH, 1895.

PRESENT : Mr. H. Caracciolo, V.P. (in the Chair), Professor Carmody, Messrs. Syl. Devenish, T. I. Potter, T. W. Carr, R. R. Mole, John Barclay, A. T. Reoch, H. J. Baldamus, J. H. Collens, E. D. Ewen, A. Taitt and F. W. Urich, Hon. Secretary. The Secretary read on behalf of Mr. C. W. Meaden a paper on "A Cattle fly" giving a description of the fly, the damages it does and the remedies to be applied to infected animals. Mr. Mole read a note on "The Dimensions of Animals." Mr. Caracciolo exhibited a large collared owl (*Pulsatria torquata* Daud.) and a Mygale spider. Mr. Urich showed some bats *Chilonycteris rubiginosa,* Wagn from the Oropouche caves, which, according to Mr. Oldfield Thomas, were rather uncommon and only represented in the national collection by one specimen from Trinidad. A number of specimens of *Peripatus trinitatis* from St. Anns were also on view and the Secretary informed the meeting that he had recently secured large numbers of specimens which evidently represented at least two species. The reason why these comparatively rare creatures had been recently found in such abundance was the exceptional dryness of the season which had caused them to congregate in the ravines for the sake of the little moisture to be found in them. Upwards of 100 specimens had been secured in this manner. The specimens were compared with the illustrations in a splendid Monograph (Sedgwick) of the Peripati with a considerable degree of interest.

CLUB PAPERS.

BATS.

By Henry Caracciolo, F.E.S., M.S. Ent. de France, &c.

BY recent systematists bats have been divided into two groups, viz.: *Megachiroptera* or fruit eaters and *Microchiroptera* or insect devourers. Amongst the former we have the larger bats, the insectivorous ones being generally small. According to Dobson's Catalogue (1878) there are on an average about 100 fruit-eaters known, whilst 300, more or less, are recognized as insectivores. These have been arranged in 80 genera and six families, but since 1878 many species have been added and the number now described must be very nearly 500. These creatures at one time were supposed by some naturalists to have evolved from terrestrial mammals, which gradually developed wings to adapt themselves to circumstances, but remains of bats have been found in the Eocene deposits and even further back, which proves that they are a distinct class, though closely allied to the insectivorous mammals, and, more wonderful still, the sloths. The distinctions between the insectivorous and frugivorous species are as follows. The former have their teeth cusped or sharp in the form of a **W** and the molars are divided transversely, whilst the teeth of the fruit eaters are flat and longitudinally grooved. The index finger of the insect eaters never has more than two joints, but usually only one, and it never terminates in a claw as it frequently does in the fruit bats. Then again the two margins of the conch of the ear arise from separate points instead of being joined together and the tail is joined to the membrane and appears always on the upper surface. Though there are many other distinctive marks characterising the two groups, those I have given are sufficient to identify them. The wings of bats consist of a membrane which has its origin on the sides of the neck. It was once believed that this membrane is of a callous nature but experiments made by Spallanzanni and others have proved that it is a very sensitive organ. I had an Artibeus alive for some time in a large cage, the bat being apparently asleep, I took a thread and merely touched this membrane, this excited nervous movements and the animal was evidently uneasy for some time. The fingers in bats are developed to an extraordinary degree and they thus serve to stretch the flying membranes, when in use, just as the ribs of an umbrella keep its covering open. The hinder limbs can scarcely be called organs of locomotion for the creature is exceedingly awkward when placed on the ground, walking or creeping very much like a sloth when in similar circumstances. In regard to their geographical distribution bats are found almost all over the world but principally in the tropical zone, where the largest number of species are known, in fact they are supposed to be the characteristic type of mammals in the tropics, only a few species inhabiting northern climes. The distinctive appearance of the two groups, as given above, enables us to decide off hand to which of the two great sub-orders any particular individual may belong, yet the observations made by naturalists show that a few species deviate in their habits from the rule indicated by these distinctive characteristics. For instance, in the case of *Noctilio leporinus* the food has varied considerably, probably owing to a gradual adaptation of habits to altered conditions of life. The species, which is widely distributed in Central and South America and is rather common in Trinidad, lives under roofs of houses but principally in caves. This bat is stated by no less an authority than Linnæus to be a fruit-eating bat, and is referred to by him as *Victilaus fructibus arborum* though its characteristic dentition and other peculiarities place it among the insect eaters. Dobson found some fruit seeds in the stomach of a specimen sent to the British Museum, apparently the seed of *Morus tinctoria*. But other naturalists (Gosse and von Tschudi) remarked that it lived in hollow trees and feeds on beetles; I have examined many stomachs myself and often found the tibiæ of beetles in them. But now, strange to say, the species has shown a still more remarkable deviation from its original habits. It has given up its diet of fruit and beetles and has taken to fishing in salt and fresh water. The first account of this new custom is that of Mr. Fraser who observed it in Ecuador. There he saw the bats skimming the bank of the river at Esmeralda, every now and then making a dash along and actually striking the water, catching in their swift career minute fish which were incautiously swimming near the surface. Some time in January 1892 I accompanied the late Mr.

McCarthy to the Caves at Gasparillo, where these bats live in great numbers, for the purpose of ascertaining whether this fishing habit was a fact and the precise manner in which the bats went about it. We arrived late, after 7 p.m., on a beautiful moonlight night and we observed the bats leaving their caves and performing their fishing operations. They generally flew near shore, say about ten feet from the rocks and one foot above the surface of the water, going to and fro, until they perceived a fish, then suddenly they made a swoop, striking the water with a swish and off again. At times this manœuvre is made very quickly and at others the animal seems to be checked a little, no doubt the prey in this instance giving a little more trouble. I could see the interfemoral membrane, *i.e.* the membrane between the thighs, close in the form of a bag, but this I believe is done to check their flight which is very rapid, and not in my opinion for the purpose of collecting fish. I have also examined their deposits collected from the caves at Gasparillo and from an old gateway, which once faced the wharf, and all the deposits contained remains of fish and beetles. These bats are also to be seen fishing in the drains and gutters in Tranquillity and may be observed in this interesting operation at dusk any evening and also on moonlight nights. Mr. McCarthy suggested at the time that the feet might be used for capturing their prey but my opinion is that there is no reason for believing so. The claws are only used to hang by, which is the reason for their great development. This bat has a large expanse of wing. The specimen before you measures, $6\frac{1}{4}$ inches from tip of nose to the end of the extended interfemoral membrane ; the length of head and body is $3\frac{1}{2}$ inches ; length of fore arm 3 inches; hind foot $1\frac{1}{2}$ inches and expanse of wing 23 inches. The colour is orange red. I have found it living in company with a smaller species *Mormops megalophylla*, the first specimen of which was sent to England by Sir William Robinson in 1889. The bats of Trinidad belong to three families, viz.: Vespertilionidæ, Emballoneuridæ and Phyllostomatidæ. Of the first family we have up to now only discovered two species— *Vespertilio nigricans* and *Thyroptera tricolor* though it is very probable, according to Mr. Oldfield Thomas, that there are more. These are distinguished by having no exfoliations or nose leaves on the muzzle and by having the tail contained in the interfemoral membrane, the tip slightly projecting beyond it. *Thyroptera tricolor* is by no means common ; it is a small bat with a spread of wing of about 9 inches. It is chiefly remarkable on account of an adhesive circular pad at the base of the thumb and on the foot, by which it adheres, to the various objects on which it may settle ; it is an insect eater. In the next family Emballoneuridæ the members are characterized by the absence of a nose leaf, and as a rule they have a small tragus in the ear and only a single pair of upper incisor teeth which incline towards one another. The tail is partially free and perforates the interfemoral membrane, appearing on the upper surface or produced far beyond the posterior margin. This family is largely represented in the Neo-Tropical Region in which Trinidad is situated, and contains fruit eaters, insect eaters as well as salt and fresh water fishers. *Noctilio leporinus* belongs to this family. They are also commonly called mastiff bats on account of the supposed resemblance of their broad, wide-mouthed muzzle, to the head of a mastiff and are the first representatives of the above family. They are characterized by the thickness of their tails which are prolonged beyond the outer margin of the interfemoral membrane. Of all our bats they are the ones that can best crawl, on account of a peculiar corn-like callosity at the base of the thumb. The hind legs are free from the wing membrane and their wings are long and slender. This great length and narrowness of the wings indicates as Dr. Dobson observes, rapid flight, and since they have the power of varying the great length of the membrane between the legs by a sort of reefing process they must have great dexterity in suddenly changing their direction. It is curious to note *Molossus (rufus and obscurus)* flying at dusk between the Zamand trees in Queen's Park, catching insects. They fly high and very much like swallows, their flight is exceedingly rapid and it is interesting to note them in the act of capturing their prey, their power of sight must be intensely developed for, notwithstanding the rapidity of their flight every now and then they will dart side ways, and sometimes swoop down a distance of at least 15 feet to pick up an insect which flies about one foot from the surface of the savannah. *Molossus obscurus* is not so common in town, though I have captured them in the roofs of houses. They are smaller than their ally *M. rufus* and insinuate themselves between the galvanized sheeting and the wood-work of roofs, their deposits often collect in the gutterings and is a source of danger to those who use rain water for drinking purposes as they pollute the water and it may become injurious. This bat however is more often found in the outskirts of the town and lives in decayed stumps of trees. An interesting little bat is the *Saccopteryx bilineata*, a pretty little creature of a brownish colour with two longitudinal white lines running down the back. We possess three members of this genus *S. bilineata S. leptura* and *S. canina* which are remarkable for their pouched wings and have derived their name from this peculiarity. The number of incisors is 1-3 and the cheek teeth 5-5 on each side. These pouches are well developed in the males and almost rudimentary in the females, and are supposed to be a means of sexual attraction on account of a red coloured strong smelling substance contained therein. In *Saccopteryx leptura* the pouches are very much larger than in the others. A curious little bat of this family is *Rhynchonycteris naso* Wied. They are often met with on the banks of the Caroni, in groups, resting on stumps and trunks of trees, especially those over-hanging the water. This is our smallest species and has a spread of wing of about 8 to 9 inches. Its colour is brownish gray, very closely imitating the tree trunks, they are also very often found under the loose bark, which no doubt they seek for protection against the hawks and other birds which feed on them. The muzzle is rather elongated and fine, the tail perforates the interfemoral membrane and appears on its surface. The next and last family of Trinidad bats is the Phyllostomatidæ— largely represented here and in America, and differing considerably in size though all agreeing in the possession of the nose leaf, which in some instances is longer than the head. They are collectively styled Vampires and are exclusively confined to Central and South America and the West Indies. We have up to the present discovered 17 species in this island and may probably, very shortly add several new ones to the list. They are characterized by the presence of three bony joints in the third or middle finger. A very curious species inhabiting this island is *Glossophaga soricina* and represents with *Anoura geoffroyii* a group of bats characterized by their long tongues which project considerably from their mouths and can be pulled out twice the length of the head. The tongue ends in a mass of papillæ and in shape somewhat resembles a tooth brush. This peculiar modification it was once thought was used for abrading the skin of animals, previously to sucking their blood, but it is now known to be used for licking up the pulpy substance of fruit. The lower lip is deeply channelled, the incisors enlarged, the medians larger than the laterals and projecting with a fine cutting edge. This small and harmless bat is thought by many persons to be the blood sucker, no doubt owing to the length of its tongue, in the very same way that 99 out of 100 persons not only in Town but in the country (where the people have the facility of acquiring better information) believe that snakes sting with the filamentous process which they project from their mouths and which is nothing else but their tongues. These bats simply use their remarkable tongues for licking up the pulp of fruits and the honey of flowers. Again we possess in this family the largest bat of the New World, in fact the largest in the world, except the great fox bat of India. The specimen on view, *Vampyrus spectrum,* with the skull attached will show you what a formidable creature it apparently is. It belongs to a group of this family in which the tail, when present, perforates the membrane between the legs ; the nose leaf is spear-shaped. It is a most horror-striking animal when viewed from in front, with its ears standing erect from the sides and the top of the head, its spear-shaped nose-leaf and its powerful canines. No wonder it was imagined that this animal must be the terrible blood sucking vampire, but such is not the case, for it is a most harmless bat and is reported to feed principally on fruits. This species has no tail. *Chilonycteris davyi* is a small bat closely allied to *Chilonycteris rubiyinosa.* I have caught them both in the gardens. The former is characterized by its wings being attached to the centre of the back instead of the sides of the body and its back is entirely naked, a peculiarity not yet accounted for. This genus differs from the other members of the family in the absence of a nose leaf the function of which is performed by folds of skin depending from the chin. It is usually of a dull brown colour. Of the fruit eating species belonging to this family, *Artibeus*, four species of which up to date have been discovered, the most con· spicuous are *A. perspicillatus, A. planirostris, A. quadrivittatus* and *A hartii.* They are easily recognizable by their blunt faces and the interfemoral membrane, concave behind, no tail being present. The largest species, *A. planirostris*, has a large and fleshy nose leaf, the sides of the process being rounded and not produced upward into lobes. They are particularly fond of sapodillas, mangoes, pomme-malac and guavas and begin to fly very early in the evening. They seem to be very strong and muscular, often making several darts at a fruit but eventually picking it, however difficult it may be ; after picking the fruit the bat resorts to another tree, sometimes a long distance off, there hangs by its hind feet and clasping the fruit with its thumbs eats it leisurely ; they do not seem to be always able to discriminate between ripe and unripe fruits, and often pick green ones which they drop ; this accounts for the quantity of fruit one finds under these trees. *A. perspicillatus* is very little smaller and can be at once recognized by the two longitudinal white lines on its head ; these two species are often found under the eaves of houses. *A. bilobatus* is smaller yet and is distinguished by the edges of the basal part of the nose leaf being produced upwards into lobes ; it is much less common and more often found in the country districts than in Town. Its habits are identical with *A. planirostris.* I may here add that the latter was supposed by Charles Waterton to be the blood-sucking bat. We come now to the true blood-sucker, *Desmodus rufus.* The canines are as sharp and pointed as needles, it has no true molars and no spur on the ankle for the support of the membrane between the legs. Darwin, up to date was the only person said to have had the good fortune of seeing one caught in the act of sucking the blood of its victim, but Mr. Libert a zealous member of the club reports having personally caught one in the act.   I had heard it said that some persons are proof against bat-bites and having been several times exposed to their attacks without being bitten I very nearly believed it, until one night I determined to give them all facility.   I slept in a hammock at Gasparee exposing my foot, next morning I was quite surprised to find myself bathed in blood, the result of a bat-bite.   They seem to delight in poultry and attack their birds in the neck.   I once saw an unfortunate hen which had been bitten in the night so weak she could not stand and in nine cases out of ten the feathered victim dies.   Some country people pretend that these bats go out twice a night in search of prey.   It may be they are misled by the fruit eating species who appear at dusk and late during moonlight nights; but since we have had electric lighting the insectivores have quite a good time of it and feast greedily up to early morning.

4th May, 1894.

NOTES ON SOME TRINIDAD BUTTERFLIES.

By Lechmere Guppy, Jr.

THE following notes give an account of the striking resemblance between some members of the genus Danais in one or more of their three stages; at the same time showing the strong difference between them in other stages; as for instance while the larvæ are very much alike, the pupæ differ very much, and then the imagos are again alike and *vice versa*.

Drawings and specimens of the undermentioned are shown.

I have also added some notes on *Ageronia feronia*, and *Gynaecia dirce*.

Tithorea flavescens and *Sais eurymedia*.—The eggs of these butterflies are laid singly on the underside of the leaf of a very well-known creeper here, Echites sp., which is very common in Queen's Park (referred to later).   The former species *T. flavescens*, often chooses those vines which are in open situations while the latter only deposit on the small ones in the depths of woods, or in very shady, obscure, moist and protected places suited to the frail imago; these vines are extremely pretty when sheltered from the sun, and the venation of the leaves is beautifully developed, showing up clearly against the dark green; they seldom, however, grow in such situations to a height of more than 1 to 2 feet and the supply of leaves is only sufficient to support one or two larvæ, more often than not only one is found on each plant under a leaf.

In the open, exposed to the sun, like the vines in the Queen's Park, so much frequented by metallic tinted wasps, the leaves lose their beauty completely, but whereas the plants that grow in the open are luxuriant in foliage, the others that grow in the shade are deficient in this respect.

I have found several larvæ of *T. flavescens* on some creepers and they are very easy to rear in captivity.

In the larval stage *T. flavescens* and *S. eurymedia* are almost identical.

The larvæ of the former are black and white with about eleven latitudinal bands or belts, marked with black and white, regularly, like a chequer-work pattern, on close examination, appearing like so many links of a chain, as in *S. eurymedia*, except that the latter are reddish about the feet and under the body, and when mature are just about half the size of *T. flavescens*—remarkable, thin, tapering horns rise from the second segment, and are slightly curved at the tips and are vibrated rapidly when disturbed—the movement is not so rapid as that in the larvæ of *Lycorea* sp. mentioned later on.

The pupæ of both are again similar, the splendid burnished golden appearance, except where the thorax, eyes, wings, and a few specks, marked in black indicating the venation and other parts mentioned, most distinctly and clearly. In some of the pupæ the dorsal depression on the ridge between the head and abdomen is more developed than others but it is always conspicuous, which seems to cause the protuberance of the thorax as in *Callidryas*. The resemblance to *Mechanitis* sp. in this stage is marked and I refer to it later on. The imago of *T. flavescens* described by Kirby, (see F. N. C. Journal) No. 3, August 1892) and *S. eurymedia* are very different, but in the early stages these insects are very much alike. The delicate, frail, *Sais eurymedia* with its soft, transparent, pale greenish yellow wings, margined with black, the fore and hind wings rounded at the tips, flies low and slowly along, while *T. flavescens* is tawny yellow and black, strong, and jerky in its flight, has pointed forewings and is very different in this stage.

There are two types of larvæ of this insect which, however, produce very little variety in the imago,—that is, some few of them are much darker than the usual type met with. They are almost black, the white marks showing very faintly, the pupa is more marked with black in the venation of the thorax, etc., but this species is liable to such changes, and not constant in its markings, though according to Kirby more so than some other members of the group.

A comparison of the above with *Lycorea* is interesting; the eggs are deposited in the same manner on the plant which is a species of *Oleander*, and the larvæ very much alike, merely the markings being different, it is striped with broad white and black belts, (with the exception of the first four segments which are black and pale yellow), the white belts are, however, curved up and pointed just above the feet, resembling hooks. The horns are longer and vibrate most rapidly whenever the insect is disturbed, the pupa is however quite different, being yellow almost transparent and with no marked dorsal depression, there are black spots on the thorax and a stigmatal and dorsal row, otherwise it is simply smooth, without the striking appearance of *T. flavescens*.

The imago is again so different, resembling more the style of *Mechanitis*, which however, differs from it in every other respect, as the following account shows :—The eggs of *Mechanitis egansis* are laid in clusters on the upperside of the leaves of the cockroach plant—(*Solanum* sp.) they are white conical ribbed longitudinally, the larvæ are gregarious—in these respects differing quite from *Tithorea*, *Lycorea* and *Sais*, although so similar to the last mentioned in its perfect state.

The pupa is very much like *T. flavescens* and *S. eurymedia* and has the same beautiful appearance, but the dorsal depression is not so pronounced. In the imago the likeness to *Lycorea* appears and this is the only stage in which there is any resemblance, this is striking enough to lead one to suppose there would be some characteristics in the two previous stages, both in the habits of the larva and form of pupa which would give one some clue to the relationship to *Lycorea* sp.

These insects may all be procured near the same spot and are usually seen together. They are protected in the same manner as *Heliconius* by a peculiarly offensive odor very much like the smell of the oil in the cashew-nut. *Mechanitis eyænsis* is one of the most easily captured butterflies, from its fondness for a certain weed, (I am hoping to find out its name) which grows to about five feet in height, with clusters of small white flowers, somewhat like a thistle in shape, and very common in some localities. As many as thirty or forty of them may be captured with the fingers, when these plants are in flower. From a good sized plant I have picked them off in this manner, one by one for my amusement. When liberated they will fly round and return again ; so great is their fondness for the flowers. Apparently there is no external difference between the male and female, but the sex may be. at once discovered by lifting the fore wing a little, and if there is any hair on the costal nervure or vein of the hind wing we know at once that it is a male specimen. The same distinction of sex is noticeable in

Tithorea flavescens and *Sais eurymedia*. Specimens may be captured all the year round and their metamorphoses are undergone in the short period of about a month.

NOTES ON *"Ageronia Feronia"* AND *"Gynæcia Dirce."*

The larvæ of the former are solitary and at first they are green. Head in proportion, body blackish covered with branched spines, there are two horns on the head (which is minutely spined) slightly curved, black, and clubbed at the tips with a few small spines. On the first segment a pair of long spines branched at the top and covered with minute spines to the root. On all other segments these spines are branched but there are no minute spines at their sides, and they are slightly shorter. There is a lateral row of spines, small and finely branched rising from lilac spaces with three intermediate white spots. On close examination there are dark and light brown spaces, not sufficiently prominent to be noticed otherwise. I have found the larvæ in October. The imagos appeared on the 27th. The pupæ are remarkable, the two horn-like projections on the head are flat and slightly waved. From a side view there are to be seen two dorsal humps, the upper one rounded, and the lower more angular, sloping down gradually, and forming a depression which again slopes up causing a slight thickening at the end of the abdomen ; the color is dark green with lighter tints of the same hue, suffused here and there with yellowish tints, and marked with brown. The manner in which the crackling noise is produced by the perfect insects, and the peculiar manner in which they stick close against the trunks of trees afford a great deal of interest, and deserve attention as striking peculiarities ; the former habit was noticed by Darwin in his book "The Naturalist's Voyage," page 33, in which work they are called *Papilio feronia*. " This is the only butterfly I have seen which uses its legs for running. They are very active, flying round and round in chase of one another and crackling loudly when at close quarters, frequently choosing the Palmiste, on the side of which they rest, about ten or twelve feet from the ground, against the trunk, and they may be passed by without being noticed, protected by their striking resemblance to the discoloured spots where some fungus growth or whitish portion conceals them. They stick close against the trunk in the invariable position, head downwards, except when about to lay their eggs, on the low creeping vines ; then they may be seen flying alone, with a series of sharp, regular, jerks, gliding along, and at this period, I suppose it is about the only time they ever alight on anything green against which they would offer a striking mark." According to Mr. Doubleday the noise they make is produced as follows :—" They are remarkable for having a sort of drum at the base of the fore wings, between the costal nervure and the sub-costal. The two nervures moreover, have a peculiar screw-like diaphragm or vessel in the interior." Darwin describes the noise as similar to that produced by a toothed wheel passing under a spring catch.

Gynæcia dirce is somewhat similar in its habits to *Ageronia feronia.* The female lays her eggs on the Trumpet tree under the fresh leaves of the very young plants ; strange to say, they never seem to choose the full grown trees, I have only found them on trees two, three, or four feet high. I can at once always detect when there are any larvæ on the plant, by the appearance of the leaf they are on, which has the principal veins cut so that it forms a tent, the palmate portions drooping around ; they can thus feed securely from their enemies (except those that know it) although already protected by the formidable spines with which they are covered. Sometimes as many as three or four may be found under a leaf, however, more often, only one is found. They are not active, but remain under one leaf as long as the supply of food is sufficient, and when disturbed they keep perfectly still, in the manner of the larvæ of *Ageronia feronia.* The larvæ are formidable in their appearance which is as follows :—Their general colour is black, covered with white and yellow branched spines. Head and body velvety black. There are six branched spines on every segment. From the second segment the spines have four branches as finely pointed as a needle, pale yellow and white. On the first segment the spines are simple. On the head are a pair of projections like horns, slightly curved back, and clubbed at the tips, spined and irregularly notched. There is a lateral row of yellow spaces or spots, which show up very prominently against the black ground. They remain in the pupæ state for about ten days ; in August 1893 they were plentiful at San Fernando, when I found a great many larvæ. The pupa is remarkable in appearance (and can be best judged by the imperfect illustration), it may with some reason be said to closely resemble a bit of decayed wood, being the same dull brown colour with paler patches of the same. The dorsal ridge is covered with notches and pointed prominences which are most highly developed on the abdomen, there is a noticeable rounded hump on the thoracic portion and the head terminates in two points like a **V.** There are just faint indications of the presence of wings on the thorax. The imago is very lively in its movements, alighting in the same manner as *A. feronia,* except that the wings are closed vertically, and it confines itself to shady places. The marking of the under wings has gained for it the popular name of "Zebra."

6th July, 1894.

NOTES ON SOME FUNGUS GROWING ANTS OF TRINIDAD.

By F. W. Urich, F. E. S.

IN his second paper on the *Œcodoma cephalotes*, Mr. Tanner stated that there were two distinct species of Parasol Ants in Trinidad and he was inclined to believe that there was also a third. That he was not wrong the results of my collecting have fully proved. During the past eighteen months I have collected seven distinct species, of fungus growing ants, of which no less than four belong to the genus *Atta* or "Parasol Ants." The specimens have been named by Prof. Dr. A Forel whom I desire to thank for the readiness with which he helped me in studying the species referred to. These fungus or mushroom growing ants are already well known, so far as their ravages on cultivated plants are concerned, but until lately no one seemed to know exactly what they did with the leaves they carried into their nests. Our late member, Mr. Tanner, was the first to give us an insight into their domestic economy, but Belt was the first to suggest the existence of a fungus by the passage " I believe . . . . that they are in reality mushroom growers and eaters." That this careful observer was not wrong Dr. Moller has ably proved in his " Pilzgärten einiger Sudamerikanischer Ameisen " on the fungus of the Brazilian species of *Atta* and *Apterostigma*, a most fascinating work, teeming with interesting information for the Mycologist as well as the Ant student. As I shall have to refer frequently to this valuable work in the course these notes, all quotations not otherwise indicated are taken from it. As Mr. Tanner has so ably given us a description of the manner in which the Ants prepare leaves for their "mushroom gardens " and the time of their development, I shall confine myself to describing the Ants themselves, their nests and the parts of the gardens they make use of.

Genus : Atta, Fabr

= Cephalotes, Latr

= Œcodoma, Latr

1. Sub-genus : Atta, sens strict, Fahr.

A. sexdens, Linn.

This species is very much like the following one, *A. Cephalotes*, but does not seem to be so common, I only found it once on the Tucuché and perhaps it may be confined to the higher parts of the island.

A. cephalotes, Linn.

This species is very common all over the Island, on flat lands as well as on the heights. Its ravages are well known to the cocoa

proprietors who spend large sums of money to get rid of it.   The
nests of these ants vary in size, a single colony occupying not more
than about a cubic foot of space and externally is not very conspicuous in the long grass of the cocoa estates, but in the woods I have
seen nests covering an area of about 2,500 square feet and the
earth and used parts of plants, brought up from the interior, formed
regular little hillocks fully 4 to 5 feet high.   These large nests
consist of hundreds of colonies each of about 1 cubic foot, all
are connected with each other by subterranean passages, but as a
rule each colony has an exit hole to itself which seems to be
exclusively used for carrying out the refuse of the garden and
other waste products.   The nests are made in soft clayey soil,
and the reason why these large nests are divided into several
smaller ones is perhaps because the ants are taught by nature
the danger of constructing a large nest which may cave in or be
destroyed by man or animals and the colony exterminated
entirely   The fact of there being several cavities, each with its
garden, increase the chances of one or two nests escaping the
general disaster and thus the ants have a greater facility for
perpetuating their species.   But whatever the reason be the
ants prefer to make another cavity a little way off instead of
enlarging the original nest to a size corresponding with the
increase of the fungus gardens and number of the inhabitants.
In an artificial nest of this species when all the available space had
been filled I noticed that the workers carried fungus, eggs,
and larvæ to a dark corner of the room and proceeded to found
another nest.   "The mushroom gardens consist of a gray spongy
"mass full of chambers like a coarse sponge, containing the
"larvæ, pupæ and workers.   This mass never touches the sides of
"the cavity, and there is always a clear space of about a
"finger's width between the garden and the wall of the cavity.
"The older part is differently coloured, but not very sharply
"marked off from more recent growth.   The older part is
"yellowish red in colour ; newly built portions, forming the
"surface of the garden are of a blue black colour.   The round
"particles are penetrated by and enveloped in white fungus
"hyphæ which hold the particles together.   Strewn thickly
"upon the surface of the garden are seen round white bodies
"about 0° 25 m.m. in diameter ; they always occur in the nests
"except in the very young portion of the gardens.   They consist
"of aggregatons of peculiar swollen hyphæ and are termed by
"Möller the "Kohlrabbi" clump."   These form the principal
food of the Ants.   On the authority of Kew the fungus cultivated by these ants is the same as that found by Möller in Brazil
viz : *Rozites gongylophora* Möller.   What plants are used by the
ants would be hard to tell as they use so many, but they seem
to have a special liking for introduced ones, and are particularly

fond of our cultivated Cocoa. On some nests I have seen a large quantity of freshly cut leaves while on others there is no such deposit. In an artificial colony the ants deposited leaves outside the nest, as long as there was not any available room inside, and used them by degrees. Some weeks later the nest diminished considerably and the ants no longer left leaves outside, as there was room in the interior, where they were deposited and used up by degrees. The roads these ants make in the woods are well known and have been ably described by Bates, Belt and other travellers in Tropical America. Now as to the inhabitants of the gardens ; there are six distinct kinds of workers or neuters the smallest being 2 m.m. in length and the largest, or soldiers, 15 mm. The smallest workers perform the duties of nurses to the larvæ and also weed the garden "and this is so well done "that a portion of it removed and grown in a nutrient solution " gives a perfectly pure culture, not even containing bacteria ; " they also go out with the leaf cutting gangs and are carried home on the pieces of leaves. They do not seem to take any part in cutting the leaves and I cannot find out why they go, unless it is because the ant doctor orders the nurses sometimes to leave their troublesome charges and take the ant equivalent of carriage or equestrian exercise. Those varying from three to six mm. are the foragers which do us the damage by cutting the leaves. Those from six to fifteen mm. are the soldiers or fighting ants. Their only use is the defence of the nest, and for this purpose they have strongly developed pairs of mandibles and a very large head. They cannnot sting, but draw blood at each bite. They accompany the foragers or leaf cutters, marching on the flanks of the column. Their mandibles close like scissors and are toothed, but these teeth, when viewed under a high power do not appear to be pointed but are more like the edge of the new fashioned bread knives which are so largely advertised just now. The tips of the mandibles overlap when closed. Of all the *Atta* species in Trinidad *cephalotes* is the most energetic and in the woods work day and night, but in the cocoa groves they only sally out after dark. Winged forms are found in the nest from May to July and this is the best time for killing them.

Subgenus : ACROMYRMEX, Forel

A. octospinosa, Reich.
= *A. Güntheri*, Forel.

In the same way that *Atta cephalotes* is the terror of the country people this species is a regular plague to gardens, even those situated in the heart of the town. They also occur in the country districts. They do not form such large colonies as do the preceding ones, but they form small nests from ½ to 1 cubic foot in clayey soil which they excavate, but they readily avail themselves of any suitable crevices in the masonry of buildings, under woodwork and when once lodged there are very difficult to get rid of. Their mushroom garden is constructed in the same way as *A. cephalotes*, but presents a more compact appearance. The "kohlrabbi" clumps are identical with those of *cephalotes*, and are found in the gardens in the same way. This species attack many garden plants, but seem to be very fond of roses, which suffer very much from their forays. The workers vary from $2\frac{1}{2}$ to 8 mm. The smaller ones acting as nurses in the nest and not cutting leaves. The large ones seem capable of laying eggs, as an artificial nest I have containing no queen or female ant, contains newly hatched larvæ although they were separated from the mother nest, six months ago. This is the species referred to by Mr. Tanner, who has made the same observation on the egg laying capacity of these large workers. Dr. Fritz Müller ascertained, many years ago, that the workers of species belonging to this genus, are capable of producing at least one fertile egg. Another curious thing about this ant is that on two occasions at different places I have seen fertilized females working just as hard and engaged in the same occupation as the neuters, viz : cutting leaves and carrying them to the nest. They all issued from the same nest and therefore could not have been mothers of new colonies. Any one who studies ants knows that queens will not work if they can help it and this proceeding is interesting especially as there were workers in the nests who could do the work. I also noticed that several females lost their wings in nests without any marital flight, although a few weeks later the winged ones swarmed out in the usual way on a damp evening. The larvæ of these ants are covered by a white shining skin, studded over with short stiff hairs which, at their ends, are shaped like an anchor. When a nest is disturbed small workers can be seen carrying larvæ much larger than themselves. And as it is quite impossible for these to open their mandibles wide enough to grasp the large larvæ firmly, they probably take hold of these hairs and it would seem that they are there for this purpose. Moller also records these hairs on the larvæ of *Acromyrmex discigera* from Brazil.

4. Subgenus : TRACHYMYRMEX, Forel.

T. Urichii, Forel, n. sp.

The nest of this species is excavated in clayey soils and never anywhere else. It consists of one chamber at about the depth of a foot and is never directly under the entrance hole, but always on one side at right angles and about 9 inches away from it. It has a habit of carrying the particles of earth which result from it's mining operations a little way from the entrance

hole, say about a foot, and deposits them in a small conical heap.
This seems to be a protective measure, for other ants always
have their rubbish heaps and middens immediately near the
entrance and naturally anyone in search of the ants would look
near the heap of débris.   I was at first deceived in this way,
until I noticed their dodge in a clear piece of ground.   These
ants also cultivate a fungus and if it is not *Rozites gongylophora*
it is very much like it.   As in the two preceding species their
"kohlrabbi" clumps consist of an aggregate of hyphæ with large
spherical swellings on their ends.   Any roots of plants going
through the ants chamber are not cut away, but are made use of
to suspend their mushroom garden to which are in their case
regular hanging gardens.   Unlike the other species mentioned
there is only one class of worker in these nests, measuring between
4 and 4½ mm., and of a brownish colour with a hairy pubescence.
They are nocturnal in their habits and when disturbed sham death.
I have often found them on lawns in gardens and they seem to
be rather common about St. Ann's.   They are not so dis-
tructive as the preceeding species, and seem to like small
fallen flowers and the fruit of various kinds of plants to be found
in gardens, but at the same time they do not despise rose plants
especially the young and tender shoots.   They are not at all
energetic and are very slow in their movements.

Genus : SERICOMYRMEX, Mayr.

S. *opacus*, Mayr.

The nests of this species are constructed in clayey soil with
raised entrances, consising of a narrow cylinder composed of the
particles of earth resulting from their mining operations and used
up portions of the mushroom garden.   Internally the nest consists of a sort of antechamber in which I have often found
flowers, leaves &c., brought in for use.   Small entrances leading
into the excavated cell open on this antechamber.   They are
about two to three cubic inches in size and like the preceding
species they readily make use of any roots growing through their
chambers to hang their gardens on, which is exactly like the
T. urichii with the same kind of "kohlrabbi."   The larvæ are
white and when viewed under the microscope are sparsely
covered with long tapering hairs.   There is only one kind of
worker, never exceeding 4 mm. in length.   Unlike *T. urichii*
they work during the day and it seems to me more so than
during the night, judging from observation of an artificial nest.
They also sham death when disturbed, but on the whole are
more active.   They are quite harmless and like fallen fruit better
than anything else, small flowers are also readily used for their
mushroom gardens.

Genus : APTEROSTIGMA, Mayr.

A. urichii, Forel, n. sp.

Möller suggests the name of "Hairy Ants" for this genus on account of the long hairs covering their bodies. Unlike *Atta* this species does not excavate its nests but builds them in rotten trunks of trees in any dark crevice it can find. I succeeded in securing several nests intact without destroying their structure. They are built in a hanging position *i.e.* the ants start working from the top, but never let the nest touch the bottom of the cavity. Unless the garden is quite recent and small, it is always enclosed in a delicate white covering, which on first sight looks like fine cobweb, with an exit hole at the bottom. The nests therefore look like a more or less round ball and are never larger than an apple. On breaking away this delicate covering a small mushroom garden is found consisting of irregular cells in which the ants, larvæ and pupæ, are scattered. It is at once apparent that the inhabitants are also mushroom growers and eaters, and a microscopical examination reveals the fact that they have a fungus like *Atta*, belonging apparently, according to Möller, to a different genus of *Rozites*. The chief point of interest however is that they have not cultivated and selected the "kohlrabbi" to the same degree as *Atta*. Their "kohlrabbi" is of lower type, the hyphæ being only slightly swollen into a club shape instead of being large and spherical, and they are not aggregated into groups. An examination of the covering shows it to be a fabric woven with the mycelium of the same fungus which holds the cells together. The ants, no doubt, arrange and regulate the growth of this covering, with their legs and antennæ. As I said before the gardens are always found under rotten wood and the ants invariably use the excrementa of wood-boring insects as a medium for growing their fungus on. I have kept these ants in captivity, but never with much success as they will not touch fruit, flowers or leaves. They took excrementa of wood-boring insects readily and cassava farine sparingly. From an economic point of view, they are therefore quite harmless. The colonies of these ants are small, not numbering more than 20 or 30 dark brown workers, all of about the same size viz. 6–6½ mm. and with abnormally long legs which measure 7–7½ mm. without the hip. They are of nocturnal habits.

A. mayri, Forel, n. sp.

This species can be readily recognized from *A. urichii* by its smaller size. The workers measure 3–3½ mm., but are of the same colour and also covered with hairs. They are mushroom growers and eaters and construct hanging gardens, not only under rotten wood but also under large stones in suitable dark cavities. They also have a covering to their garden like *A. urichii*, though I have not found it in such perfection as with that species, but they have the same kind of fungus and their "kohlrabbi" seems to be of a finer quality, the *hyphæ* are swollen into a club shape like *A. urichii*, but they are aggregated into regular groups, which are plainly visible. It would thus appear that *mayri* is more advanced than *urichii* in the cultivation of the fungus. Like the preceding species it takes the excrementa of wood-boring insects to construct its garden with, but is also fond of fruits and even parts of flowers. They would not take cassava farine when I offered it to them. They are of nocturnal habits, and when disturbed sham death for many seconds.

Genus : CYPHOMYRMEX, Mayr.

C. rimosus, Spinola.

This is a common little brown ant found under the bark of trees and in damp spots about walls. It does not cultivate any fungus, but two species of the same genus found in Brazil are mushroom growers and eaters, as Möller has undoubtedly proved.

ECONOMIC NOTES.

Of the eight species mentioned above, the last five may be considered harmless, at least the *Apterostigma* and *Cyphomyrmex* species are quite so and a *Trachymyrmex* are a pest in a very slight degree, but as the nests they make are not large they can very easily be disposed of. Of the remaining species *Atta cephalotes* is very destructive in the country and *A. octospinosa* destructive in gardens in town. *Cephalotes* is always engaging the attention of cultivators and a few years ago an ordinance was drafted to help the planters against their tiny foes, by giving them power to enter their neighbours' lands, to destroy nests, if the owners or occupiers would not do so themselves. *Cephalotes* do not seem to have any natural enemies through whose agency they could be considerably diminished, and this is not to be wondered at as they maintain a complete military department and a huge standing army of veritable giants who, like all soldiers, are remarkable for their prowess in the feeding and fighting lines. Unlike the armies of European powers they are composed, as is the army of Dahomey, of ferocious females. Their courage and daring are indomitable, for they rush on the foe without a moment's hesitation in huge swarms and grasping any portion of him they can lay hold of never let go until their mandibles have met. If they can then release themselves they immediately make another bite and this they continue to do until their bodies have been torn from their terrible heads.

These armies are of great service to the community, making its existence pretty safe. Those ferocious hunting ants *Eciton foreli*, Mayr, which are so common in the forests and render the Trinidad Agriculturist such good services by destroying much vermin on the Cocoa estates, are powerless against *Atta*, whose nests they seem to take no notice of, although they do not spare some other species of ants such as *Campanotus* and *Dolichoderus*. Even if they invade the nests, they are soon repulsed (as my friend Mr. Albert Carr has seen in the woods). The Superintendent of the Botanic Gardens in a recent Bulletin calls attention to *Amphisbænæ* and *Typhlops* as natural enemies to parasol ants, and recommends preserving them. But these reptiles do not seem to do the ants any harm at all. As a rule *Amphisbænæ* are not found in every nest dug up, they only turn up now and then and many of them are very badly bitten and show numerous marks of the soldiers' mandibles. It seems pretty certain that their occurrence in the nests is accidental and this also applies to the smaller *Typhlops*, which are also found there. Neither *Typhlops* (nor his congener *Glauconia*) would be much use if he encountered a Parasol Amazon, for the formidable jaws would speedily shear him clean in two. What seems to attract these two reptiles is the number of worms in the loose soil brought up by the ants which contains much vegetable matter rejected from the mushroom garden and which is always infested by worms and insects. This of course only applies to very large nests, and the country people all say that "Two-headed snakes" as the *Amphisbænæ* are called, are only found in very large nests. The only way to deal with these ants is to poison them. Many substances have been suggested, such as Cyanide of Potassium, Bisulphide of Carbon, Coal-tar, but none of these have proved very successful. The method in general use all over the Island is that known as puddling, but it is very expensive, especially in districts where there is a want of water. Mr. Ewen patented a process a few years ago, the principle of which is driving into the nest a poisonous fume which acts instantaneously on all the ants and larvæ, with which it comes into contact. This process was very successful at some experimental trials, but does not seem to have been universally adopted, and agriculturists are still trying to find out something which would rid them of their chief enemy in a cheap and effectual manner. With regard to *A. octospinosa* in gardens, it is easy to dispose of nests built in the ground with cyanide, coal-tar or very little puddling, but when they are in masonry as is often the case, or in the roots of trees, the task of disposing of them is difficult, and some poisonous gas must be driven in with a suitable apparatus.

6th October, 1894.

NOTES ON TICKS.

(Concluded.)

SO much for the naming.　The apparent complete localisation of the Gold Tick in Antigua is a very curious circumstance ; and would seem incredible considering the interchange in cattle which has taken place since the parasite has been known here.　For many reasons it is desirable to learn if it has spread to other islands of the group—in particular as bearing on the possible connection between it and the Antigua cattle disease.　Similar large ticks are well known in Jamaica, the mainland, besides South Africa and different parts of Asia, but they would appear to belong to different species.　While collecting the specimens for Mr. Michael, I instituted a series of experiments on the Gold Tick with a view to determining its fertility, its early life, and what became of it during the winter months.　It is a matter of common experience that these ticks cease to be troublesome about September, and reappear in the following May : what becomes of them in the interval ?　This is a matter of great importance, when it is considered that any really effective action must be taken before the parasites appear upon the Cattle.

It is generally acknowledged that ticks in their earlier stages are dwellers on plants.　It would appear further from the accounts of explorers, who invariably seem to suffer from their attacks, that great multitudes pass their entire existence in regions where it is next to impossible for them to meet with the animal host which they so readily attack.　Sir Joseph Hooker in a well known passage in his Himalayan Journals, writes as follows :—

" The path was soon obstructed, and we had to tear and cut our way from 6,000 to 10,000 feet, which took two days very hard work.　Ticks swarmed in the small bamboo jungle, and my body was covered with these loathsome insects, which got into my bed and hair, and even attached themselves to my eyelids during the night when the constant annoyance and irritation completely banished sleep.　In the day-time they penetrated my trousers, piercing to my body in many places, so that I repeatedly took off as many as twelve at one time.　It is indeed marvellous how so large an insect can painlessly insert a stout barbed proboscis, which requires great force to extract it, and causes severe smarting in the operation.　What the ticks feed upon in these humid forests is a perfect mystery to me, for from 6,000 to 9,000 feet they literally swarmed, where there was neither path nor animal life."

It should not then be regarded as improbable that the ticks spend their winter months upon the pastures, either in the protective egg-shell or in a semi-dormant youthful condition.　To this I shall return directly, indicating that the latter is probably the case.

I obtained several well-filled female ticks, and placed them in glass boxes. The largest of these weighed °17 oz. and was nearly an inch in length. All soon began to lay their eggs. These were deposited in a heap as large or larger than the parent tick, immediately in front of the mouth. In reality the genital opening is situated very far forward, and each egg, as laid, is passed on under the proboscis. I calculated the number of eggs in several cases. It was so large as to render it almost impossible to count them directly; and I accordingly adopted the plan of weighing them. First of all I selected a small portion of the heap and counted them carefully—usually about 600. I then weighed these accurately on a delicate chemical balance. The whole mass of eggs was then weighed and the numbers easily calculated. For two of the ticks the results were 17,000 and over 20,000 respectively (the latter from the large tick referred to above.)

I was not surprised to learn that the time taken in depositing all these eggs is considerable, especially as the oviposition is a rather complicated act in these creatures. The larger tick commenced to lay on July 30. By August 30 it had laid 13,000; and by September 10, when it finished laying, a further 7,000 had been deposited. In this case then the total period of laying was exactly six weeks. At the commencement the tick was fully distended, but as the time proceeded a gradual shrivelling took place, so that at the end the animal would not have been recognised but for the legs which stood out more prominently every day. All the female ticks died shortly after the deposition of the last eggs.

Almost the whole of the 20,000 eggs hatched. To determine the period of incubation was difficult, and the results varied a good deal. In two cases I noted the day on which the first eggs were laid and that on which the first young tick emerged. The intervals were 51 and 23 days respectively. In one case again the date on which the last egg was laid was noted and compared with that on which the last ticks were hatched. The period in this case was 32 days. From this it would appear that the time spent in the egg varies, possibly according to the condition of the female at the time of removal from the animal. In no case was there any indication that the ticks passed the winter in in the egg condition. Those eggs which were not hatched shrivelled up.

With regard to the remaining experiments, they were rendered difficult by constant absences from the laboratory. On one occasion the pots were left untended for 6 weeks, and on another for two months. A series of closed glass vessels were placed over flower pots in which plants were growing. These were in the form, of large lamp chimneys whose bases were sunk in the earth, while their tops were sealed by plates of glass lined with cotton wool. It was found that the young ticks had extraordinary powers of insinuating themselves through the cotton wool, and heavy weights had to be placed on the glass plates before the chambers were properly secured. From the causes already mentioned, and the absence of anyone in charge of my laboratory, all the plants died long before the conclusion of the experiments ; but this seemed to make little difference to the life of the young ticks.

Having prepared the chambers, small lots of eggs were inserted in test-tubes so that the date of emergence could be observed. Control lots of eggs were placed in test-tubes and plugged with cotton-wool. The length of life of the young ticks in these latter tubes was usually about a fortnight, after which they died and shrivelled up. In the closed chambers the ticks, upon hatching, immediately emerged from the open test-tubes, spread in all directions, and the cotton-wool seemed to have a special attraction for them. After a while however they congregated in small crowds, usually in the axil of a leaf or some place where they were not likely to be disturbed. One of these groups contained I suppose about 200 ticks in a state of rest, immediately moving if molested. They remained thus huddled together on the leaf of a *Tradescantia* for weeks.

The last ticks emerged from the eggs on October 3, and on March 4 following, or just five months later there were still some ticks living. No appreciable growth had taken place in these ; they seem merely to have been quiescent during the time, probably imbibing small quantities of vegetable matter from the surrounding grass debris in which they had wintered. The living ticks were found in dried up Bahamas grass.

From these and other observations, interrupted as they were, I am led to conclude.

(1). That the common, Creole or blue tick is probably the *Rhipicephalus concinna* of Koch, and widely distributed in these islands.

(2). That the Gold Tick of Antigua, wrongly called the St. Kitts Tick, is the *Hyalomma venustum* of Koch.

(3.) That it is not indigenous in this Island, but was introduced about thirty years ago from the West Coast of Africa.

(4). That its young are first noted in quantities upon the cattle about May, and that the ticks cause great inconvenience till about the month of September after which they disappear.

(5) That the full females lay from 15,000 to 20,000 eggs, all of which may be fertile.

The time of oviposition varies with the number laid. In one case it was six weeks, the number of eggs being very large.

The period of incubation varies. In cases noted it may be put down as between 23 and 51 days.

(6). All the good eggs are hatched by about October, but the young do not increase greatly in size at first. The period of vitality of the eggs is short, and any remaining unhatched speedily dry up.

(7) The young ticks born in the autumn spend the winter months huddled together in the vegetation of the pasture, probably at the roots of the old dead grass.

It will naturally follow that the keeping of the pastures clean, either by burning or, much better, by cutting or feeding, will aim a severe blow at the young ticks, but a consideration of remedies I propose to leave, for want of space, to a later number of this journal, when the very similar large ticks of the Cape, South America, India and elsewhere will be referred to.—C. A. BARBER, M.A., in Leeward Islands' *Agricultural Journal*, July, 1894.

Communications and Exchanges intended for the
should be addressed to the Honorary Secretary,
Port-of-Spain, Trinidad, B.W.I.

Price 6d.   Annual Subscription, 3/.

Vol. 2.   JUNE, 1895.   No. 8

*J'engage donc tous à éviter dans leurs écrits toute personnalité,
toute allusion dépassant les limites de la discussion la plus
sincère et la plus courtoise.*—LABOULBENE

Trinidad·Field·Naturalists'·Club.

NATURA MAXIME MIRANDA IN MINIMIS.

Publication Committee:

President.

H. CARACCIOLO, F.E.S., *V.P.*:   P. CARMODY, F.I.C., F.C.S.
SYL. DEVENISH, M.A.,   R. R. MOLE.
F. W. URICH, F.E.S., *Hon. Secretary.*

CONTENTS :—

Report of Club Meetings :	
April	187
Caudal Appendage in C. Horridus	189
Dimensions of Animals...	191
New Scale Insect from Grenada	194
New Mealy-Bug on Sugar Cane	195
New Trinidad Flounder	196
Ticks, their effect on Animals...	197
Economic Uses of the Compositæ	204

FIRE! FIRE!!
FIRE!

ACCIDENTS WILL HAPPEN

Insure your Houses and Property

of every description against loss by Fire in the

NORWICH UNION

Fire Insurance Society

one of the Oldest English Companies
ESTABLISHED 1797.

Head Offices: London and Norwich.

DIRECTORS.

President: H. S. Patterson, Esq.,

Vice-President: Major-General James Cockburn; Major F. Astley Cubitt; S. Gurney Buxton, Esq; F. Elwin Watson, Esq; William Foster Bart: I. B. Coaks, Esq; R. A, Gorell, Esq; George Forrester, Esq; William Birkbeck, Esq; Edward Palgrave Simpson, Esq; G. Hunter, Esq; —— Tuck, Esq.

☞ CLAIMS PROMPTLY AND LIBERALLY SETTLED.

Trinidad Agent,

RANDOLPH RUST,
Marine Square.

Sub-Agent,

L. W. BONYUN,
San Fernando.

☞ All Communications and Exchanges intended for the Club
should be addressed to the Honorary Secretary, Port-of-
Spain, Trinidad, B.W.I.

JOURNAL

OF THE

Field Naturalists' Club.

VOL. II.   JUNE, 1895.   No. 8.

REPORT OF CLUB MEETINGS.

19TH APRIL.

PRESENT: His Honour Sir John Goldney, V.P. (in the
chair), Hon. Lieut. Col. D. Wilson, C.M.G., V.D., Professor
Carmody, Messrs. W. S. Tucker, T. W. Carr, J. Barclay, Thos.
I. Potter, E. D. Ewen, J. H. Collens and F. W. Urich, Hon.
Secretary. The Secretary was directed to send a letter condoling
with Mr. H. Caracciolo in his recent bereavement.—The Secretary
stated that the Victoria Institute authorities had, subsequently
to the failure of the negociations for the amalgamation of the
Club with the Institute in January, met and elected all the
members of the Club to membership in the Victoria Institute,
without exacting any membership fees and subscriptions and it
would now be necessary to rearrange the meetings. He would
propose that they should alternate their meetings with those of
the Victoria Institute. It was decided that the meetings of
the Club should take place on the second Tuesday evening of
every alternate month.—Mr. Urich read a paper on a recent
visit he had made to the Guacharo Caves of Oropouche and
described the road, the caves and the animals he saw. Colonel
Wilson, who has always been a pioneer to the opening up of
some of the most fertile districts of the Island, supplemented
Mr. Urich's paper by some valuable information, he having previously visited the district a short time before. The gallant Colonel
referred to the rich cocoa soil to be had there, the healthiness
of the district, the beauty of the scenery, the beautiful rivers,
which he said, by the way, were not correctly given in the map
of Trinidad. He then went on to describe how General Mendoza

came to settle on this spot and finally said that he was sure that if the place were opened up it would become one of the important cocoa districts of the Island.—Mr. Ewen said : About a fortnight ago, in the *Port-of-Spain Gazette* there appeared a small extract from some English paper (the name of which I don't think was given) referring to an efficacious fever medicine given to the members of Sir Walter Raleigh's expedition, 300 years ago, by the Indians "out of hatred to the Spaniards;" when many of them were down with fever on the Caroni.   The extract goes on to say that if anyone now-a-days could discover those efficacious roots and leaves, there would be many people in certain yellow fever haunted localities, disposed to barter rather more than their weight in dollars for them.   I have pleasure in stating that I think I have discovered them, and if the Club can, and is inclined to barter them for dollars, I wish it all possible luck in that line.   The plant is a sedge called *Adrue* (in Jamaica English) and *Cyperus articulatus* L., botanically.   I am not prepared to say that that particular sedge is to be found on the Caroni, but I have no hesitation in stating—and our friend Mr. Carr, can no doubt corroborate me,—that there are many members of that genus to be found here, while it is quite reasonable to infer that the reputed curative properties by the *Articulatus* exist in several of them.—Professor Carmody exhibited a young ant-eater *Cyclothurus didactylus*, which he had obtained from an adult female which he had been trying to keep alive, but without success.—A letter was read from Mr. Caracciolo expressing his regret at not been able to attend the meeting and forwarding a monstrosity presented by the Hon. W. Howatson and a *Peripatus trinitatis* found in the Botanic Gardens.—A letter was also read from Mr. Oldfield Thomas, thanking Mr. Caracciolo for some rodents he had sent to the British Museum.   They were in all twelve specimens of the following species : *Echimys trinitatis, Heteromys anomalus, Holochilus palmipes, Oryzomes brevicauda* and *Mus rattus* and Mr. Thomas added that they were the first he had seen of the new species found by Mr. Chapman.—In another letter Professor Riley gave the determination of two *Hemiptera Homoptera* found by Mr. Caracciolo, one belonging to the family *Membracidœ* of the genus *Publilia*, the other to the South American genus *Darnis*.   These were found on cocoa trees and the Professor added : "In case they are numerous the best thing is spraying the trees with kerosine emulsion, at the season when the young ones are hatched out." There was also a communication read from Mr. Henry Prestoe, late Government Botanist.—The Secretary showed on behalf of Mr. R. R. Mole a land *Planaria*, a genus of worms found commonly about the island, which prey on the contents of land-

shells. Mr. Potter showed an insect he found in a decayed
gru-gru palm on the heights of Santa Cruz, which he said
resembled the *Scutigera* Lamarck. Mr. Urich showed a spider
belonging to the *Attidœ* which almost exactly resembled an ant,
specimens of which were also shown for comparison. At 9.30 a
most interesting meeting came to a conclusion.

CLUB PAPERS.

ON THE FORMATION AND DISINTEGRATION OF SEGMENTS OF CAUDAL APPENDAGE IN *C. HORRIDUS*.

By R. R. Mole.

IN the *Journal* of February 1894, Mr. Urich and myself
published a short paper embodying our observations on a South
American Rattlesnake (*Crotalus horridus*). A great deal more
importance has been attached to that article in Europe than ever
we ourselves credited it with and it has been translated into
many German scientific papers and magazines solely because of
certain references to the formation of the caudal appendage.
The paper in question was dated the 15th of February
1894, just a year ago. The snake had then eight (8)
segments to the rattle having lost three on April 21st,
1893, when sloughing. At the present date he has eleven
(11) and the remains of the one which was number eight on February 15th, last year. During the year the snake has cast his
skin on several occasions but a precise record has not been kept.
It has been noticed, however, that the portion (from an inch to
half an inch long) nearest the rattle does not come away so
readily as formerly and is only cast after a time—although turned
back (reversed) like the rest of the skin, the point of attachment
being the juncture of the tail with the rattle. At the moment
of writing there is still a ring of sloughed skin round the
base of the rattle left from the last sloughing, about a
fortnight ago. The snake is confined in a box with glass
sides and top which are protected from breakage on
the outside by galvanized wire netting. The ends are perforated
zinc. The wooden frames in which the glass is set have
squared and not rounded corners; the bottom of the cage is
rough unplaned wood with two planed cross bars. There is
consequently very little to give rise to friction in the rattle, yet
the segment which was number eight on February 15 (1894) has
worn almost completely away, only the edges remain and in the

centre can be seen the terminal portion of the seventh segment which will eventually become the last one. It has also been observed that great holes have been formed in several of the rattles within the last few months; especially is this the case on the under surface and there are general symptoms of disintegration in that the older pieces and those furthest removed from the tail are much looser than those which are the result of recent exuviations. The snake has a splendid appetite and gets plenty of food, consequently observations on the growth and decay of the rattle are conducted under most favourable conditions. It is noticeable that as the period of sloughing draws near the new segment which is of a bluish black colour, can be seen growing out from underneath the scales and pushing the last formed rattle out. In this connection it may be interesting to note that a fortnight ago I received a specimen of another crotaline—*Lachesis muta*. This snake is now about to change her skin and I observe that the base of the caudal spine is dark coloured just as it is in the basal segment of the rattle of *C. horridus*. As the time approaches when the snake becomes temporarily blind the new rattle is pushed right out and the scales which covered it can be seen attached to it, presenting a withered, yellowy-whitish appearance, their free points being somewhat raised. These disappear in time and on the last occasion I searched the false bottom of the box in order to recover them if they had dropped. I only found two portions of scales and I could not say positively whether these had fallen from the new segment of the rattle or not. After the skin is cast the new segment gradually assumes a pale, yellow colour which subsequently changes into the general hue of the majority of the segments. About the 20th October last the snake appeared to be going to change, and the eyes became pale blue as they always do at this period, but no sloughing took place. Again at Christmas time were similar indications observed, but were followed by no exuviation. The change, however, was effected about the first of the present month and now there is not only the new segment but another one which is forming and pushing the new one out. The box is kept near a window and people are constantly passing, but it is very rarely that the snake springs his rattle, in fact it is only after considerable teasing that he can be induced to do so at all. The time when he most uses it, is when he is swallowing a large rat, when, if disturbed, he keeps up a continuous and very loud whirring noise. The reason for this, I imagine, is that when the rat is half swallowed the snake feels that he is in a defenceless position as he could not readily either disgorge or swallow the rat and while his mouth is so full he would be utterly unable to use his fangs. He therefore springs his rattle as much to say : "look here, you know I am a terrible fellow and I am very cross to-day, so don't come near me or you'll repent it." He blusters because he is helpless. Some time ago a very small snake with about two segments of a rattle was brought to me, I put it in a box which I placed on top of the large snake's cage. The tiny snakeling made a very weak (almost inaudible) sound with his puny rattle. The result was that the large snake at once sounded his and kept it up for some minutes. I cannot, however, say that he has since then given any further indication of recognition of a fellow prisoner. Although it has no direct connection with the subject I may mention that when I received the Lachesis I referred to a few minutes ago I had no suitable cage to put her in and consequently I was obliged to get her to chum in with the rattle snake. The latter at once gave undoubted evidences of great excitement but it was very noticeable that in paying his attentions to the visitor (who so far as I am aware disaproved of them) he did not once use his rattle. The snake almost invariably sounds his alarm when one suddenly approaches his box in the dark. The sudden flash of a lamp will often produce a warning note. With reference to the material of which the rattle is composed it appears to me to be simply thin horn very much like the material composing one's finger nails. As the various pieces become detached from the end of the snake's tail they are naturally cut off from vital sources and simply become dead horny segments, attached to each other by a purely mechanical arrangement, the older ones at the extremity are therefore constantly wearing out and the rattle is as constantly being renewed from the base.

15th February, 1895.

ADDITIONAL NOTE.—Since the above was written this snake has lost eight segments of his rattle and has now (having added one) only four. The portion broken off in sloughing has several turned back scales attached to it in fact seven of the segments have each one, and there is a large one to the third. The process of gradual disintegration by friction is well exemplified in this rattle.

THE DIMENSIONS OF ANIMALS.

BY R. R. MOLE.

THE object of these few notes is simply to point out how exceedingly unreliable are the statements which the majority of people make about animals with which they are not familiar, of which they only obtain a glimpse, or have seen dead. As a rule no attempt is made at arriving at the correct dimensions of a dead animal until the skin is removed and then it is the skin and not the animal itself which is measured, and the consequence is we are, in after years, asked to believe the most astonishing tales as to the size of these wonderful creatures. Old men tell us frequently that in their young days the deer were much larger and the peccaries fiercer than they are in these degenerate times and it seems as if in the majority of cases time has not only mellowed but exaggerated their reminiscences. Probably, if the real facts could be ascertained, we should find that the tremendous animals which were killed by our fore-fathers were very ordinary brutes after all.   It is my intention to record a few examples of this tendency to exaggeration which I have come across, not only in books but under my own observation.   One of the first instances of this was the remark-able story told me by some school fellows of the repulse of a fierce yellow animal with gleaming teeth and arched back, of the dimensions of a small donkey, which attacked them in a country road.   A further search resulted in the death of the animal under the stick of an excited school-boy and the monster turned out to be a poor ferret which had escaped from a rabbitting party and finding that it was impossible to gain a livelihood with a sewn-up mouth, had on hearing human voices, come out of its hiding place to dumbly ask relief.   But my boy friends were not a whit more prone to exaggeration than are many travellers and hunters, grown men though they be.   Lions have been said to measure 14 and 15 feet in length.   Mr. Selous says six full grown male lions' skins pegged out measured 10 ft. 3 in.; 10 ft. 6 in.; 10 ft. 9 in.; 10 ft. 10 in.; 9 ft. 7 in. ; and 11 ft. 1 in.; but the biggest lion before it was skinned measured barely 10 ft. Mr. Blanford states that male tigers rarely measure more than $6\frac{1}{2}$ feet from the tip of the nose to the root of the tail.   But there are unreliable records of tigers of 14 and even 15 feet. The elephant is another animal about which there is a great deal of exaggeration.   It is said that some of these creatures reach the height of 12 feet, and there is no doubt some ground for the assertion but sportsmen are fond of estimating the size of elephants in their native forests as being as a rule about 12 feet. It has, however, been conclusively proved of late years that 9 feet is the average height of male and 8 feet for female Indian elephants.   The African species, however, attains a rather more imposing average.   With regard to the size of reptiles and especially those with which I am most familiar the stories are no less fabulous.   A few days ago a snake was killed at Maraval and the newspapers gravely informed us that it was 16 feet in length.   I went to see this wonderful snake and as I went I confess I positively mourned because it had not been my lot to get possession of it alive and unhurt.   But I found that the skin stretched to its utmost extent only measured 18 spans ; a

span I ascertained to be 8½ inches and deducting the ½ inches
for roughness in measurement that would make it about 12 feet,
which is a very respectable size indeed for a *Boa constrictor*
(macajuel) although popular belief is they reach something like
18 or 20 feet.   I referred to my notes at home and this is what
I found : "April 9th 1894, went to see Mr. Cumberland's *Boa
constrictor* and measured it alive and found it after very careful observation to be 9 feet 8 inches from the tip of the nose to
tip of tail ; head 4 inches, tail 11½ inches   We then posioned the
snake and after it was dead and all movement had ceased hung
it up and measured it.   The result was 10 feet 2½ inches over
all.   Tail 11¼ and the head 4 inches. Mr. Cumberland skinned
and stuffed this snake and it is now in this museum and measures
12 feet 2 inches.   This fact is worth noting as it shews that the
16 foot monster of the newspapers was in reality somewhat
under 10 feet in length. Sometime ago I was invited to
see an eleven foot snake which turned out to only measure
six feet. The following are the newest authentic data of
the size of various serpents, and is taken from a work just
published by the most eminent living authority on the ophidia,
Mr. Boulenger of the British Museum.   *Python amethystinus,*
which is a large serpent found in the Moluccas and Northern
Queensland, attains a length of 11 feet. *Boa constrictor* (macajuel),
a species with a large geographical distribution ranging
from Panama to Buenos Ayres including Tobago and
Trinidad, is represented in the British Museum by a very fine
specimen 11 feet 9¼ inches long and they are credited with
attaining 12 feet. So far as my experience goes the average
length of an adult is 10 feet. *Python sebœ,* a snake found on
the West Coast of Africa attains to 23 feet. *Python molurus,* an
Indian species, reaches 30 feet and a similar length is attained by
one of the handsomest of the serpent family, the reticulated
python of Burma and Indo-China.   Probably the largest snake in
the world is what we call in Trinidad the Huillia or Huile,
Eunectes murinus, which is found from Panama to Brazil and
also in certain localities in Trinidad, of which there is on record
one specimen 22 feet in length, and which is credited, according to
Mr. Boulenger, with attaining, when living unmolested in its native
swamps far from the haunts of man, the enormous length of 33
feet.   Dr. Mitchell says that in 1810 or 1812, his god-father,
General David Stuart, was on a walking tour with the French
General, des Sources, and they found and killed near Irois a large
snake which was 32 feet long in its skin.   This snake, adds
Dr. Mitchell (who is well known as an observer of the ophidia in
the field) could only have been a Huillia.   Such a monster I
should think would be very difficult to take alive.   Imagine
him lazily leaving the stagnant waters and mud of the lagoon

and slowly arranging his mighty coils on some huge log or stool of mangroves. It would be next to impossible to capture it unhurt. It would need at least six strong fearless men to grapple with it ere it could be secured. The young of this species have a habit of constricting the hands the moment they are caught and should an adult act in this manner almost certain death would result to the unlucky person it managed to envelop in its folds. I doubt whether Huillias live long enough in Trinidad to attain such proportions; those killed of late years are seldom more than 18 feet, though I am quite satisfied that very large ones may be found in the remote swamps of the mainland of South America. But I am straying from my point. In bringing these facts before you with regard to the measurement of animals, my object has been to endeavour to impress upon members the necessity of being very careful in making statements with regard to the dimensions of animals which may come under their notice. In conclusion I may remark that I note in the *Field* of Dec. 15, Jamrach the famous wild animal dealer of London offers to pay any one who will bring him a snake 30 feet long £1,000, and for one 40 feet long (he has evidently the Sea Serpent in his mind's eye) the nice little sum of £10,000. The prices are handsome, but I am afraid Jamrach is well aware of the absolute impossibility of the capture of the one and the improbability of the existence, much less the securing of the other.

15th March, 1895.

A NEW SCALE-INSECT FROM GRENADA.

By T. D. A. COCKERELL, Entomologist of the New Mexico (U.S.A.) Agricultural Experiment Station.

I HAVE just received from Mr. Urich some Coccidæ collected by Mr. W. E. Broadway on *Citrus medica* var. *acida* in the Botanic Garden at Grenada. One of the species is *Aspidiotus articulatus*, Morgan, as already identified by Mr. Urich; the other is a very interesting new *Lecanium*.

Lecanium punctatum, n. sp.

Female scale hemispherical, moderately shiny, not ridged, long. 3½, lat 3, alt. 2 mm.; ground-color pale ochreous, peppered with brown, but largely excluded by reddish-brown blotches with blackish spots in their centres. The whole effect is rather that of a chequered pattern of ochreous and brown, but close examination shows that the brown blotches are in four dorsal longitudinal rows, besides, more irregularly, occupying the margin, which is deeply but not closely pitted.

Examined by transmitted light, it is seen that the blackish spots in the middle of the brown patches are opaque, the ochreous portions semi-opaque, and the brown patches and dots semi-transparent. Derm not reticulate, orange brown, with numerous large gland-pits, and oval or circular large recticulated patches, as in the Brazilian *L. monile*, Ckll. ined.

Anal plates ordinary, anterior external side longer than posterior external side. Margin with very small spines.

Lateral incisions with strongly thickened edges, and short stout blunt spines, which sometimes show a tendency to be bifid.

Immature females are flatter, and more or less covered by a thin glassy or waxy secretion

Hab. on *Citrus*, Grenada, as above cited. Mixed with them are larvæ of some *Orthezia*.

The specimens had been attacked by a predaceous lepidopterous larva, in consequence of which I had only shells to study, with no legs or antennæ. Yet I have no doubt that the species is easily recognisable from the details given. It is, in fact a very remarkable little form, belonging to a neo-tropical series which consist of several species known to me, but mostly unpublished at the time of writing. Out of the neo-tropical region, it seems allied most nearly to the Australian *L. scrobiculatum*, Maskell.

A NEW MEALY-BUG ON SUGAR CANE.

By T. D. A. Cockerell, New Mexico Agricultural Exper. Station.

I HAVE received the following new species from Mr. Urich, who found it in St. Ann's, Trinidad, under the leaf-axils of sugar-cane, not numerous and apparently not seriously harmful.

Dactylopius sacchari, n. sp.

Female (in alcohol) pale olivaceous or pinkish, sparsely mealy, plump, length 4, breadth 2 mm, segmentation distinct.

Antennæ 7-jointed, sometimes seeming 6-jointed from the obscurity of the joint between 2 and 3 Joint 7 much longest; 3, 4, 5, 6, shortest and subequal. 2 distinctly longer than 3. 7 a little longer than $4+5$. Joints with sparse whorls of hairs.

Antennæ pale brown, small Legs small. Trochanter with 3 bristles; a very long hair at its tip, and a short spine behind. Femur not swollen, longer than tibia. Tarsus about 2-3 length of tibia. Femur and tibia with only a few bristles. Claw large, curved, without any denticle on inner side. Digitules of claw filiform. Tarsal digitules ordinary.

Posterior tubercles not noticeable. Mentum dimerous. Anogenital ring with 6 hairs.

When I received this I expected it would prove to be *D. calceolariæ*, Maskell, but it is clearly different. The *D. calceolariæ* found on sugar-cane in Jamaica had 8-joined antennæ; it was pale brown, becoming crimson in caustic soda. At the same time, Mr. Urich sent me *Phenacoccus yuccæ* var. *barberi*, Ckll. ined., on orange at St. Ann's, new to Trinidad. This variety, recognised by its pale legs and antennæ, was discovered by Mr. Barber in Antigua and St. Kitts. The typical form of the species is common in Mexico.

A NEW TRINIDAD FLOUNDER.

ON Page 39, No. 2, Vol. I of the *Journal* (June 1892) appears a communication from the late Dr. J. F. Chittenden with reference to two new Trinidad fishes. The Hon. Secretary of the Club has now received the following description of one of these fishes which has been named in honour of its finder, the late Dr. Chittenden. The following is the description by Mr. Barton A. Bean which appears in the proceedings of the United States National Museum. Vol. xvii.—No. 1030. In the original the description is acccompanied by a wood-cut.

CYCLOPSETTA CHITTENDENI, Bean, n. sp.

On April 4, 1892, the Museum received from Dr. John F. Chittenden, of the Victoria Institute, Port-of-Spain, Trinidad Island, a single specimen of the species here described as new and named in his honour.

Diagnosis.—A single specimen. Extreme length, 197 mm. (7¾ inches.) Greatest depth of body, not including vertical fins, 76 mm. (3 inches.) D. 82; A. 62. Scales ca., 90 Gill-rakers 8 + 3 to 4, very short, tubercular, almost as broad as long.

The length of the head is contained three and one-half times in that of the body, and the depth of the body two and one-fifth times in its length, without caudal. The diameter of the eye is contained five times in the head's length. The mouth is widely cleft, oblique, the jaws curved. The cleft of the mouth is contained less than twice in the length of head. The teeth of both jaws in a single series, those of the lower jaw are strong and sharp, curved inward and backward; those of the upper jaw are not so large, and are very irregular in size.

The ventral fins are well developed, that of the eyed side being on the abdominal ridge, and about three-fourths as long as the pectoral. The pectorals are half as long as the head, their length equalling a little more than one-third of the body depth; posterior margin oblique.

Color brown; fins lighter, marked with blackish. Three small faint blotches of black on the first half of the dorsal fin, and three rather distinct blotches on the second half, the last blotch extending to the caudal peduncle. Anal fin with three black blotches situated as and similar to those of the dorsal fin. The ventral of the eyed side is blackish, that of the blind side pale. Caudal fin with three black spots at its extremity. Pectoral fin of colored side blackish; quite a large blotch of black on body under this fin.

Type.—No. 44100, U.S.N.M.

THE EFFECTS PRODUCED BY TICKS UPON THEIR HOSTS.

By C. A. BARBER, M.A.,

UNTIL within the last year or so there was nothing definitely known about the ticks of Jamaica. For many years it was felt that their attacks upon cattle were so severe that something ought to be done; but it was not even known whether these ticks belonged to one and the same species, or to widely different families. They might belong to forms which in other countries are invariably accompanied by grave diseases, or to the comparatively harmless pests found in all parts of the world where domestic animals are reared.

At the instance of the Curator of the Institute of Jamaica, collections were made, and already there are known to be five species in that island, one of which appears, according to recent investigations, to be closely connected with the dreaded Texan cattle fever of the Southern States.

In our own Islands only two have been determined, and one of these is almost if not entirely confined to Antigua. It has been the custom in the latter island to put down our troubles to the larger form—the Gold Tick—but very probably a little study would reveal the fact that there are other species upon our cattle and domestic animals. It would be important to establish whether the Texan cattle tick—*Boophilus bovis*—which has a wide range in our geographical region, is present in this Group. So far, none of the Jamaican species has been found in our islands—a somewhat anomalous fact. It is highly probable that some of the Jamaican ticks are present in the Leeward Islands, and it will be a useful piece of work to send any collections of ticks from different parts of the Colony to the Agricultural Department for determination. (They will travel well in any weak spirit.)

Let us first consider the effects produced upon cattle by ticks. We shall then be able more clearly to recognise the extreme importance of being on our guard against these troublesome but inconspicuous pests.

It is a well known fact in natural history that the female tick may, in a few days, from being an insignificant flat-bodied little creature, swell to a comparatively enormous extent, solely by the ingestion of the blood of its host. An observation made by Leidy, as to the relative weights of adult female ticks before and after being filled with blood, gave a rather startling result. The latter were found to be more than 100 times as heavy as the former. It will be at once seen what a great drain upon the animal this abstraction of blood must be.

This habit of gorging with blood is confined to the female ticks, being probably required for the maturing of the enormous number of eggs, sometimes exceeding 20,000, which each one will lay. The male, on the other hand, is less rigidly fixed to one spot; he roams about in search of the female, and, having found her, does not increase greatly in size. But it is also necessary for him to have sustenance, especially on account of his greater activity. Further, it has been justly pointed out that the fertilisation of such an enormous number of eggs is of itself a very exhausting process ; and it is highly probable that the neglected male ticks are also a serious nuisance to the animals on which they are found.

With us, animals are occasionally covered with ticks ; but it is almost certain that our cattle do not get covered to such an extent as they do in other countries. I select a few cases illustrating the degree to which these parasites may multiply.

The following account is given of some cattle which were driven along a certain route in Queensland, leaving their marks of ticks as they went. "A mob of store cattle travelling from Westmoreland station bound for Townsville were observed to be infested with ticks, and a considerable number of deaths occurred in the mob from this cause...... Since the passage of these cattle, ticks of the same description have attacked the cattle on several runs along the line of route of the infested cattle, and a considerable mortality has been the result. On one or two stations the death-rate from that cause has been heavy. On one station it is reported that while the stockmen were branding the calves they picked off the ticks in handfuls, and the ticks had to be actually burnt through before the branding iron reached the skin."

Some cattle examined in Texas were found to be so covered with the great cattle tick that it was impossible to place a silver dollar upon the skin free from them. The same observer states that he received 100 full grown ticks from each ear of one pony, and that there were besides many immature ones in the same place.

But it is needless to multiply cases of this kind. Enough has been said to show into what a state animals left for a time in the bush may get; and taking into consideration the increase of weight mentioned above, it is very easy to understand how cattle, and especially calves, may be literally sucked dry.

There are probably no countries, where cattle and sheep are kept, which do not also possess their species of ticks ; while, once established, their multiplication does not seem to be dependent upon the continued presence of their hosts.

In a previous article some account was given of the immense numbers of ticks found by Hooker in the pathless frontier region between Nepaul and Sikkim on the Himalayas. Probably, originally introduced by cattle, they are now multiplying in a region destitute of mammals, and are therefore dependent for their nutrition upon the plant world.

The naturalist Bates thus refers to their presence in the Amazon region of South America. "The higher and drier land is everywhere sandy, and tall grasses line the borders of the broad alleys which have been cut through the second-growth woods. These places swarm with carapatos, ugly ticks belonging to the genus *Ixodes* which mount to the tips of blades of grass and attach themselves to the clothes of the passers-by. They are a great annoyance. It occupied me a full hour daily to pick them off my flesh after my diurnal ramble."

Mr. Belt, in his interesting work on Central America, discusses ticks thus :—" No one who has not lived and moved amongst the bush of the Tropics can appreciate what a torment the different species of acarina or ticks are. On my first journey in Northern Brazil I had my legs inflamed and ulcerated from the ankles to knees, from the irritation produced by a minute red tick that is brushed off the low shrubs and attaches itself to the passer-by. (This is probably the *bête rouge* of these Islands or some allied form. Ed.) . . . . Through all tropical America during the dry season a brown tick, *Ixodes bovis*, varying in size from a pin's head to a pea, abounds. In Nicaragua in April they are very small and swarm upon the plains, so that the traveller often gets covered with them . . . . They abound on all the large mammals and on many of the smaller ones. Sick and weak animals are particularly affected by them . . . . and they must often hasten, if they do not cause their death. The herdsmen or "vacqueros" keep a ball of soft wax at their houses which they rub over their skins when they come from the plains, the small garrapatas sticking to it while the larger ones are picked off."

Further south in America, where vast herds of cattle are driven over the great plains, a tick similar to our Gold tick infests them. It has been repeatedly stated that in these regions great numbers of cattle are annually destroyed by these pests, and even that a scarcity of food is caused thereby.

The following refers to a visit to one of the great cattle estancias near Buenos Ayres. There were reported to be 25–50 cattle dying every day : the animals were in a fearful state of emaciation : there were 50,000 head of cattle on the estancia : there had been a recent drought : the land was hard and the grasses parched. It was noted that the garrapatos were associated with the poverty of the animals on which they lived. The natives ascribed their presence to the existence of long, coarse grasses.

Now the presence of these ticks on the poorer cattle in numbers was very naturally put down as the cause of their emaciation. But, upon considering the matter carefully, it seems quite as likely that the animals were really dying of starvation, and the ticks, according to their well known habit, congregated upon the animals about to succumb.

Similar accounts might be quoted of the cattle in South Africa, and large ticks of almost identical appearance and size with our Gold tick are found there (*Amblyomma hebræum.*)

Although, then, it is perhaps much too readily assumed that dying cattle covered with ticks are succumbing from that cause, enough has been said to show that ticks constitute one of the most formidable enemies of the stock-breeder.

The infested animals are undoubtedly much weakened by the constant drain upon their blood supply. The irritation and consequent lack of rest caused by the presence of ticks may well be held responsible for bringing the creatures to a low condition. There are, however, other and graver charges to be laid to the door of these parasites.

We read the following in a report by one of the Cape veterinary surgeons :—" Ticks have some curious effects on and sustain relationships to the animals they infest which are but little understood.

"A single tick will paralyse a lamb, or a kid. so that it appears as if its spine were injured, and if not relieved will die. And the curious thing is that if relieved of the vermin the lamb will in a few minutes after appear nearly as well as ever."

A friend took a horse to the sea : in a few days the horse fell dead lame, and hobbled about in a useless condition for two or three weeks. Presently it was found that a male bont tick (similar to our gold tick Ed.) was under its fore-arm, or in the arm pit. It was taken away and the horse ridden the next day as well as ever."

In certain cases at least it has been now definitely established that ticks are immediately concerned in the spread of deadly disease. The idea that there is a close connection between ticks and various forms of animal disease is by no means new, having occurred to stock-owners long before the tedious and difficult scientific proof was forth-coming.

We thus read of a disease among lambs called "heart-water" at the Cape. This is regarded by many as connected with the presence of ticks. When the lambs are infested with ticks in this region they very soon die of the disease. If on the other hand they are dipped early, they do not suffer from "heart-water" nor are they troubled by ticks.

A good deal again has been written about a disease among sheep in Scotland called "louping-ill " or " trembling." There appear to be grave reasons for assuming that the disease is connected with the presence of ticks upon the animals. Where "louping-ill" is found there the ticks are always present : where ticks do not exist there is no disease ; the sheep are attacked at the season when the ticks appear after their winter's rest : and, lastly, treating the pasture for the destruction of ticks not only gets rid of these pests but is found to stamp out "louping-ill." These general deductions seem to place the matter in a very clear light ; the weakest part of the evidence is the vague character of the scientific demonstration of the manner in which the ticks are connected with the disease.

But this latter charge cannot be brought against the work of the United States Bureau of Animal Industry, dealing with the suspected connection between the great cattle tick of the Southern States and the "Texan cattle fever."

The report of that Department for the year 1891–2 leaves no doubt about this connection. There is, namely, present in the blood of the cattle suffering from disease an infusorian which quickly destroys the red blood corpuscles. This minute organism has also been detected in the body of the tick. It has been again and again transmitted from diseased animals to healthy ones by means of the tick and tick alone. The presence of this infusorian in the blood is regarded as diagnostic of the disease: and the effect of its corpuscle-destroying powers is seen all over the body as well as in the red-coloured urine, which has won for the disease the Colonial name of "red-water."

Here then, undoubtedly, the tick is the bearer of a deadly disease. And it is worth while remembering that nowhere else has the question of the action of the ticks upon cattle been thus fully and faithfully worked out. There is no reason why the Southern States should be singled out for this particular display of the hidden relations between parasite and host. Already much attention is being devoted to the Cape ticks where red-water is prevalent. It is quite possible that some of the obscure cattle diseases in other parts of the world are caused by ticks, and that other countries will, in their turn, be forced to face this problem.

It is a matter of common experience that it is the poor cattle of the flock which are covered with ticks. The strong and sleek members will have ticks but usually scattered or isolated ones : it is the weak one of the herd which sickens us with its covering of ugly parasites.

This fact has been explained in various ways. It has been asserted that the weak animals are unable to rub off and get rid of the ticks by forcible means, but I can hardly think that this is a correct explanation. The insertion of the proboscis of the tick is said by travellers to be painless, and the gradual pumping of the blood is probably little felt and less localised by the animal. If the cattle rubbed these ticks off the irritation would be vastly increased, and the injury would be much greater from the broken mouth-parts remaining in the wounds.

One is tempted, for the sake of comparison, to refer to the analogous case of scale insects attacking plants. These parasites attach themselves to plants, and, under the protection of a scaly covering, suck the juices of their host. It is again the weakly plant which is attacked, and forms such a capital multiplying ground for the parasites, that it becomes a scaly mass. How do we prevent scale from attacking plants? There is no better preventive that I know than keeping the plant well supplied with good soil, and, in dry weather, with water sufficient. The scales do not seem to be able to make headway on a healthy plant. They are frequently there, but they do not multiply; they are often attacked by diseases which decimate them and show that they, in turn, are suffering from unsuitable conditions.

Looked at from this point of view, there is strong ground for adopting the suggestion advanced by Dr. Cooper Curtice, lately of the United States Bureau of Animal Industry. In his pamphlet on the "Cattle Tick," referring especially to the form determined as connected with the Texas fever, he writes:— "Almost every farmer in the tick area will ask the investigator why it is that some cattle are literally laden with ticks, while others are free or nearly so. It is usually the fat cattle which enjoy the most immunity. Whether the ticks thrive better on poor cattle or having attacked an animal cause it to become poor is a debateable question. I believe that each proposition is true, and that the fat in an animal's skin or the oily condition of the hide has much to do in protecting it against extensive invasion by ticks. It follows therefore that cattle should be kept in good order to resist these pests as will as other diseases."

Whether for purely mechanical reasons or from the presence of the obnoxious fat, the health of the animal is supposed to react upon the tick. This view seems to be most reasonable; and, as will be seen later, by far the best way of approaching the problem of the extermination of ticks is to give the animal something in its food which will escape from the pores of its skin and render this objectionable to the ticks.

The physiological action of the animal's food upon the ticks will explain the fact that cattle taken from poor pastures to rich ones are known to drop their parasites in a very short space of time.

Mr. Hutcheon. of the Cape Department of Agriculture, remarks: "It is apparent that one of the causes of the prevalence of ticks on cattle on the coast districts is the deficiency of certain food constituents in the sour vegetation there which are necessary to the healthy nutrition of the animal body. This is well illustrated by the fact that when cattle which are grazing on the coast veldt and literally covered with ticks are removed up to the sweet grass veldt and on good Karoo, the ticks drop off in a few days, often principally from the change in their nourishment."

The improvement of pastures and the assurance to the cattle of abundance of grass of the *best quality* will undoubtedly aim a great blow at their infestation by ticks.

Dr. Marx, who has made a special study of the ticks of the world, asserts that these creatures can live for their entire cycle of existence upon vegetable food, and that even after having been gorged with blood, they will return to vegetable food. It has been already pointed out that our ticks can live for a considerable time upon dried up and decaying vegetable matter, and it is probable that the young of the gold tick spend the winter months upon the dried up grasses of ill-kept pastures.

The following details are given of a sucessful attempt to stamp out ticks and the disease of sheep (louping-ill) already referred to, in Scotland ; and, although the methods may be regarded as expensive and inapplicable to our conditions, they are worthy of our consideration.

"Mr. Nicholl stated that he entered the farm of New-borough twelve years ago, and that the "louping-ill" was then very bad. Between Whitsunday term (May 26th) and August he lost 10 per cent of his lambs—32 out of sixteen score—besides several ewes. He thought the farm was not worth having. The parks were all in rough grass. Next year he bought some cattle, and far more the following year ; ate, cut down, and even burnt the old grasses ; drained and limed the parks, using six tons of lime per acre to the light and eight to the heavy land. He continued, to eat, cut and burn down the grasses as bare as he could, and both the louping-ill and the ticks became scarcer and scarcer . . . . In six years both louping-ill and ticks had almost entirely disappeared—an odd one now and then—and he has scarcely had any since . . . . During our visit we found the grasses were more luxuriant and of much better quality than on any neighbouring land. When he first entered the farm the land kept 23 score of sheep ; it now keeps 30 score, besides the extra cattle."

The many dangers to which our cattle are exposed by the unrestricted presence of skin parasites have lead the writer to approach the question of remedies from all points of view. It is easy enough to give a few well-known recipes for cleaning infested cattle. These are very important adjuncts in fighting the ticks, and it is proposed to give the formulæ which appear to receive the best support from authorities in veterinary science in the next article.

As already stated, however, on the principle that prevention is better than cure, preference will be given to such methods as do away with the laborious external application of insecticides, and the accompanying throwing down of cattle.—Leeward Islands' *Agricultural Journal*, October, 1894.

NOTES ON THE ECONOMIC USES OF THE COMPOSITÆ.

Part I

By E. D. Ewen.

WHEN Mr. Broadway was writing his excellent paper on the Compositæ in last volume, I promised to supplement it by some notes on the economic uses of the plants of that order, which I now have the pleasure to do.

It is remarkable that so vast an order should yield so few economic plants, and those, nearly all, medicinal drugs of a bitter nature, with a few oils. It is very probable, however, that many other of the plants mentioned have other local uses, and that many plants of the order are put to use and their local names and value known but to few. If any of our members can help me by information on those points, I shall be glad some day to give you a second chapter of these—at present too scanty —notes.

Many of these notes are translated from "El Medico Botanico Criollo" by the late Dr. De Grosourdy, and the rest gathered from various sources.

Wormwood. *Ambrosia artemesifolia. L.*
Artemisse (f.) Artemisa (sp.)

Is stated to be of use against dropsy and gouty disorders, and is also said to have been used with good results in diseases resulting from gravel or calculi. A bitter aromatic infusion made with ½ to 1 handful of the leaves and flowering tops to a pint of boiling water may be given by the cupful, sweetened to taste, as a daily drink in cases requiring it. Thus taken it is a good tonic stimulant, very useful in indigestion. A medicinal wine may be prepared by macerating 4 tablespoonfuls of the dried leaves in a pint of wine for 3 to 4 days in a closely corked bottle and then filtering for use. Dose 1 to 5 tablespoonfuls a day. The leaves softened in a little water make very good resolutive poultices, or they may be used raw and bruised. The plant is in·much repute as a *vermifuge*, and for this purpose an infusion is made with a handful of the leaves and tops to 1 pint boiling water, which is given, sweetened by cupfuls, in the course of a day. (De Grosourdy).

Wormwood. _Artemisia absinthium_ L.

Absinthe commune (f.) Ajenjo (sp.)

A well-known *bitter*, much used as a *stomatic* in various and other infusions. The plant is a powerful ingredient in anti-septic fomentations. Bundles of it dried are used among clothes to *prevent moths* &c. It is the principal ingredient in the well-known liqueurs, absinthe and Vermouth. According to Fortune, the Chinese use fumigations of the dry *A. indica* in *gathering honey* from their *bee hives*. The smoke appears to intoxicate the bees without injuring them, and prevents them stinging. It is also an ingredient in the "joss-sticks" known as *Mosquito Tobacco*, which are used to fumigate mosquito haunted rooms in China. The leaves and flowers of *A. absinthium* by aqueous distillation give a dark green oil with odour and flavour of the plant, sp. gr. 0·973 boiling at 205° C. soluable in alcohol. *A. cana* and *A. lippii*, similarly treated yield a colourless or yellowish oil with odour of the plant and a burning aromatic flavour sp. gr. 0·925-0·945 boiling at 175° C. The root of *A. vulgaris* (or Mugwort) distilled with water gives a thick grease of pale greenish yellow colour, peculiar odour and nauseous bitter flavour and soluable in alcohol. The leaves of Tarragon (*A. dracunculus*) besides being a favourite ingredient in salad vinegars, are cultivated in the South of France for the sake of their oil used as a flavouring in liqueurs, &c. The plant is cropped twice yearly (July and October) and distilled with water, yielding 1 lb. oil (sp. gr. 0·935 boiling at 200°-206° C) from 300-500 lbs. of plant according to season and locality. (Spon.)

Milfoils. *Achillea millefolium.*

Various parts of common Milfoil yield essential oils by aqueous distillation. That of the flowers is dark blue Sp. gr. 0·92. That of the herb is blue, of buttery consistence when cold, with a strong odour and slightly burning flavour of the herb, sp. gr. 0·852-0·917. That from the fruits is greenish. That of the roots is colourless or slightly yellow, with a dis-agreeable odour, and unpleasant flavour. By aqueous distillation of the whole plant, and from Showy Milfoil(*A. nobilis*) is obtained a thick, pale yellow oil, of aromatic camphoraceous odour and bitter flavour sp.gr. 0·97-0·98 soluable in alcohol. *A. moschata*, on being similiarly treated·before flowering yields Iva oil which is clear yellowish, of a pleasant etherous odour, and warm bitter flavour. Boiling at 130° to 210° C. (Spon.)

Bideus leucanthus W.

Correopside odorante des antilles (f)
Coreopsida de flores blancas (sp.)
Used as an *emenagogue*. v. W. I. Camomille.
W. I. Camomille, *Chrysanthellum procumbeus* Rich.
Camomille des Antilles (fr.)
Manzanilla de America (sp.)

With a handful of this plant to one pint boiling water is prepared a useful *Emenagogue Infusion*. With the flowering summits can be prepared a good *Emenagogue Wine* analagous to that of Wormwood (q.v.) and which may be used in the same circumstances and doses (De Grosourdy). The dried and pounded flowering summits might be useful as an insect powder. *C. parthenium* by aqueous distillation yields a greenish oil, depositing stearoptene by keeping.

Cows' Tongue. *Elephantopus scaber* L.
Lengua de Vaca (sp.)

The decoction of one handful of this bush in a pint of water, administered by cupfuls is much used as a mild *astringent*.

A good vulnerary (Sloan)
Cudweed. *Guapholium americanum* Milt.
Vira-vira (sp.)

This plant (and others of the same family) is very *astringent*. Barham recommends a syrup made with the flowering tops against *Coughs and Catarrhs*. In Venezuela, according to De Grosourdy, it is in repute both as a *sudorific* and an *emenagogue*. They give a decoction of one handful of it to one quart water, sweetened, in the course of a day.

Dandylion *Leria albicaus* D.C.
Is credited with *Diuretic and Sudorific* properties.

Eclipta alba Hassk. Yerba de tajo (sp.)

Is celebrated in Venezuela as a *Vulnerary*. The fresh leaves are pounded and applied as a poultice twice a day. (De G.)

Cigar Bush (Jma.) *Critonea Dalea* D.C.
Eupatoire de la Jamaique (fr.) Guerrera (sp.)

The dried leaves of Cigar bush have a strong odor of Vanilla (and yield a strongly scented *essential oil*). They are much used in Puerto Rico to *aromatise* tobacco. In that island a tincture of three handfuls of the dried leaves in a quart of strong rum has much repute as a *liniment to cure rheumatic pains*. An infusion of a handful of the dried leaves in a quart of boiling water taken internally is a good *sudorific*.

Eupatorium odoratum L.

Eupatoire odorante (fr.) Eupatorio oleroso (sp.)
E. repandum W.

Eupatoire à feuilles d'arroche (fr.) Eupatorio de hojas escotadas (sp.)

These plants are *bitter, aromatic, stimulant, stomatic tonics,* prepared and used as wormwood (q.v.)

Ayapana (sp.) *E. ayapana* Vent. Zerbe à femme. (f.)

This plant has a repute similar to that of guaco as a *remedy* against *snake poison.* An infusion of two handfuls of the plant to one pint boiling water is given, sweetened to taste, as a strong *sudorific.* The dried leaves have a bitter and astringent taste and an odor recalling that of Tonka beans, its properties appear similar to those of Tea and it is often used in place of the latter. In Porto Rico the pounded fresh plant or its juice are applied to cure the wounds of fighting cocks, and *promote rapid cicatrization:* it doubtless might be used on men with good results.

Christmas Bush. *E. conyzoides* Vent. Herbe chatte (Trd. f.)

A tisane of this plant is given to women after child-birth, and to bring on menstruation.

Guaco (sp.) *Mikanea guaco* Desc.

Herbe aux serpents (fr.) Guaco rojo ó Bejuco de aradores (sp. Prco.)

Guaco, white. *M. orinocensis* Kth. Guaco Rebalsero (Gua.)

Guaco is in much repute as an *antidote and preservative against snake poisons.* Its most certain effects are said to be obtained by inoculating the body in six different places with the dried and powdered plant, and taking for nine days three table-spoonfuls of the expressed juice per day. Besides which it is thought prudent to drink the like doses for five to six days each month. This fortifies the body; but in case of a snake bite besides giving the juice internally the fresh pounded plant is applied as a poultice to the wound after cupping, &c.

On the Spanish Main, in addition to the above use, the pounded fresh plant is used as a *vulnerary* to apply in falls and contusions, and to *apply to hard and scrofulous ulcers.* According to Dr. Almodova of Porto Rico the very strong decoction of these plants is a *powerful emenagogue,* which he gives in doses of about three tumblerfuls per day sweetened to taste, during the three to four days preceding the period. The very strong decoction has been applied with great benefit as a dressing to venereal and slow ulcers and fistulas causing a prompt diminution of pain and suppuration, and rapid cicatrization (De. G.)

Halbert weed.   *Neuroloena lobata* R. Br.

Herbe à piques (fr.)   Conisa de hojas lobadas (sp.)

Geurit tout (fr.)   *Pluchea odorata* Cass.   Conisa oleroso (sp.)

P. purpurescens D.C.

Conyse purprée (fr.)   Salvia colorada (sp.)

The Halbert weed as well as the two latter plants, have the same *tonic properties* as the real sages and may be used in the same manner and circumstances. Barham classes Halbert weed as "a most noble wound herb" and recommends it outwardly for *application to sores and ulcers*, and the decoction inwardly to *act on the kidneys and to stop fluxes*.

Halbert weed has a bitter and aromatic flavor similar to and almost as strong as that of Gentian in place of which it may be used. It may be employed with advantage in *nervous* affections, chlorosis, and in want of appetite produced by inter-mittents. A decoction of half to one handful of the leaves and flowering summits (before the flowers open) in a quart of water is given between the attacks in cases of mild fever. (De G.) Two leaves macerated in a bottle of strong rum make a valuable stomatic bitter, and locally a liqueur glassful 3 times a day is accounted a good remedy in *Dysentery and Diarrhœa*.

Flea bane.   *Veronia Arboresenes* Sw.

According to Piso, the bruised leaves are good against *pains and inflammation of the eyes*, and that the aromatic leaves are good in *baths*, and to put amongst clothes to *keep away moths*.

Golden Cudweed.   *Ptercaulon virgatum* D.C.

"The whole plant is drying and restringent which makes it good against all sorts of *fluxes, catarrhs, quinsies and ulcers*." (Barham).

Rolandra Agentea Rottb.   Zerbe equilles (fr. Trd.)

The whole plant is very bitter, and an infusion of half a handful of it to one quarter of rum is used as a *stomatic tonic*.

Lettuce.   *Lætuca sativa* L.

Laitue cultivé (fr.)   Lechuga de hortaleza (sp.)

This well known salad vegetable has long been celebrated for its cooling and wholesome properties. A *mild sedative* drug called "Lactuary" is prepared in Europe from the milky juice of the flower stalks of some varieties. A transparent sweet oil is said to be extracted from the seed in India.

From the flowering summits is prepared a distilled water useful as a vehicle in the composition of sedative potions. Lettuce seeds serve to prepare an emulsion much appreciated as a *refreshing and calming drink* and very *useful in inflamations of the chest.* (De G.)

Wild Lettuce. *Sonchus palidus.*
Chicorée Blanche (fr.) Lechuga del Pais (Sp.)
Sow Thistle. *S. oleraceus* L.

Wild Lettuce is used in the French West Indies as a *salad vegetable,* and the milky juice of Sow Thistle *contains* about ·27 per cent. of *India rubber.*

Porophyllum ruderale Cass. Namú ó Cas. (sp.)

This plant is used on the Spanish Main as an *antispasmodic* and *sudorific.* With the leaves and flowers which exhale a strong disagreeable distinctive smell they make *vapor and common baths,* on coming out of which a cupful of the infusion made with a handful of the substance to 1 quart boiling water is given. The spasms disappear immediately and the body is bathed in perspiration. (De G.)

Pui Madame Ligne (Trd. f.) *Oliganthes condensata* Schultz.

Found on the Bocas Islands. The only one of the Compositæ yielding a Timber in this country. It gives pieces of 8"—10" x 12'—15' and is valued for its durability and strength (Crüger).

Spilanthes Uliginosa Sw.
Cresson de Para (p.) Chimapaga (Sp. Peco.)

This plant (and others of the same family common in the W.I.) is used as a masticory to *alleviate toothache.* (De G.) It is also used as a *vermifuge,* prepared and used like Wormwood, &c.

Tussilago. *Leria Nutans* D.C.
Soldat de Napoléon (fr.) Bretónica (sp.)

This common herb (as well as others of the same family) is considered a good *resolutive* especially *in opthalmia.* Half a handful of the herb is boiled in a tumblerful and a half of water, till the liquid is reduced to one tumblerful : after straining it is used in lotions and local baths. The flowers are an ingredient in *Pectoral tisanes* (De G.)

Wild Wormwood. *Parthenium Hysterophorus* L.

Absinthe bâtarde des antilles (fr.) Escoba amarga (sp.)

The bruised leaves are used to apply to *ulcers*. The whole plant is in great repute for making baths and fomentations for *ulcers and skin diseases*. It may be used in other respects the same as wormwood *q.v.* and it is the source of the *alkaloid Parthenine*, a reputed remedy for neuralgia.

Arnica. *Arnica montanum.*

A tincture is prepared from the roots for application to bruises &c. It is sometimes taken internally as a stimulant and diaphoretic. The flowers yield a blue or brownish essential oil, and the roots a brownish yellow one.

Sun flower. *Hilianthus annus.*

Girasol (sp. & fr.) Surúj-mukhi (H.)

Has long been grown for its oil-seeds and other products in Russia and India. In India it is considered a very exhausting crop for the land, which is allowed to fallow after every second crop. Amount of seed to sow an acre, 4–6lbs. The average yield (in Mysore) 11½ cwts. per acre. The seeds are threshed out of the dried heads with flails, and when quite dry are hulled in horse grinding mills. Continued pressure extracts the best oil, which is considered equal to olive oil when purified. It is pale yellow in colour, thicker than Hempseed oil, of 0·926 sp. gr., dries slowly, and solidifies at 160°c. The yield of the oil is 16 to 28 per cent. of the seed. The chief uses of the unpurified oil are for woollen-dressing, lighting and candle and soap manufacture, for the latter purpose it is superior to most oils. The leaves are used for *fodder*, to make *asthma cigarettes*, and after being steeped in Tobacco water, to adulterate tobacco. The meal after expression of the oil is excellent food for fowls, cattle and horses. The stalks make good fuel and yield much potash. They are said also to yield a valuable fibre.

☞ All Communications and Exchanges intended for the Club
should be addressed to the Honorary Secretary, Port-of-
Spain, Trinidad, B.W.I.

JOURNAL

OF THE

Field Naturalists' Club.

VOL. II.   AUGUST, 1895.   No. 9.

REPORT OF CLUB MEETING.

11TH JUNE

PRESENT : Sir John Goldney, V.P. (in the chair), Professor
Carmody, Dr. Rodriguez, Messrs. H. Caracciolo, W. S.
Tucker, J. R. Llanos, John Barclay, T. I. Potter, R. R, Mole,
E. D. Ewen, F. Hernandez, and F. W. Urich, Hon. Secretary. The following gentlemen were elected Town Members of
the Club : Mr. W. H. Ince, Ph.D., F.I. C., and Mr. J. B. Inniss.
Mr. Urich read a paper on the web insects *Embia urichiana*,
N. SP., de Saussure in litt., giving a short description of the insects
together with notes on the life history, parasites and methods of
extermination. The paper was illustrated by specimens of the
insects and their webs.—Mr. Caracciolo gave a description of a
grotto in the island of Gasparee which he had visited, and shewed
specimens of lime stone formation found there.—Mr. Mole called
attention to two specimens and pointed out the difference in the
young and adult of the snake *Scytale coronatum.* A specimen
of *Uraniscodon plica*, a lizard locally called "Old Man," was
exhibited.—The Secretary showed a specimen of the uncommon
burrowing snake *Glauconia albifrons* and said that some people
regarded it as a lizard and thought it a very important natural
enemy to Parasol ants, in reality it was quite incapable of doing
the ants any harm and stood a very poor chance of escaping the
mandibles of the large soldier ants.—A specimen of the green
tree frog *Phylomedusa burmeisteri* was also on view. This frog
has the peculiar habit of making a nest of leaves and laying
its eggs in it instead of in water, as is the case with other species.
—Mr. Ewen made some remarks on the jigger and the means of

getting rid of it.—Mr. Caracciolo laid on the table specimens of a Capsid bug he had received from Father Clunes which were found at Carriacou, Grenada. They were known there as cotton flies for they swarm on the trees to get at the cotton seeds in the pods. They assemble round these trees by the thousand, but are perfectly harmless (*sic*).—At 9 15 the meeting adjourned.

NEW TRINIDAD SPIDERS OF THE FAMILY · ATTIDÆ.

BY GEORGE W. AND ELIZABETH G. PECKHAM.

THE descriptions of the following species were published in the "Occasional Papers" of the Natural History Society of Wisconsin Vol 2, No. 2, 1894 and Vol 2, No. 3 1895, from specimens collected by Mr. W. E. Broadway during his residence in this Island. The original descriptions are accompanied by well executed plates.

The measurements are given in millimeters.

Marptusa broadwayi, sp. nov.

Male : Length, 9. Length of cephalothorax, 3; width of cephalothorax, 2.

Female : Length, 8. Length of cephalothorax, 4; width of cephalothorax, 2°8.

Legs, male, 1423 ; female, 4132 ; first pair stoutest in both sexes.

The cephalothorax of the female is widened out more behind the eyes than that of the male. The quadrangle of the eyes is one-fourth wider than long, is wider in front than behind, and occupies two-fifths of the length of the cephalothorax. The first row is a little curved, with the eyes all separated, the middle being less than twice as large as the lateral. The second row is about half-way between the first and the third row. The third row is much narrower than the cephalothorax at that place. The falces are vertical and moderately long and stout.

The male has the cephalothorax dark reddish-brown, with the eye region almost black. The hairs are all rubbed off, excepting a small bunch in the post-ocular depression and a few on the anterior sides. The abdomen is dark brown, with the upper central part occupied with a lighter colored herring bone stripe, which is outlined in white. The legs and palpi are dark brown, the first and second pairs being almost black ; they all have short, white hairs, and the first and second pairs have long black spines   The palpi have short, white hairs on the patallæ and black hairs on the tarsi.   The under surface is dark brown.

The female resembles the male in markings, but is much lighter in colour.   The cephalothorax is almost covered with a mixture of white and rufous hairs, and there are many long white hairs around the eyes, and hanging down over the dark, iridescent falces.   The abdomen is hairier than that of the male and has a decidedly rufous tinge.   The upper surface and the lower sides do not differ much in colour, but the herring-bone stripe is marked by a black line with a white line inside it.   The palpi are light brown, covered with long, white hairs.   The legs are light brown.   The mouth parts are dark brown, almost black, but the rest of the under surface is light brown.

Of this species we have three females and two males, sent to us from the west coast of Trinidad, by Mr. W. E. Broadway.

Breda milvina, C. Koch.

Females: Length, 13.   Length of cephalothorax, 5°8; width of cephalothorax, 4.   Some females are only 8 or 10 mm. long.

Legs, 4132; the first, second and third are nearly equal in length; the first pair is the stoutest, the second next.

The quadrangle of the eyes is equally wide in front and behind.   The middle eyes, of the first row are fully twice as large as the lateral and stand out very prominently.   The labium .s as wide as long and is one-half as long as the maxillæ.

The cephalothorax is dark brown, with the eye region black. In the post-ocular depression and extending backward from it are some white hairs; and there is a narrow white band around the lower border.   Around the eyes of the first row are some stiff, upright, black hairs.   On the clypeus are some long hairs, which are white, tinged with copper color.   The falces are dark red-brown, with a few white hairs.   The palpi and legs are brown and hairy, the first pair being the darkest.   The abdomen is brown and hairy.   The sides are covered with white dots in the anterior part and with copper-colored dots behind.   The upper dorsal surface has a longitudinal, angular band which is white in front and copper color behind, where it sends out some projections towards the sides.   Underneath, near the spinnerets, there is, on each side, a distinct, round, white or copper-colored spot.

In the young male the brown on the abdomen is replaced by black; there are no dots on the sides, and instead of copper-color, the posterior part of the dorsal band is brilliant orange-red.   The palpi are pale.   C. Koch had this species from Bahia.   We have numerous females from Santarem and two immature males from Trinidad, one from Port-of-Spain and one from the East Coast.

Epinga barbarica, sp. nov.

Females: Length, 15. Length of cephalothorax, 5°5; width of cephalothoiax, 3°7·

Legs, 4123; first pair plainly stoutest, second next. This is a large heavy spider.

The quadrangle of the eyes is one-sixth wider than long, is almost equally wide in front and behind, and occupies two-fifths of the cephalothorax. The cephalothorax widens out considerably in the thoracic part. The sternum is long, the anterior coxæ are almost touching. The labium is longer than wide. The abdomen is large and rounded, more like *Bavia* than *Marptusa*.

The cephalothorax is dark brown. The cephalic plate is black with green and purple reflections.

The middle of the abdomen is of a rather bright olive-green with a few iridescent scales—all that are left of the original covering. On each side is a longitudinal, brilliant red band, and, below this, a band of snowy white. The first legs and the falces are black and somewhat iridescent, like the cephalic plate. The falces are well covered with black hairs, and, as they project a little forward, they are visible from above. The palpi and the second, third and fourth pairs of legs are light brownish-yellow, with white hairs. The mouth parts are black; the sternum coxæ and the venter are light brown, the venter being marked, more or less distinctly, with three darker longitudinal lines.

We have three females from Port of Spain, Trinidad sent to us by Mr. W. E. Broadway. In shape of cephalothorax it is much like *Epinga chapoda,* although the slope from the dorsal eyes to the posterior border is not so unbroken and gradual a curve, as the second half falls a little more abruptly than the first half, looking something like *Breda.* In the *ornata* the curve is smooth and unbroken, but is shorter than in *chapoda* or *barbarica.*

Deza sumptuosa Perty.

Male: Length, 11. Length of cephalothorax, 5; width of cephalothorax, 2°5.

Female: Length, 11°5. Length of cephalothorax, 4; width of cephalothorax, 2°7·

Relative length of legs, of female, 1423; of male 1423; first pair stoutest, especially in the female.

The coloration exceeds in brilliancy that of any other species of this group. In the male the cephalothorax has for a ground color a covering of highly iridescent purple scales, on which is an elaborate pattern in iridescent silvery scales, with a prevailing tint of exquisite light green. This pattern, which will be best. understood by a reference to the figure consists of a large central spot just above the anterior eyes, which is connected on each side with a spot between the second and third rows of eyes; of a curved band behind each dorsal eye; and of a band which encircles the cephalo- thorax, beginning on the face at each side of the anterior middle eyes. These silvery scales, which in some lights show reflections of pink, blue and violet as well as of green, are found also on the fourth face of the falces, on the palpi and on the first pair of legs, which are darker in color, than the other pairs. The abdomen has four transverse bands of these scales, which here, however, have rather a golden than a silvery tinge, alternating with four bands of bright red. The posterior one of the iridescent bands is extended backward, in the middle line, to the spinnerets. The cephalothorax of the female is brown, with a lighter spot on the anterior thoracic part, and seems to have been covered with iridescent violet scales. The palpi and the second, third and fourth pairs of legs are light brown, the palpi having a bunch of white hairs on the tarsus. The first legs are bright brown, with some short, snow-white hairs. The abdomen is like that of the male

We have one female from Santarem, Brazil, in the Smith collection, and one male from Port-of-Spain, Trinidad, sent us by Mr. W. E. Broadway.

Anoka parallela, sp. nov.

Male: Length, 5°5. Length of cephalothorax, 2; width of cephalothorax, 2.

Legs: 1423; first much the stoutest and longest.

This spider is a little below medium size, with long first legs, which are darker in color than the others.

The cephalothorax is short, wide, and moderately high. The sides round out from the anterior end to the widest point, just behind the dorsal eyes, and then approach each other again. The eyes differ a little from those of *grenada* and *moneagua* the first row being slightly curved, with the middle eyes twice as large as the lateral. The second row is nearer the first than the third, and the third row is not so wide as the cepholothorax at that place. The falces are stout and are directed obliquely forward.

The whole body is of a mahogany-red colour, growing paler toward the posterior end of the abdomen. Encircling the sides of the cephalothorax and extending throughout the length of the abdomen are two wide bands of pure silvery-white hair; the space between these bands, and the lower sides of the abdomen are covered with small, yellow scales. Under alcohol, an in- distinct, blackish chevron and a pair of black spots are visible on the anterior part of the abdomen. Around the eyes of the first row are brilliant red hairs, and the clypeus is covered with glistening, silvery-white hairs. The first legs are bright mahogany red above, shading into black beneath. The other legs are of a pale yellowish color. The falces are of a bright reddish-brown. The mouth parts are almost black.

One male, from Port-of-Spain, collected by Mr. W. E. Broadway.

Anamosa inconcinna, sp. nov.

Female: Length, 6·5. Length of cephalothorax, 2·4 ; width of cephalothorax, 2·6.

Legs, 1423 ; first pair much the stoutest. The quadrangle of the eyes is one-fourth wider behind than in front. The first row of the eyes is straight. The middle eyes are less than twice as large as the lateral, and are a little separated. The lateral are well separated from the middle and are placed further back. The second row is narrower than the first and is twice as far from the third as from the first row. The dorsal eyes are large and are placed on the sides of the head. The falces are stout and vertical. The maxillæ are broad and heavy ; the labium is rounded. The sternum is narrower in front, wide and rounded behind.

The whole aspect of the spider is clumsy and thick, the cephalothorax being short and broad, while the abdomen is very large, much wider than the cephalothorax, which it overlaps.

The general color effect is dark brown. Looked at closely the cephalothorax is seen to be black, with the sides covered with rather bright yellow hairs, and two narrow lines of yellow hairs down the middle. The abdomen is dark brown and seems to have been covered with yellow hairs, which, in this specimen, are largely rubbed off. Under alcohol two or three pairs of black dots appear on the dorsum. The falces, palpi and legs are dark brown, with a good many yellowish-white hairs. The first legs have the femoral and tibial joints considerably enlarged. The second, third and fourth pairs have the proximal end of the metatarsus light yellow and translucent, while the tarsi are black.

Although we have put this species into the genus *Anamosa*, it does not agree, so far as the general shape of the cephalothorax is concerned, with the other species, *callosa*. In *inconcinna* the sides are gently rounded, the anterior and posterior ends being about equally wide, while in *callosa* the anterior end is much narrower, as may be seen by the drawing.

We have a single female from Port of-Spain, Trinidad, sent to us by Mr. W. E. Broadway.

ADDITIONS TO THE REPTILIAN AND BATRACHIAN FAUNA OF TRINIDAD.

THE following notes are based on a small collection of Reptiles and Batrachians sent by Mr. R. R. Mole and Mr. F. W. Urich, to Prof. Dr. O. Boettger, for examination.

REPTILIA.

ORDER: CHELONIA.

Family. TESTUDINIDÆ.

Nicoria punctularia, (Daud). typ.

Boulenger, Cat. Chelon., 1889, page 123.

A full grown specimen from Caparo, Trinidad. Altogether characteristic in form and colour.

Family: CINOSTERNIDÆ.

Cinosternum scorpioides (L.)

Boulenger, Cat. Chelon., 1889, page 41.

Greatest length of carapace 13 c.m., Caparo, Trinidad.

New to the Island. Up to now only known from Guiana (especially Cayenne) and North Brazil.

ORDER: OPHIDIA.

Phrynonax fasciatus (Pts.)

Boulenger, Cat. Snakes. Brit. Mus., vol. ii., page 21.

New to the Island.

Carenage, Trinidad.—One specimen (young). Internasals shorter than præfrontals; præoculars in contact with the frontals. 8 supralabials, of which 4, 5 and 6 reach to the eye; 7 infralabials in contact with the anterior sub-mentals; temporals 2 + 2. The five middle rows of scales slightly keeled; the ventrals angulated on the sides. In the young specimen a dark spot under the eye on the light coloured upper lip is specially conspicuous, under which a similar spot is to be found on the lower lip. Up to now this species was only known from Guiana and from the Amazons. The specimens at the British Museum are from British Guiana, Surinam, and the upper Amazon in tropical Peru.

Streptophorus atratus (Hall.) var. lansbergi, D.B.

Boulenger, Cat. Snakes., vol. i., page 294.

From Oropouche, Trinidad, found under rotten stems of trees. New to the Island. This variety was previously only known from Ecuador, U.S. of Columbia and Venezuela.

Squ. 19, G. 0, V. 143, A. 1, Sc. 62/ 62 + 1.
„ 19, „ 1, „ 147, „ 1, „ 53/ 53 + 1.

Herpetodryas carinatus (L.) var. macrophthalma, Jan.

Jan. Incon. Ophid., part 31, plate 2, fig. 2, (spec).

This variety is above all characterized by the large number of sub-candals—170–173. The temporals stand $1 + 2$; of the supra-labials the 4, 5 and 6 or the 5 and 6th only reach the eye; ventrals 167–174.

Squ. 12, G. 2/1, V. 167, A. 1/1, Sc. 170/ 170 + 1.
„ 12, „ 2/2, „ 167, „ 1/1, „ 173/ 173 + 1.
„ 12, „ 1/1, „ 174, „ 1/1, „   ?

Rhadinæa cobella (L.) typ.

Liophis of former lists.

Boulenger, l.c. vol. ii., page 166.

Squ. 17, G. 3/3, V. 161, A. 1/1, Sc. 62/ 62 + 1.

Tantilla melanocephala, (L.)

Homalocranium of our former lists.

Spilotes pullatus (L.) var. B., "Black Tigre."

Boulenger. l.c., vol. ii., page 24.

Var. B. "nearly entirely black, only the anterior ventrals being partly yellow.

New to the Island. Up to the present without any fixed locality.

Squ. 16. G. 1/1, V. 212, A. 1, Sc. 114/ 114 + 1.

Elaps corallinus (L.) var. *circinalis* D.B.

Jan. Incon. Ophid., part 41., plate 6, fig. i.

New to the Island.

Squ. 15, G. 3, V. 181, A. 1/1, Sc. 44/ 44 + 1.

The white ring at the back of the head is slightly interrupted immediately behind the parietals, otherwise the coloration is typical.

Epicrates cenchris (L.) typ.

Squ. 49, G. 11/10, V. 227, A 1, Sc. 53.
  „ 49, „ 11/11, „ 228, „ 1, „ 53.

Corallus cookei, Gray typ.

Boulenger, l.c. vol. i., page 100.

Squ. 43, G. 14/15, V. 247, A. 1, Sc. 103.

Coloration typical. The number of ventrals 247 is less than that of the smallest known number 253. It varies from 253-285.

BATRACHIA.

Hyla maxima. Laur.

Boulenger, Cat. Batr. Sal. page 349.

From Oropouche and Port-of-Spain. New to the Island. Previously recorded from Guiana (especially British Guiana) and North Brazil.

The specimen, an adult male, has longer legs than usual, for when the hind limb is carried forward, the tibio-tarsal articulation surpasses the snout slightly. The granulation of the back is not very distinct. Length of head and body about 90 mm.

DESCRIPTION OF A NEW LECANIUM FROM TRINIDAD.

By T. D. A. COCKERELL.

(From the Entomological News, 1894 page 203-4.)

Lecanium urichi, n. sp.

Red-brown, very shiny, nearly circular, moderately convex, the segments marked on upper surface by black or blackish transverse lines interrupted at regular intervals. Posterior incision nearly 1 mm. long, with contiguous sides. The two anal plates together forming a practically equilateral diamond-shaped quadrangle length $4\frac{2}{3}$ mm., breadth $4\frac{1}{4}$ mm., alt nearly $2\frac{1}{4}$ mm

I found no trace of legs or antennæ after boiling in soda, nor could I detect them by examining the underside of an untreated specimen with a lens. The mouth parts were distinct, as usual. The lateral incisions are very deep and large, bulbous, . with the margin of the bulb thickened and appearing dark

brown. This appearance is very different from that in any
other *Lecanium* I have seen. Margin with extremely small,
but moderately stout spines, fairly numerous. Derm not
tesselate, but crowded with very large gland pits, which by
transmitted light appear dark brown on a light brown ground.

Hab :—Trinidad, West Indies, in a nest of the ant
Cremastogaster brevispinosa, Mayr var., sent by Mr. Urich,
September, 1893.

I was not able to study this species at the time it was
received, and until I made a careful examination of it recently,
I did not realize its interesting character. It is the first *Lecanium*
known to live in ants' nests, and is, besides, a very peculiar
form, perhaps subgenerically distinct from the rest of the genus.
Its shape and appearance somewhat recalls *L. begoniæ* Douglas,
but it is widely different from that in important characters. It
is also apparently the first Coccid found in the nest of any
Cremastogaster, the species of that genus being, according to my
experience, arboreal.

[Since the above was written the species was found in
Brazil at Sao Paulo by Dr. von Ihering].

DESCRIPTION OF A NEW SPECIES OF TELENOMUS BRED BY MR. F. W. URICH, FROM A COCCID.

By WILLIAM H. ASHMEAD.

Telenomus minutissimus (sp. n.)

Female.—Length 0·5 mm. Polished black, sparsely sericious ;
sutures of trochanters, narrow annulus at base of tibiæ and the
anterior tibiæ beneath light brown ; tarsi, except the two last
joints, pale or whitish ; tegulæ black ; wings hyaline, ciliated, the
venation light brown, the marginal vein being as long as the
stigmal. The head is broad, much wider than the thorax, about
$3\frac{1}{2}$ times as wide as thick antero-posteriorly, the frons subconvex, smooth, highly polished, the lateral ocelli close to the eye
margin, with a short, oblique grooved line behind, the eyes oval
sparsely pilose ; antennæ 11 jointed, when extended as long as
the body, the scape slender, slightly bent and not quite as long
as the flagellum, without the pedicel, the pedicel conical as long as
the first three funicle joints united and much stouter ; funicle
5-jointed, the joints all very minute, rounded, the first and fourth
the smallest, the fifth a little transverse ; club 4-jointed, the
joints 1-3 transverse-quadrate, the last ovate. Thorax smooth,

convex, but under a high power exhibiting some sparse microscopic punctures on the mesonotum, the scutellum highly polished, impunctate. Abdomen not longer than the thorax, truncate behind, or, as viewed from above, spatulate, highly polished, with some striæ in the suture between the first and second segments, the second segment being wider at apex than long.

Hab.—Trinidad W.I.

Types in National Museum. Described from 3. female specimens received by Mr. L. O. Howard from Mr. F. W. Urich, who reports having reared them from a Coccid, *Dactylopius* sp.

This species belongs to that section of the genus, in which the pedicel is longer than the first joint of the funicle, and it is quite distinct from the several species described by the writer in his recent "Report upon the Parasitic Hymenoptera of the Island of St. Vincent (Journ. Linn. Soc. Zool. vol. XXV., 1894, pp. 201-210) or in his Monograph of the North American Proctotrypidæ (Bull. 45 U.S. Natl. Mus. pp. 143 *et seqq.*)

In general appearance it comes nearest to my *Telenomus minimus* taken at Arlington, Virginia, agreeing with it in venation, but it is smaller and is readily separated from it by the much wider head and the relative porportion of the antennal joints.

NOTES ON SCALE INSECTS. I.

By F. W. URICH, F.E.S.

Dactylopius citri, Boisduval.

„   *destructor*, Comstock.

This species of mealy-bug is new to the fauna of the Island. Under the name of *D. destructor* Professor Comstock gives the following description of this insect "*Adult female*—Length, " 3. 5 mm. to 4 mm. width 2 mm. Color dull brownish yellow, " somewhat darker than with *D. longifilis*; legs and antennæ " concolorous with body. The lateral appendages (seventeen on " each side) are short and inconspicuous and are subequal in " length. Upon the surface of the body the powdery secretion " is very slight. In spite of the small size of the filaments, the " spinnerets and the supporting hairs are as numerous and as " prominent, or nearly so, as in *D. longifilis*; those upon the " anal lobes being especially long. Antennæ 8 jointed ; joint 8 " is the longest and is twice as long as the next in length, joint " 3, after joint 3, joints 2 and 7, subequal, then 5 and 6, joint

" 4 being the shortest. The tarsi are a little more than half the
" length of the tibiæ and the digitules are as in the preceding
" species; claws strong."

" *Male.*—Length 0.87 mm. expanse of wings, 2·5 mm., color
" light olive brown, lighter than in following species; legs
" concolorous with body, antennæ reddish; eyes dark red; bands
" darker brown than the general color; anterior edge of
" mesoscutum and posterior edge of scutellum darker brown.
" Body, as will be seen from measurements, rather small and
" delicate compared with size of the wings; head small, with
" almost no hair; antennæ 10 jointed, joints 3 and 10 longest
" and equal; joints 2, 6, 7, 8, and 9 nearly equal and considerably
" shorter than 3 and 10, joints 3 and 4 subequal and a trifle
" shorter than the following joints. The lateral ocelli are each
" just laterad of the centre of the eye, and not at its posterior
" border, as in the following species (this however, is a character
" which will not hold with specimens long mounted). Prothorax
" short; legs sparsely covered with hairs, tarsal digitules extremely
" delicate, and the button is very difficult to distinguish; we
" have been unable to discover a trace of the hair belonging to
" the claw. The anal filaments and the supporting hairs are
" similar to those of the following species.

" This species is readily distinguished from *D. longifilis* by
" the shortness of the lateral and anal filaments in the female.
" Indeed, for convenience's sake, we have been in the habit of
" distinguishing them as the mealy bug, with short threads and
" and the one with long. The life history of this species differs
" quite decidedly from that of *D. longifilis.* in that true eggs,
" which occupy quite a long time in hatching, are deposited.
" The female begins laying her eggs in a cottony mass at the
" extremity of her abdomen, some time before attaining full
" growth and the egg mass increases with her own increase,
" gradually forcing the posterior end of the body upwards until
" she frequently seems to be almost standing on her head. The
" young larvæ soon after hatching spread in all directions and
" settle—preferably along the mid rib on the under side of the
" leaves or in the forks of the young twigs, where they form
" large colonies closely packed together, as mentioned in the
" description; they are only slightly covered with the white
" powder, and many seem to be entirely bare with the exception
" of the lateral threads.

" *Habitat.* This species is very abundant upon almost
" every variety of house plant in the Department green houses,
" but especially so upon the Arabian and Liberian coffee plants.
" On these plants they were found curiously enough, in small pits
" or glands on the under side of the leaf, along the mid rib almost
" every pit, of which there is one at the origin of each main vein,

"contained one or more mealy bugs and the larger ones whole
"colonies. The name *destructor* is however, proposed for this
"insect from the damage done by it to orange trees in Florida,
"especially at Jacksonville and Micanopy, where it is the most
"serious insect pest to the orange.'

This species has been recorded from Jamaica by Mr,
Cockerell "from a few examples in the Jamaica Museum con-
"tained in a tube marked "Scale insect on Crosset [? Gogset]
Coffee. *D. citri* is said by Penzig to be one of the worst of the
orange enemies in Italy.

So far as Trinidad is concerned it is found singly on a variety
of plants about gardens and sometimes in large numbers. In the
country districts I have seen it very numerous on cocoa pods, in
fact some pods were nothing else but a mass of mealy bugs and
white secretion, in which quite a small insect fauna could be
observed, such as several species of parasitic hymenoptera, lady
birds, lace wing flies, aphis flies, *Baccha* sp. in all stages, and
mites. On the Cocoa the young lice settled in the fork of the
young trees or in the grooves of the pod. I have also observed
them on Sugar cane in the leaf axils. Oranges are also subject to their attacks. They are not regarded as injurious by
planters, but I certainly think that their presence ought not in
any way to be neglected.

Several parasites were bred, which will be described on
another occasion.

Dactylopius virgatus, Ckll.

var. *ceriferus* (Newstead).

This coccid is also new to the fauna of the Island, Mr.
Cockerell to whom I referred it for determination writes as
follows :—"Curiously it agrees with *Dactylopius ceriferus*,
"Newstead, Ind. Mus. Notes, vol. iii., found in Madras on
"Croton, by Miss L. E. Tomlin. Unfortunately, however, it
"is also most dreadfully close to my *D. virgatus*, and especially
"to that form of it named var. *farinosus*. The typical true
"*virgatus* has the back with dark spots and bands, but in var.
"*farinosus* the back is very mealy and the banding is practi-
"cally lost.

"Singularly, in his description, Newstead also compares his
"*ceriferus* with *filamentosus ;* though as I wrote you, it has
"nothing to do with that insect. In my original description of
"*D. virgatus*, I described four varieties which I had at first
"taken to be distinct species or sub-species. The insect is in
"fact a general feeder, and very variable. According to my
"present judgment, your croton species is most assuredly a form

" of *virgatus* but it is also apparently Newstead's *D. ceriferus.*
" The creature is a very obnoxious one in gardens. *D. virgatus*
" is one of the worst pests in Jamaica."

A technical description of the life history of this insect is
being worked out but for the present the following will do.

Adult female :—Length 3½ to 4 mm., width 2 mm. Colour
dull brown yellow, legs and antennæ a trifle lighter. Young
larvæ light yellow. Body covered with a white secretion. Long
fine hairs from all parts of the body take the place of lateral
appendages, very conspicuous in young specimens, but gradually
lost as the adult stage advances. Two caudal filaments comparatively thick, half the length of the body.·

Male :—Wing expanse 2 mm. Length of body 1–0·99 mm.
Colour light brown ; thorax antennæ and legs darker brown.
Eyes dark claret colour. Wings iridescent. Body long and
stout very little haired. Head large almost no hair. Antennæ
and legs hairy. Antennæ 10 jointed, joint 3 longest, 10 next
longest.

Life history :—This species is common about gardens in
Port-of-Spain and is found on quite a variety of plants, croton,
orange, guava, foliage plants and nettle. I have found it
abundant on guava and croton. The female is active when
disturbed. She surrounds herself with a cottony mass, in which
the eggs are laid and the young remain under the parent for
some time. Growth is rapid and if left to themselves the adults
form a thick cottony mass on the leaves and twigs of the food
plants. In this mass on which a smut fungus soon grows are to
be found quite a number of predaceous insects, such as larvæ of
aphis flies, lady birds and lacewing flies. The male larva forms
a cottony sac, in which it undergoes its transformations.

This mealy bug should not be neglected when observed, as
owing to its rapid multiplication it soon becomes a pest.

Natural enemies.—From a mass of adult females a small
brown parasite was reared which Mr. L. O. Howard kindly
determined as *Aphycus* n. sp. remarking in his letter that he had
never seen a *Dactylopius* parasite like this one. From a mass
on croton the two following species emerged ; one *Telenomus
minutissimus* n. sp. Ashmead, a description of which will be
found on page 220, and the other an *Encyrtus* (fam. CHALCI-
DIDÆ), which could not be described from want of material.
With regards to the first-named Mr. Howard writes " The
" occurrence of three specimens of *Telenomus* among the seven
" specimens sent is interesting for this reason : the species of
" this genus are invariably parasitic upon the eggs of other
" insects. It seems to me that this form is almost too large to
" have come from the eggs of Dactylopius and I wonder perhaps,
" if there were not the eggs of some other insect in the mass of

" *Dactylopius.* In other words, I doubt the parasitism of
" *Telenomus* upon *Dactylopius.* The *Encyrtus*, however, is
" without much doubt properly considered a parasite of the
" coccid."

Mr. Howard is undoubtedly right, but it is hard to tell
who really was the host of this parasite as these masses of
Dactylopius present quite a little insect fauna of their own,
which I hope on some future occasion to give a list of. In
conclusion I desire to thank Messrs. Howard, Cockerell and
Ashmead for their kindness in helping me.

NOTES ON THE ECONOMIC USES OF THE COMPOSITÆ.

PART II.

THE following is the conclusion of Mr. Ewen's excellent
article begun in last number.

Jerusalem artichoke. *H. Tuberosus.*

This plant is said to attain the fullest perfection in Brazil.
Its tubers are excellent vegetables, and are the basis of the well
known Palestine soup. The stems are said to afford much
valuable textile fibre. The plant requires a soil rich in Potash.
The root contains Inuline (a sort of starch) and uncrystalizable
sugar.

Safflower. *Carthamus Tinctorius.* Carthame (fr.)

Is cultivated for its Petals and oil-seed in France, Spain,
Italy, Egypt, China and India. The coloring matter (Carthamic
acid) ($C14. H16. O14.$) is extracted from the petals of the flowers.
It is used as a red, rose and plush tint dye for silks and satins.
Safflower extract, dried and mixed with ground talc forms the
popular cosmetic called Rouge. The seeds yield about 20 per
cent of fatty oil, called "Curdie" oil. There are two ways of
extracting it. (*a*) The clean seeds are crushed and pressed, producing about 25 per cent. of light colored oil of good burning
qualities. (*b*.) The natives in India prepare a black tarry oil from
the seed by a rude sort of destructive distillation *per desensum.*
It is useless for burning but is highly esteemed by them for
dressing and preserving leather vessels exposed to water, such as
the "Masaks" of the water-carriers.

Benné seed or Ginjelly. *Sesanum Orientalé* L.

Sésame (fr.) Alégria (Sp.) Tili (H.)

The seed of this plant (which has found its way to the West Indies from Africa or still further east) is a well known ingredient of sweetmeats in Trinidad. In India the cultivators recognise two sorts, white (Saféd-til) and black (kála-til). It is grown in sandy land, and is rarely irrigated. It is reckoned that an acre requires one-sixteenth bushel of seed and yields 1½–2 bushels and occupies the land for three to four months. The seed contains oil to the extent of 56 per cent. but the yield of the usual pressure processes is 45 to 50 per cent. The oil has a sp. gr. of 0·923, Boiling point 100° C. congealing point 5° C. When well prepared it is said to keep for years without rancidity. Its uses are cooking, medicine anointing the body and hair, absorption of the odors of flowers in perfume making, illumination and the adulteration of olive oil. It is one of the most valuable articles of export from India, and there is no reason why it should not become a much more extensive and valuable minor industry in Trinidad.

The fresh leaves chopped and beaten in cold water, make a cooling and refreshing drink, taken sweetened. If hot water is used instead the liquid it becomes very mucilaginous, and then is useful in all inflammatory intestinal disorders, especially dysentery. Dr. Moringlane of Porto Rico cured with it cases of dysentery which had resisted all other remedies, giving the drink daily for a time, as well as daily enemas of a very stong infusion. If a quantity of the chopped leaves are boiled in water and then allowed to stand near the fire for some hours, the resulting liquid has the appearance and consistency of white of egg and has similar properties. (De G.) The marc or oil-cake is excellent food for animals, and according to Soubeiran and Girard 11 per cent. of nitrogen is found in it.

Niger-seed. *Guizotia oleifera.*

Sirgújia (H.)

Much grown in some parts of India and Africa. It yields about 35 per cent. limpid, sweet flavoured oil, much used as an edible and lamp oil by the poorer classes in India. It is much inferior to Gingelly oil, and is used to adulterate it.

Active Principles of the Compositæ.

The following list of the Active Principles of the Compositæ (extracted from Sohn's Dictionary of the Active Principles

of Plants, London /93) will be of assistance in the investigation
of local species. (Contractions. A = Alkaloid. G. Glucoside.
B. Bitter Principle.)

Achillea millefolium (milfoil).
A—— *moschatus* (Iva) (whole plants).
   Achillein G., (C2. H38. N2. O15.) von Planta.
   Ivain B., (C8. H14. O.)   do.
   Moschatine A., (C21. H27. NO7.) do.

Anthemis nobilis (flowers).
   Anthemine A. Pattone.

Arnica montana (Leopard's bane).
   Arnicin B. (C20. H30. O4. (?) Walz.

Artemisia absynthium (Wormwood).
A —— *abrotanum* (Southern wood).
A —— *maritima* (Wormwood).
A —— *Herba alba* (Asso.) does not contain c.
A —— *Silberi* (Bess.) does not contain c.
   (*a.*) Absynthin B. or G. (C40. H28. O8. Senger.
   (*b.*) Abrotine A. (C21. H22. N2. O2.) Giacosa.
   (*c.*) Santonin B. (C15. H18. O3.)

Atractylis Gumnifera.
   Viscin G. (?) (C20. H48. O8.) Rheinsch.

Bacharis Cordifolia.
   Bacharine A. P. Arata.

Calendula officinalis (Marygold).
   Calendulin (Neut. Prin.) Geiger.

Chrysanthemum Tanacetum (a. and *b.*)
C —— *Cinerariæfolium (c.* and *d.*).
   (*a.*) Tanacetin B (C. 61·46 % H. 7 7 %) Leppig.
   (*b.*) Tanacetic Acid (uncertain).
   (*c.*) Chrysanthemine A. (C14. H28. N2. O3).
     M. Zuco.
   (*d.*) Pyrethrosin B. (C34. H44. O10).

Chichorium Intybus (Chicory).
   Chicorin G. R. Nietzki.

Cnicus Benedictus.
   Cnicin B. (C42. H56. O15). (?Dragendorf.)

Crepis foetida.
   Crepin B. Walz.

Eupatorium Cannabinum.
   Eupatorine A. (C20. H25. O36. HNO. 3). Shamel.
   Eupatorin G. G. Latin.

Lætuce Sativa (Lettuce).

Lactuca Virosa.

L —— *Altissima.*
    Lactucin B.   (C22. H18. O7).   Kromayer.
    Lactucopicrin B.   (C. 44, H. 32, O. 21).   Kromayer.
    Lactucon.   (C16. H. 26O.).   Flückiger.
    Gallactucon.   Franchemont.
        from *L. Altissima.*
    Lactucerin.   Hesse.

Mikanea Guaco. (Leaves).
    Guacin B.   Faure.

Leontodon Taraxacum.   (Dandy lion.)   (Juice.)
    Taraxacin B.   Kromayer.

Vernonia Nigritana.   (Batjentjor).
    Vernonin G.   Heckel & Schlagdenhauffen.   (a cardiac
    poison).

Xanthium Strumaria (seeds).
    Xanthostrumarin G.   A. Zander.   (Poisonous).

THE MEANS OF FREEING CATTLE OF TICKS.

IN previous papers on this subject, sufficient stress has been laid on the importance of keeping the cattle in as healthy a condition as possible.   Whether or not the ticks select the poor cattle for attack, the fact remains that well-fed sleek animals are not usually much troubled by these parasites.

The same applies to the necessity of keeping the pastures in as good order as possible.   It is not easy in tropical countries to ensure the growth of the right kinds of grasses and shrubs ; but this is certainly no reason why the task should be altogether avoided.   A little care would at any rate materially improve the character of the herbage.   The fencing off of portions to be eaten down in rotation is not an expensive matter, and this is the most important item in the improvement of pastures.

Such measures as burning, draining and liming should be adopted where necessary, but while the latter is a much valued application for all grass lands, the former will not often be of use in these islands.   For the rest, there seems no need of insisting that clean pastures will produce clean animals.

The most direct and primitive method of the removal of ticks is that of simple extraction.

One frequently hears of scraping the cattle by means of a blunt knife, and thus removing the ticks. But to what pain and danger does this expose the animal! The bodies of the ticks will assuredly be removed, but what of the deeply sunk and invisible mouth parts? Even if the creatures were willing it would be impossible for them to with-draw the buried parts in time. It is asserted by those who have studied the subject that the mouth parts of the ticks if left in a wound exert an influence altogether disproportionate to the simple mechanical irritation. There is, according to these observers, something very like a poisonous reaction upon the animal. Forcible removal of ticks seems then rather to court skin diseases than to be a wise method of protection of the animals.

The question has again and again been broached as to whether ticks are capable of their own free will of detaching themselves from their host, some asserting that so firmly is the barbed proboscis fixed and anchored to the animal's flesh that it would be an impossibility to withdraw it unbroken. The consensus of opinion among those who have made the life of the tick their special study is however that the parasite can voluntarily let go its hold. Certain it is that the egg-laying of the female does not usually take place upon the host, while it is not uncommon to meet the males roaming about in places where cattle are used to congregate. The males are provided with an apparatus as formiable as that of the female, and yet it is everywhere agreed that he passes from one place to another with impunity.

It is thus a matter of importance to apply if possible substances so offensive to the tick that it will voluntarily shift its quarters.

We should avoid the use of external caustic applications upon the body of the tick; we should avoid such mechanical injury as cutting the body of the tick with scissors; for the observed effect of such brutal methods is the deeper thrusting of the parasite's proboscis, with a wince of pain from the unfortunate host.

Finally, the danger of leaving open bleeding wounds upon the animal's skin is obvious. Such spots, unless in very healthy animals, are of necessity the points of attack of various disorders. To quote one instance, in a report on "The Texas Screw-worm," issued by the Louisiana State Experiment Station, it is pointed out that the great majority of cases of animals attacked by this plague resulted from the deposition of eggs in the vicinity of where ticks had been killed, the flies being attracted by the blood. In an island where a troublesome skin disease appears at intervals, as in Antigua, special note should be taken of facts like these

Anyone then who values his beasts will put a stop to the barbaric remedies of scraping his cattle with a knife, cutting the ticks with scissors or injuring them in such a way as to cause them to penetrate deeper into the animal's skin.

In almost all countries there are certain local applications to the skin of animals intended to free them from ticks, and other parasites. In India a vegetable substance called *Nim oil* is said to very efficacious. In Barbados various plants are used to rub the cattle and horses to produce a sleek skin and to keep off vermin. Among others the *Belly-ache bush* (*Jatropha gossypifolia*), *black sage* and *rock sage* (*Lantana* species) were mentioned to me as especially useful. The oil-glands in the former of these plants at once point out the reason of its application, while the strong scent of the latter may be objectionable to flies and ticks.

The members of the tick order can be readily placed *hors de combat* by stopping up their breathing holes, and a very harmless series of methods is thus opened up before us.

Any oily substance will be of value, and a well-known remedy used locally in Antigua and in various other parts of the world is *old axle grease* rubbed over the parts attacked. But independently of this suffocating effect of oily matters there appears, as already pointed out, to be an objection on the part of the tick to any greasy or fat skin, so it is difficult to decide how exactly the application re-acts upon the parasite.

But, as will be evident, the remedies so far suggested have the grave drawback that the cattle must be held or thrown down during treatment. The throwing-down of cattle, excepting for special operations. cannot be too strongly condemned. The ordinary estate labourers are not particularly light-handed, and the brutality sometimes inflicted upon the cattle for the purpose of efficiently smearing them with some vermin-destroying compound is very probably more harmful than the pests themselves. It has been said that the days are fast passing away when it is customary to tie up a cow " fore and aft " and throw it heavily to the ground. A little consideration will show that there is no necessity for any such proceeding.

To throw down and drench a number of cattle entails besides a great deal of labour, so much so that this method cannot be employed frequently enough to be economically efficient.

We must again appeal to the methods adopted in the cleansing of plants from infestations. There is this difference between the vegetable and animal kingdom that, while the animals have to be caught, the plants to be treated are stationary. They need not be brought into compulsory quiet, but can easily be treated at any moment. But the very great surface of foliage to be treated has long ago proved that careful washing of the

leaves as at first practised is impossible as far as utility is
concerned. And the results of experimental work of the last
few years, especially in the United States, has placed in the
hands of nurserymen a completely new set of apparatus. The
different means of thoroughly covering every part of a plant with
the minimum of waste, have reached a high state of perfection,
and it is to these means that we must turn for dealing with our
cattle.

With a powerful spraying machine, such as those now
employed in all kinds of agriculture in the States, every part of
an animal may in a few seconds be searchingly covered with any
desired substance.

To be continued.

CLUB PAPER.

A VISIT TO THE GUACHARO CAVE OF OROPOUCHE.

By F. W. Urich, F.E.S.

ON Saturday the 2nd March, at 12 o'clock my small party
consisting of myself, a huntsman called Ped and a boy, left
Mr. Wallen's, Valencia, (where I had spent the night) for the
Cavestero or Cerro of Oropouche, where the Guacharo Caves
are situated. These hills form part of the Northern range and
according to Messrs. Wall and Sawkins belong to the Caribbean
group and rise to an elevation of about 2,000 feet. The first
part of our route lay on the Matura Road and after crossing the
Quaré River we turned into a bridle path leading to the North.
The first part of the road was of a sandy nature and did not
possess anything of an interesting character although there
were numerous butterflies flitting about, many of them old friends
of the St. Ann's Valley and a few, which did not seem familiar to
me. We followed the winding valley of the Barro River and crossed
it three times. Here we commenced a series of ascents and des-
cents (a good locality for a switch back railway) until we reached
the Turure River. It was here that the grand scenery of the
high woods and refreshing coolness of the streams peculiar to the
Northern Range of hills was first noticeable. My clumsy words could
give you but a very poor description of the beauty of the stream
on the banks of which we stood. The clear crystal water
rippling over white quartz pebbles was broken into miniature
rapids and whirlpools by great boulders which lay about in wild
confusion. On the banks, a rich vegetation of tall trees with

lianes hanging down in festoons made a deep shade into which the light could hardly penetrate from above, the banks were covered with a beautiful carpet composed of ferns, balisier, arums, and many other small plants.   Leaving the river we began to ascend the Turure Hill by a winding road from the top of which we got glimpses at intervals of beautiful pictures, the frames of which were tall trees and the subjects fine views of the surrounding country.   To the South we saw the Montserrat hills in the distance ; Mount Tamana the centre of the Island and at its foot the plain of Arima with its cocoa estates, all marked out by the flaming red flowers of the Immortel trees.   Later our attention was attracted by a beautiful view of the Matura cavestero to the East.   From here to our destination the scenery was beautiful. It would take a Kingsley to describe it.   The road goes up and down hill, one moment you are on the spur of the hill, the next you find yourself in a shady valley, again, in another, you are looking down into a shady glen from which the murmur of a small stream rises, and what with the bell birds tolling or I should rather say hammering, and the choruses of the other forest birds, in which the whistling frogs and the insects join, one does not get tired and the journey is agreeably shortened. At a place called Cumaque, where a small stream crosses the road and then loses itself down a beautiful and shady valley we made a halt. So far as natural history was concerned, our interest was always taken up by something.   On the Turure hill a beautiful machete escaped us, on another part of the road we caught a little snake *Tantilla melanocephala* L, trying to swallow a small centipede which it had already killed.   Huge nests of parasol ants were found all along the road, the inmates of which were all at work in spite of the day-light.   When near the Cerro Oropouche we found a fine specimen of coral snake bathing, (or was he fishing?) in a small pool near the road. After passing a place called La Peila, where I found the first shell we emerged into a clearing in the forest, a natural amphitheatre, all surrounded with high hills covered with rich vegetation, where on gently rising ground General Mendoza's house is situated, which, through the General's kindness, was to be my abode for the next two days. It was about five o'clock and with the help of the Overseer of the estate we set about cooking our dinner.   The night was delightfully cool and made a blanket very acceptable.   At daybreak on Sunday we rose and I wanted to at once explore the Guacharo caves ; but Ped, animated no doubt by the superstitious ideas of the natives of Venezuela, who, according to Humboldt, believe that the souls of their ancestors sojourn in the deep recesses of caverns thought that the Guacharo cave was too black to enter on a Sunday.   We went instead to the Oropouche River. We waded down the stream which takes its rise in the Guacharo caves, until we came to the Oropouche.   On our way down we saw several traces of otters, and at one place we found the remnants of one of these gentlemen's dinners in the shape of fish scales, which it is well known otters do not eat. The number and beauty of the Micro-lepidoptera was conspicuous, and it would well repay the trouble to spend a fortnight in this locality to collect them. The scenery of the Oropouche river which we ascended was grand. Rich vegetation on the bank, changing every now and then into perpendicular walls of crystalline limestone, made the scenery uncommonly beautiful, the only thing wanting perhaps being the graceful bamboos of our suburbs. The river is about the size of the St. Joseph River with more water at this time of the year and with beautiful basins. After enjoying a bath in a large pool we bent our way homewards through the rich vega, spending a long time collecting shells. The forest is here very dense and damp and right through the woods are scattered upheavals of tremendous boulders of crystalline limestone on which were to be found the shells *Cylindrella trinitaria* by the hundred hanging to the dry sides of the boulders. Among the vegetable debris that had accumulated in the fissures and depressions of the boulders *Cistula aripensis* was common, also other species, but the latter were fewer, in numbers. There were many interesting insects and spiders, and *Polydesmidœ* to be seen, but all my time was devoted to shells. After a rich harvest of the latter we returned to the house and rested the remainder of the day. Early on Monday morning we commenced making preparations to enter the caves. Before leaving the house our first care was to make torches or as the Spanish peons call them pavils, they consist of old cloth well soaked in melted wax of wild bees belonging to *Trigona*, a species found commonly about the woods in the hollow trunks of trees. When dry we twisted the cloth into long strips. These the Spaniards say are the only safe lights to enter the caves with, as they cannot be put out by the wind, on the contrary the stronger the latter the better they burn. After a good breakfast we started for the caves, which are situated about ten minutes walk from the General's house. Passing through the cocoa and ascending the river which takes its rise in the caves we were soon at our destination, making use of a trace that had been cut through gigantic wild tanias growing on the banks of the river. The Guacharo cave is pierced in the vertical side of a wall of crystalline limestone a hundred feet high, the entrance is about 25 feet in height and about 15 feet broad. Strewn about it are some huge boulders. The first thing to greet us was the smell of the birds and their deposits. They were of a decided "cockroachy" aroma. There are numerous rocky ledges and cavities in the sides. It is difficult for me to judge the distance I penetrated as progress was very slow, but

I should say that about 150 yards is not over estimated. In the bottom of the cave is a clear running stream in which we waded, reaching as a rule to above the ankle and sometimes above the knee. The bottom is composed of white quartz pebbles and sand ; strewn about the bed are large boulders with jagged ends which make progress very slow. At the sides of the river, out of the current, there is a large accumulation of guano and seeds of, so far as I could judge, several kinds of palms. The roof of the cave slopes downwards and at the further end I could not stand erect, while the water was above my knees. Divesting ourselves of as much clothing as we could without running the risk of getting chilled, we each lighted a pavil and entered the cave. The scene which met our eyes and the noises we heard were of peculiar weirdness. Above our heads about a hundred birds fluttered, wheeled, darted, and screamed. The beating of their wings their shrill and piercing cries and croakings together with the rushing and murmuring of the stream created an impression which cannot be described but which was intensified by the vaulted rocks and repeated by the echoes in the depths of the cavern. Ped, by fixing a pavil at the end of a long rod bent at the end like a shepherd's crook, to which was attached a fish hook, showed me the nests of the birds some 25 feet above our heads. They were mostly in the holes and fissures of the rock, with which the sides of the cave are riddled. Feeling about the nest with the hook we managed to get two young ones, but they were very young and not yet fit for table. Most of the birds were still sitting on nests built of clay of a reddish colour. As we penetrated into the cave the noise increased, but when we got into the lower parts, no more guacharos were seen and we got into the region of the bats, which belonged to[*] one species and were in numbers, treating us to a shrill concert which was answered by the plaintive cries of the guacharos in the distance. After securing a few specimens we retraced our steps and after various tumbles over the boulders emerged into daylight and seated ourselves at the entrance on the banks of the river and rested. We were glad to leave "a place where darkness does not offer even the charm of silence and tranquillity." The guacharo is known to science as the *Steatornis caripensis* Humb., and is almost the only fruit-eating nocturnal bird known. It feeds on hard fruit and at night fall I could see them flying over the house and in the cocoa, uttering their lugubrious cries.

19th April, 1895.

---

[*] *Chilonycteris rubiginosa.*

Communications and Exchanges intended for the Club should be addressed to the Honorary Secretary, Port-of-Spain, Trinidad, B.W.I.

Price 6d. Annual Subscription, 3/.

Vol. 2. OCTOBER, 1895. No. 10

J'engage donc tous à éviter dans leurs écrits toute personnalité, toute allusion dépassant les limites de la discussion la plus sincère et la plus courtoise.—LABOULBÈNE

Trinidad·Field·Naturalists'·Club.

NATURA MAXIME MIRANDA IN MINIMIS.

Publication Committee :

His Honour Sir JOHN GOLDNEY, *President.*

P. CARMODY, F.I.C., F.C.S. *V.P.*: SYL. DEVENISH, M.A., *V.P.*

HENRY CARACCIOLO, F.E.S. R. R. MOLE.

F. W. URICH, F.E.S., *Hon. Secretary.*

CONTENTS :—

Report of Club Meetings :

August	236
Secretary's Report	236
List of Members	239
Financial Statement	241
Publication Committee's Report ...	242
Four Oncidiums of Trinidad	243

☞ All Communications and Exchanges intended for the Club should be addressed to the Honorary Secretary, Port-of-Spain, Trinidad, B.W.I.

JOURNAL

OF THE

Field Naturalists' Club.

VOL. II. OCTOBER, 1895. No. 10.

REPORT OF CLUB MEETING.

13TH AUGUST 1895.

PRESENT: Mr. H. Caracciolo V. P., in the Chair, Dr. Rodriguez, Dr. Ince, Messrs. H. J. Baldamus, W. S. Tucker, J. R. Llaños, John Barclay, J. T. Rousseau, J. B. Inniss, Charles Libert, Thos. I. Potter, L. Guppy, jnr., R. R. Mole and F. W. Urich, Hon Secretary and Treasurer. Mr. Frank E. Beddard, F.R.S., was elected an Honorary Member of the Club.—The Secretary presented the following reports.

Secretary's Report for 1894-95.

ACCORDING to Rule 7, I beg to lay before the Members of the Club, the Honorary Secretary's Fourth Annual Report. It is a *resumé* of the Club transactions for the year 1894-95.

The number of Members on the list at the close of the year 1893-94 was 124. We have elected during this year two Honorary and nine Town Members and after deducting the losses through death, resignation and other causes the total number of Members on the Roll is 118.

Amongst the Town and Country Members of the Club who have passed over to the majority are Dr. Beaven Rake, who was elected President at the last Annual Meeting, Messrs. Henry Tate, B.A., and Arthur Gaywood and the Honble. Robert Guppy, M.A., and of the Honorary Members lost through death Professor Carl Vogt, who always took a lively interest in our doings and helped us in many ways. The usual list of Members is appended.

The balance of $26.60 of last year has been raised to $110.95. The total receipts do not amount to those of last year, owing to the fact that the advertisements have not yet been collected, but when this is done the receipts will be more than those of 1893–94. There are also many Members' subscriptions still to come in. The usual statement of finances is appended.

During the year there have been nine Ordinary Meetings, at which the attendance has been on the average better than in former years, and the papers read have maintained a high standard of excellence. On the 26th October 1894, under the auspices of the Club, Mr. Mole delivered at the Victoria Institute, a lecture on Snakes which was well attended.

The *Journal* is published as usual, details of which will be found in the Publication Committee's Report.

The following is a list of the additions to the Library during the year 1894–95.

PRESENTED BY THE AUTHORS

Materials for a history of the Mackerel Fishery by G. Brown Goode and J. W. Collins.
Contributions to the Natural History of the Bermudas by J. Mathew Jones and G. Brown Goode.
Materials for a history of the Sword fishes by G. Brown Goode.
The Colour of fishes by G. Brown Goode.
A preliminary catalogue of the Reptiles, Fishes and Leptocardians of the Bermudas by G. Brown Goode.
Do Snakes swallow their young ?—by G. Brown Goode.
The Museum of the future by G Brown Goode.
A short biography of the Menhaden by G. Brown Goode.
The first Decade of the United States Fish Commission by G. Brown Goode.
The Status of the United States Fish Commission in 1884 by G. Brown Goode.
The Genesis of the United States National Museum by G. Brown Goode.
The Origin of the National Scientific and Educational Institutions of the United States by G. Brown Goode.
American Fishes by G. Brown Goode.
Miscellaneous Papers on Economic Entomology by F. M. Webster.
A Manual for the study of Insects by J. H. Comstock.
Descriptive Catalogue of the Butterflies found within fifty miles of New York by W. Beutemuller.
Descriptive Catalogue of the Orthoptera found within fifty miles of New York by W. Beutemuller.

PRESENTED.

By The Trustees of the British Museum Natural History :
Guide books to the collections (Twelve in number).
By Mr. F. A. Ganteaume :
Girad, Entomologie 2 volumes.
By Mr. William Brewster :
J. B. Smith, Catalogue of the Lepidopterous super family Noctuidae found in Boreal America.

By United States National Museum :

8 Manuals of directions for collecting and preserving the following Mammal skins, Rough skeletons, Birds, Birds eggs and nests, Reptiles and Batrachians, Insects, Mollusks, Recent and fossil plants.

By Mr. C. A. Barber, M.A. :

Miscellaneous Papers on Cane disease.

By United States Department of Agriculture, Division of Entomology :

Reports of observations and experiments in the practical work of the Division.
Experiment Station Record Vol. VI Nos. 8-10.

By Supt. Botanic Gardens, Trinidad :

Annual Report for 1894, Bulletin of Miscellaneous Information and Circular Notes.

In Exchange.

Smithsonian Reports for 1891, 1892 and 1893.
Report of the United States National Museum for 1892.
Proceedings of the United States National Museum Vol. 16—1893.
Transactions and Annual Report of the Manchester Microscopical Society for 1893.
Bulletin du Muséum d'histoire naturelle No. 1-3.
Bulletin of the American Museum, Natural History, (New York) Vol. IV. V. and VI.
Agricultural Journal of the Leeward Islands Nos. 1-3.
Insect Life Vol. VI Nos. 4 and 5 Vol. VII Nos. 1 and 4.
Annual Report of the Entomological Society of Ontario for 1894.
Canadian Entomologist.
Tuft's College Studies Nos. 1, 2 and 3.
Journal of the Institute of Jamaica Vol. II No 1.
Agricultural Lectures of the Institute of Jamaica.
Proceedings of the Entomological Society of Washington Vol. III Nos. 1-4.

Purchased.

Science Gossip.—Vol. 1.

The Club collection has progressed favourably during the year, although it does not show up so well apparently. The reason is that mostly insects were collected and owing to the quantity and difficulty in naming, these have not yet been placed on exhibition, but it is hoped that this may be done in the course of the ensuing twelve months. Through the exertions of Mr. A. B. Carr a fine specimen of *Galictis barbara* was added to the Mammals to which Mr. Caracciolo added 8 new ones viz.: 6 *Muridæ*, 1 *Didelphys trinitatis* and 1 *Cyclothurus didactylus*.

The following is a list of the donations and the names of their donors for the year. The thanks of the Club are due to those who have thus contributed to the collection.

MAMMALIA.

6 Specimens of *Muridae*,   ⎫
1 *Cyclothurus didactylus*,   ⎬ presented by Mr. H. Caracciolo.
1 *Didelphys trinitatis*,   ⎭
1 *Cyclothurus didactylus*, presented by Dr. A. W. Wight.
   A Bat presented by Mr. J. Graham Taylor.

REPTILIA AND BATRACHIA.

A collection of Snakes containing some rare types presented by Mr. J. F. Wallen, Valencia.
A *Glauconia albifrons* presented by Mr. A. B. Carr.
A crowned Snake, *Scytale coronatum*; a mangrove Snake, *Liophis cobella* presented by Mr. H. Payne.
A clouded Snake, *Leptognathus nebulatus* presented by Mr. T. I. Potter.
A Beh belle chemin, *Liophis melanotus* presented by Mr. J. T. Rousseau.
A *Boa constrictor* presented by Mr. L. de Verteuil.
A coral Snake, *Elaps riisei* presented by Mr. P. Maingot.
Two young Mapepires, *Lachesis muta* presented by Mr. H. Hutton.
A Tree frog, *Hyla crepitans* presented by Mr. J. Graham Taylor.

INSECTA.

Electric Bugs, *Nepa grandis* presented by Mr. Syl. Devenish, Mr. H. K. Collens, Mr. W. Norton, Mr. J. O. Urich.
Several Coleoptera presented by Mr. J. O. Urich.
A giant Locust presented by Master H. Kernahan.
Praying Mantis presented by Miss Fenwick and Mr. E. C. Purcell.
A chrysalis of *Tithorea flavescens* presented by Mr. W. Le Gall.
A Hag moth caterpillar presented by Mr. W. A. Ghent.
A silk moth, *Attacus hesperus* presented by Inspector A. D. P. Owen.
A butterfly, *Helicops cupido* presented by Dr. S. M. Lawrence.

MISCELLANEOUS.

A Hermit crab presented by Mr. D. Hart.
A small centipede, *Geophilus* species presented by Mr. D. F. King.
A Scorpion presented by Mr. W. J. L. Kernahan.
An abnormal egg presented by Mr. A. Lucien.
An abnormal chicken presented by Hon. W. Howatson.

In conclusion I can only congratulate the Members on the success which has attended the efforts of the Club during the past year, in spite of the heavy losses sustained at the opening of the twelve months in the deaths of Dr. Rake and Mr. Tate and the departure of Mr. Broadway, all of whom were hard workers.

F. W. URICH,
Hon. Sec. and Treasurer.

TRINIDAD FIELD NATURALISTS' CLUB.

Officers and Committees elected for the year 1895-96.

President :

Sir John T. Goldney.

Vice-Presidents :

Prof. Carmody, F.I.C., F.C.S. ; Syl. Devenish, M.A.

FINANCE COMMITTEE.

Sir John Goldney, P.	P. L. Guppy, jr.
T. I. Potter,	

BUSINESS COMMITTEE.

Sir John Goldney, P.	R. R. Mole,
Lieut.-Colonel D. Wilson,	W. S. Tucker,
H. Caracciolo,	T. I. Potter,
Syl. Devenish,	F. W. Urich.

PUBLICATION COMMITTEE.

Sir John Goldney, P.	H. Caracciolo,
Syl. Devenish, V.P.	R. R. Mole,
Prof. Carmody, V. P.	F. W. Urich.

LIST OF MEMBERS OF THE TRINIDAD FIELD NATURALISTS' CLUB.

Honorary Members.

	Date of Election.			Date of Election.
Boettger, Prof. Dr. O.	6 10 93	Morris, D., M.A., Ph. D.	6 10 93
Brewster, William	4 5 94	Riley, C. V., Ph. D.	2 3 94
Broadway, W. E.	10 7 91	Smith, G. Whitfield	13 11 91
Brown-Goode, Dr. G.	9 11 94	St. Vraz, E.	8 8 91
Crowfoot, W.M., M.B.,		Stone, Sir J. B., F.L.S.,	
F.R.C.S.	3 3 93	F.R.G.S., &c.	7 7 93
Gatty, S. H.	19 8 92	Thomas, Oldfield, F.Z.S.	6 3 92
Gunther, Dr. A., F.R.S. &c.	1 7 92	Warming, Prof. Eug.	4 12 91
Hamilton, Hon. C. B.	5 2 92	Wilson, H. F.	1 7 92
Howard, L. O.	6 7 94	Wells, S.	2 9 92
Kirby, W. F.	2 9 92	Whitehead, C. E.	7 10 92
Mitchell, P. Chalmers	2 12 92		

Corresponding Members.

Bock, E.	2 6 93	Elliot, H. V., Lieut, R.N.	3 3 3
Chapman, F. M.	5 5 93	Ganteaume, Harris	8 8 1
Cockerell, T. D. A., F.E.S.,		Jefferys, W. J.	3 3 3
F.Z.S.	5 5 93	Terry, John, F.R.G.S. &c.	5 1 94
Deyrolle, E.	7 10 92	Wright, Rev. E. Douglas	6 4 94

Town Members.

Broome, H. E. Sir. F. N., K.C.M G.	6	5	92
Agostini, Edgar	4	11	92
Agostini, J. L.	2	3	94
Alcazar, Hon. H. A.	2	2	94
Anduze, Jules	2	3	94
Archer, Julian H.	7	7	93
Arnott, T. D.	7	7	93
Bain, F. M.	7	7	93
Baldamus, H. J.	5	1	94
Barclay, John	11	1	95
Borberg, A.	7	7	93
Caracciolo, H., F.E.S.	10	7	91
Carmody, P., F.I.C., F.C.S.	8	1	92
Carr, T. W.	3	11	93
Collens, J. H.	5	2	92
Creagh-Creagh, G.	5	1	94
Cumberland, S. A.	1	9	93
Devenish, Syl., M.A.	8	1	92
Dickson, J. R., M.B., C.M.	1	6	94
Dumoret, R.	7	6	92
de Labastide, F.	5	10	94
Eagle, F.	8	1	92
Eagle, S.	2	9	92
Ewen, E.D.	2	9	92
Garcia, Hon. G. L.	2	3	94
Gerold, E.	4	3	92
Goldney, Sir J. T.	6	1	93
Gordon, Hon. W. G.	7	7	93
Guppy, P. L.	10	7	91
Harding-Finlayson, M.	5	1	94
Hernandez, F.	4	8	93
Hoadley, John	1	9	93
Holt, Rev. E. J.	6	4	94
Ince, W. H., F.I.C. Ph D.	11	6	95
Inniss, J. B.	11	6	95
Koch, W. V. M., M.D.	2	6	93
Knaggs, R. H. E., M.R.C.S.	5	5	93
Knox, C. F., M.R.C.S.	5	1	94
Llaños, J. R.	2	3	94
Libert, C.	7	10	92
Lovell, Hon. Dr. F. H., C. M. G.	2	2	94
Lota, A., M.D.	7	6	92
Maingot, F. J., F.G.S.	7	6	92
Meaden, C. W.	7	6	92
Miller, J.	6	4	94
Mole, R. R.	10	7	91
Munro, D.	9	11	94
Murray, J. R., F.B.S.S.	2	10	91
Potter, T. I.	10	7	91
Prada, E., M.R.C.S.	2	3	94
Reid, R. S.	8	2	95
Reoch, A. T.	5	10	94
Rodriguez, F. A., M.B., C.M.	11	1	95
Rousseau, J. T., M.A.	2	3	94
Saurmann, C.	2	3	94
Scott, C. W.	10	7	91
Seccombe, G. S., L.R.C.P.	2	3	94
Smith, R. H. S.	5	1	94
Smyth, Hon. J. Bell	4	11	92
Sorzano, Til	4	11	92
Spooner, F. V. H.	6	10	93
Taitt, Alfred	10	7	91
Taylor, J. G.	2	2	94
Tripp, E.	6	10	93
Tucker, W. S.	5	1	94
Urich, F. W., F.E.S.	10	7	91
Walker, C. W.,	8	8	91
Warner, A., B.A.	5	1	94
Wilson, Hon. Lieut-Col. D. C.M G., V.D.	7	10	92
Wilson, Edwin C.	2	3	94
Wilson, James (Tertius)	3	8	94

Country Members.

Carr, A. B.	19	8	92
deVerteuil, L.	7	6	92
deVerteuil, F. A., M.R.C.S.	6	10	93
Eakin, J. W., M.B.	5	1	94
Goode, John	6	1	93
Guilbert, J.	19	8	92
Greenidge, J. S.	2	9	92
Hobson, J. J.	4	12	91
Kernahan, G. J.	6	10	93
Laing, Alex. S.	5	1	94
Mahoney, F. J.	7	10	92
Mitchell, J. W. L.	2	12	92
O'Connor, P. A. T.	2	2	94
Thavenot, C. J.	6	5	92
Wight, A. W., M.R.C.S.	2	6	93
Wilson, J. S.	7	10	92
Woodlock, A., L.R.C.S.	7	10	92

TRINIDAD FIELD NATURALISTS' CLUB.

TREASURER'S STATEMENT FOR THE YEAR 1894-95, ENDING 31ST JULY, 1895.

Dr.	$ c.	Cr.	$ c.
To Balance from last year	26 60	By expenses for collections	80 24
„ Members' Subscriptions and Entrance Fees ...	410 26	„ Printing *Journal*	203 76
„ Receipts from *Journal*	52 20	„ Printing and Stationery	35 24
„ Nett proceeds of Lecture	9 76	„ Postage	25 93
		„ Expenses incurred at meetings ...	8 30
		„ Miscellaneous (Porterage, collecting Subscriptions, Attendants, &c. &c.)	34 40
		„ Balance in hand	110 95
	$498 82		$498 82

F. W. URICH,
Hon. Sec. and Treasurer.

TRINIDAD,
1st August, 1895.

Examined and found correct,
On behalf of the Finance Committee,

H. CARACCIOLO,
Vice-President.

Publication Committee's Report.

IN accordance with Rule 29, the Publication Committee have the honour to make their Annual Report on their transactions for the year 1894-95.

During the year the Committee have issued six numbers of the *Journal* containing 144 pages of Reports of Proceedings, original papers and extracts from the publications of societies engaged in Zoological research, a decrease of 34 pages upon the previous year. The Reports of Meetings are up to date and the great efforts of the Committee have been successful in reducing the number of original papers in arrear from eleven to six.

During the past 12 months 24 original papers have appeared in the *Journal*, and whereas last year the Committee were disposed to take a desponding view of the gradually accumulating mass of manuscript they are now seriously alarmed at an anticipated dearth of material—the next two *Journals* will exhaust their stock completely and the Committee therefore take this opportunity of appealing to the members, and the country members especially, to help forward the Society's work by placing their observations on record. The simplest facts in Natural History, and especially economic Natural History should be noted and forwarded to the Committee and will always find a place in the *Journal* where they cannot fail to be of service to other members. It has been strongly felt for sometime past that the members do not fully appreciate the value of the opportunities afforded them in this respect, and in this connection it is interesting to note that only 12 of the 24 papers above referred to have a distinct value as dealing with questions of economic Natural History. What a vast scope for improvement is there here ! The Committee would be glad to see the Club making more use of the pages of the *Journal* and cordially invite it to do so.

The death roll of the Club has been exceptionally long during the past year and it has been the melancholy duty of the Publication Committee to compile a corresponding number of obituary notices. The most important of these was one which the Committee feels was a very inadequate attempt to place on record in the annals of the Club the services of our late President, Dr. Rake. The capital portrait which accompanied it was obtained through the kind instrumentality of an eminent scientist who is also one of our Honorary members, Professor Dr. Boettger of Frankfurt, who has done good work for us in this and other ways and to whom our grateful acknowledgments are due.

The Committee again tender their thanks to the Editors of the daily Press for the liberality with which they have published matters in connection with the affairs of the Club.

Signed on behalf of the Committee.

HENRY CARACCIOLO.

On the motion of Mr. Potter seconded by Dr. Rodriguez the reports were unanimously adopted. The meeting then proceeded to the election of officers for the ensuing year. A ballot being taken the following was the result : President : Sir John T. Goldney ; Vice-Presidents : Professor Carmody, F.I.C., F.C.S., Mr. Syl. Devenish, M.A.

Mr. F. W. Urich was unanimously re-elected Honorary Secretary and Treasurer. The Finance, Business and Publication Committees were elected. A special vote of thanks was recorded to Mr. R. R. Mole for services rendered to the Publication Committee during the past year.

At 9 p.m. the meeting adjourned.

CLUB PAPER.

FOUR ONCIDIUMS OF TRINIDAD.

By T. I. POTTER.

OF all the numerous genera of orchids to be found in the Island of Trinidad there are none to compare with the four representatives of the genus Oncidium which form the subject of this paper.

Among our native orchids we have many beautiful species, but for beauty of bloom and duration of flowers in perfect condition none to surpass these four species, a fact which is proved by the demand for them from abroad.

There is the beautiful *Diacrum bicornutum* (known locally as the " Virgin " from a fancied resemblance in the shape of the column to a statue of The Holy Mother) with its almost white sweetly scented blossoms which in the dry season adorns the rugged and waterworn rocks of the Bocas and North Coast. Then again the curious white waxy slipper shaped pair of flowers of *Stanhopea grandiflora* relieves the gloom of the dense forest and attracts the solitary bees in large numbers by its rich aromatic perfume; and besides these two lovely species there are very many beautiful and fascinating orchids adorning our forest trees with their spikes and racemes of curious or resplendent flowers, but none are so keenly sought for nor more universally admired than the four Oncidiums I am about to attempt to describe.

Their botanical names are *Oncidium ampliatum, Oncidium lanceanum, Oncidium luridum,* and *Oncidium papilio,* but to the unscientific horticulturist they are known as "The Yellow Bee", "The Cedros Bee", "The common or brown Bee" and "The Butterfly orchid," these names being given on account of a fancied resemblance in the shape of the flowers to the insects whose name they bear.

I will deal with them in the order in which they stand above, and will therefore begin with *Oncidium ampliatum*—The Yellow Bee.

Of this species there are two representatives, viz. the type and a larger variety known as *O. ampliatum majus.* The type is an epiphytal orchid with roundish pseudo-bulbs somewhat compressed, pale green, striped and spotted with reddish brown, and bearing on each pseudo-bulb a pair of shiny rich green, oblong divaricate leaves. From the base of the pseudo-bulb, when it is fully matured, springs a bright green flower spike which terminates in an ample raceme of bright yellow flowers, almost white on the under side, minutely spotted at the base of the broad lip and on the sepals and petals with reddish brown.

The flowers last a very long time in beauty if protected from the weather and the garden syringe. It flowers in the dry season (January to May).

When in full bloom a large mass of this Oncid is a grand sight, the bright yellow colour predominating in the blossoms being beautifully relieved by the shiny dark green of the leaves.

I have had two large plants, growing in baskets, in bloom for seven weeks, and they were admired by all who saw them.

This species grows well either on the bark of living trees or in a basket with good drainage. For the latter method peat may be use as compost, but it decomposes so very rapidly in a warm climate that I have found thoroughly dried coconut husk with the outer skin peeled off and cut up in pieces of the size of a hen's egg a very good substitute if similar sized pieces of charcoal be intermixed with it. For large masses the basket treatment is preferable to blocks. In fact *Oncidium ampliatum* takes a very long time to develop into a fine plant when grown on a block or a tree, although it grows and flowers well, but the growth is slow and the flowers are generally produced from one pseudo-bulb at a time, while under the basket treatment the plant sends leads from several sides, and these when mature generally produce one, and sometimes two flower spikes which are often more vigorous than those produced by plants grown on blocks.

The flowers of *O. ampliatum* are visited by a number of yellow bodied wasps (*Polistes* sp) which with the co-operation of a small ant seem to affect the fertilisation of the flowers, as this Oncidium goes to seed very readily, and can be easily propagated from its seeds. The writer has two nice little plants raised from seeds sown naturally from an older plant in his possession. One of these seedlings will bloom next year for the first time.

To those who have the patience necessary to take an interest in raising orchids from seed, a method which requires much of this virtue and of care, the writer recommends the top of an orchid basket or pot, the top of an old block, preferably of Calabash wood, or the rough bark of a living tree as a suitable place for sowing the very fine seeds. Some time, perhaps weeks and months will elapse before the seedling makes its appearance but once it has taken a start it will, with ordinary care, soon grow into a fine little plant and in three or four years will reward its cultivator with its first bloom.

The next Oncidium with which I propose to deal is the well known, beautiful, and much desired *Oncidium lanceanum* or " Cedros Bee."

This orchid has not yet been recorded from any other part of the Island but the South western extremity known as the Cedros Ward, but it is also a native of Venezuela and British Guiana, which countries can boast of all our native species and many more besides.

This Oncidium belongs to the section which makes no pseudo bulbs, the large, thick, somewhat oval acute green leaves spring from a short knotty rootstock, which throws out a vast number of cordlike roots greyish white with an orange coloured growing point ; the leaves when exposed to sunlight are much spotted with brown and from the base of the new leaf there springs a raceme spike bearing a number of beautiful and sweetly scented flowers of which the sepals and petals are pale olive green thickly blotched or spotted with reddish brown the broad somewhat spathulate lip being of lovely violet purple at its extremity (more or less intense according to the variety) and gradually shading off to a deeper purple at the base. The column and tubercles are of a purple shade also.

The flowers measure about 2 inches in diameter when fully espanded, though in some varieties I have found flowers which measured two and a half inches across.

Although found only in one quarter of the Island, this orchid flourishes everywhere else under cultivation, but to grow it to perfection it should be put in a wooden basket. I have ound that light baskets made with rods of cedar, with plenty of charcoal for drainage and peat or coconut fibre as potting material answer extremely well, but it will also grow and flower fairly well on wooden blocks or tree fern root.

Oncidium lanceanum in its native state grows on the upper branches of very large forest trees, and an equal amount of light and shade is indispensable for its healthy existence. Too much of either will certainly do harm in the long run, either by scorching the leaves or by producing rot and blind buds, which latter is also caused by over watering.

O. lanceanum being accustomed to stand long periods of drought in its flowering season should never be watered more than once a day, or once every other day, when the flower spike is appearing. When the flowers have opened it will require the usual supply of water to maintain them in bloom, but by no means should the water, wherever it may come from, be allowed to fall on the flowers or they will soon fade and decay.

Of this beautiful orchid there are several varieties, as is easily seen when a number of them is put together. In some the flowers are large and are borne singly on the spike ; in others the flowers are of the typical size but the colour of the lip is nearly white, while in others again the sepals and petals are blotched and spotted with reddish brown and these parts of the flower are somewhat acute, wavy and more incurved than in the type.

I venture to classify them as follows :—

1. *O. lanceanum*, type.

Leaves broad, compact erect growth, flower spike erect and flowers, borne in clusters, sepals and petals spotted with brown lip violet purple.

2. *O. lanceanum majus* (?) variety.

Leaves longer than type, straggling growth, flower spike long, rather drooping, flowers very large, borne alternately on the spike, sepals and petals, much spotted with reddish brown lip &c., deep violet purple, divisions of flower very much separated from each other.

3. *O. lanceanum*, blotched sepals and petals.

Leaves almost eliptically erect, compact growth, smaller than type, flower spike erect, flowers borne in cluster on short spike, division of flower close together, sepals and petals heavily blotched with reddish brown in confluent patches, lip violet, rather pale.

4. *O. lanceanum*, almost white lip.

Leaves a little channelled horizontally spread, growth inclined to be straggling, flower spike erect, bearing as a rule, a few flowers lip pale white at tip the violet shade being more pronounced towards the base of the lip, sepals and petals, pale olive green spotted with reddish brown.

There may be other varieties besides these, but I have not been able to note them as carefully as those given above.

It is much to be regretted that this beautiful Oncidium is gradually and surely becoming scarce in the locality in this

Island where it formerly abounded. Many reasons have been given for its increasing scarcity, but there is good ground to believe that the alienation of Crown Lands and their cultivation is the chief, and perhaps the only cause of the rapid disappearance of *O. lanceanum.* When the lands are cleared by the small proprietor or planter the forest growth is cut down and burnt, and in this way hundreds of orchids of all sizes and of various species perish with their natural supports.

In this way *O. lanceanum* is being rapidly exterminated and with it another and much rarer Oncid with regard to which latter I shall make some observations later on.

Some persons are of opinion that the great demand from abroad for this *Oncidium* and the ruthless manner in which it is collected is the chief cause of its increasing scarcity but I beg to differ from this opinion, for from *actual experience* I know that unless the forest where an orchid is indigenous is destroyed it is next to impossible to exterminate it. You must annihilate its habitat in order to prevent the minute seed from springing up in a genial climate, and no amount of mere collecting can do this, as it is illegal to fell trees on Crown Lands, without a wood-cutter's license and this the orchid collector if he be only an orchid collector does not require.

It is to be hoped, however, that such plants as may be protected from destruction by having established themselves on trees on Government reserves may in the course of time disseminate their seeds around so that the larger trees of the cultivation which replaces the virgin forest, may in a few years be covered with a new generation of this lovely species and thus save this orchid from complete extermination.

Oncidium luridum. "The common or brown bee"

The next *Oncidium* of Trinidad with which my paper proposes to deal, is the well known, and much admired *O. luridum* better known to residents as the "Common or brown bee."

This orchid may be said to be almost ubiquitous in this Island, and is rather common on the Continent of South America also. But there is no doubt that it is a very beautiful orchid and were it scarcer would be as eagerly sought for as are many less beautiful but rarer orchids.

Like *O. lanceanum* (which I may call its first cousin) it belongs to the bulbless section of Oncidiums, and the long somewhat channelled acute leathery leaves of green colour sometimes finely dotted with brown, spring from a knotty rootstock not unlike that of *Oncidium lanceanum.* The roots are similar in shape and colour to those of this latter orchid except that the growing point is more of a yellow colour, and the long flower spike which springs from the front of the base of the almost matured leaf is of a reddish brown sometimes greenish brown colour, drooping, much branched and racemose.

The flowers borne in clusters on the various branches are small, perhaps about an inch in diameter, sepals, petals and lip of an olive green sometimes of a brownish green colour, mottled with a darker shade of brown, the column and tubercular process being of a pink white colour with a slight touch of yellow.

There are very many varieties of this great old orchid and in the more desirable ones the spikes are about five to six feet in length, proportionately broad, and bearing hundreds of the small but curious flowers in clusters on the several branches of thespike.

This orchid grows very well in or on anything : in fact it is very hardy indeed, but it must be treated like *O. lanceanum* to grow it to perfection.

When grown on a living tree in a favourable situation it often goes to seed and in a few years will cover the adjacent limbs of its support with its seedlings.

On account of its being found nearly everywhere in this Island it is undeservedly despised, and it is looked down upon by many, but to the true lover of flowers it is a very charming species, and a well grown plant with its flowers and spikes in full bloom is really an object of admiration.

Small ants are very fond of visiting the flowers which possess a peculiar aromatic odour.

Before leaving the bulbless Oncids it is necessary to notice the rare and most beautiful *Oncidium hæmatochilum* which has hitherto been considered a distinct species, but which certainly bears out the idea that it is a natural hybrid or cross between the two orchids last described. In a letter to the *Orchid Review* of No. 30, 1895, I gave my reasons for thinking that this native was a natural hybrid and the Editor of that Journal seemed to think that I was right (*vide Orchid Review* 1895).

This extremely rare orchid is a native of Trinidad, possibly it may occur in Venezuela and British Guiana also, in fact wherever *O. lanceanum* and *O. luridum* grow near to one another. It has, however, not yet been recorded from any part of the Island but Cedros, where *O. lanceanum* also only occurs and where *O. luridum* is often found growing upon the same bough and sometimes entangled with the former orchid.

The parent plants flower about the same time of the year. They are visited by the same insects. There are two distinct kinds of *O hæmatochilum.*

It was these facts which led me to believe that *O. hæmatochilum,* a bulbless Oncid with leathery green and brown spotted leaves sometimes like those of *O. luridum* and at other times resembling those of *O. lanceanum,* with a reddish brown spike of medium length somewhat drooping, bearing a considerable number of flowers, in shape like those of *O. luridum*, in size like those of *O. lanceanum*, and in colour, olive green or yellowish green, spotted with chocolate brown, or reddish brown, as regards the sepals and petals, but blood red and bright yellow as regards the lip, with purplish violet tubercular processes and column, was the result of a natural cross between the two species above mentioned.

In one of the two kinds, the lip is blood red with a narrow border of yellow in which a few dots and dashes of blood red appear.

In the other, and perhaps the more beautiful kind, the blood red blotch is very small, and the yellow border predominates, the column and operculum with the rather pronounced tubercular processes being of a violet or purplish violet colour.

This natural hybrid, grows easily and to perfection if treated similarly to *O. lanceanum*. It is very free flowering if exposed to strong diffused daylight, and the flowers last about three weeks in beauty.

It is, however, extremely rare, and as it closely resembles both *O. luridum* and *O. lanceanum*, it is very difficult, for an inexperienced person to distinguish it from them when it is out of bloom, and hence many amateurs have met with sad disappointments in attempting to obtain a specimen. The advice I would therefore give is : never buy a plant out of bloom, unless guaranteed by a respectable person.

The three Oncidiums just described though very easily grown in this their native colony with ordinary care are nevertheless subject to the ravages of an insect, a species of *Capsus*, one of the plant bugs, which if not speedily exterminated, raises a large family on the under sides of the leaves and by sucking the sap from these parts they dessicate the leaves and thus destroy the healthiest plant.

This *Capsus* is particularly fond of thick or coriaceous leaved orchids and affects not only these *Oncidiums* but *Epidendrums* also including the *Cattleya* genus and last year it nearly destroyed a very fine plant of *Diacrium bicornutum* in my collection, just as the plant was sending forth its flower spikes.

Through the kind assistance of Prof. Howard of the United States' Department of Agriculture, an honorary member of our Club, I was able to get the genus of this orchid pest determined, and he suggests the application of a dilute solution of kerosine emulsion as an effective remedy. My own method of getting rid of them was to seek them out and destroy the whole brood, and this I did with such good results that on the receipt of Prof. Howard's remedy I was not able to find a single plant in my fairly large collection with *Capsus* on it, and therefore I am not able to say what effect it has on them, though I feel certain it is an excellent remedy.

Oncidium papilio, " The butterfly."

This old and exquisite orchid is almost world-renowned. It occurs in many parts of Trinidad and has been found in great quantities at the North Western extremity.

It is an Oncidium with pseudo-bulbs. These are almost circular, rather compressed and somewhat wrinkled, and bearing a short elliptical tough leaf with a well-defined midrib or nerve down its centre. In colour, the leaves are sometimes dull green, but where the plant grows in a good light, they are of a green colour prettily marked with marbling of reddish brown. The roots are rather fine, of a brown colour, fleshy, with a green growing point.

The flower spike springs, as is usual with orchids of this genus, from the base of the last matured pseudo-bulb, and is about 30 inches in length, jointed, and bearing small bracts at the joints. It is somewhat compressed or flattened, as it were, and from its apex the single flower is produced, to be succeeded by another as soon as it fades, the plant continuing to flower in this manner for as many years as it is kept in health.

The flowers are large and spreading, too well known to need description ; the three sepals are narrow, erect, and reddish brown in colour, the petals are broader, drooping, tapered, wavy, and barred with yellow and cinnamon brown, the large, spreading, almost circular lip is yellow with a border of cinnamon brown, the depth of this border varying with the variety of the species. In the best variety it is very deep, the yellow centre being small in comparison. Of the varieties of this orchid there are many, but the most desirable is *O. papilio majus.* In this the flowers are nearly double the size of the type, and the lip is conspicuous by its brown border and yellow centre.

There is said to be another and exceedingly rare variety of *O. papilio* for the possesion of which fortunes have been offered and vigorous attempts have been made to find it, but in vain.

This variety is described as *O. papilio albiflorum,* having *pure uhite* flowers, and it is said that a specimen was *once* found at Caracas, Venezuela.

There is no doubt but that if such a variety does exist it would be a most valuable orchid, but I am inclined to think that *O. papilio albiflorum* is a myth, as the Venezuelan peons, or labourers, call such orchids as *Epidendrum atropurpureum, Brassavola, nodosa* or *venosa* and even *Diacrum bicornutum.* " *Mariposa blanca,*" (white butterfly) and it is my idea that this is how the story of a white butterfly has originated.

Moreover, except *O. jonesianum,* which posesses a white lip, there is not a *pure white Oncidium* known, and as yellow or some shade of yellow and brown seems to be the predominating colours of the majority of *Oncidiums* it is hardly possible that a "white butterfly" or pure white variety of a typical yellow-brown coloured orchid could exist in accordance with nature's laws.

The one who discovers this most wonderful curiosity (if it does exist) would without question be entitled to the fortune which would certainly be the result of such a discovery, but the search for it would be more fruitless than that of Raleigh for El Dorado, or the efforts to reach the poles of the world.

Oncidium papilio grows best in this climate on a block in a situation, where it gets a fair amount of light and shade and is screened from the direct rays of the sun during the warmer hours of the day. The best wood for blocks is calabash (*Crescentia cujete*) with the bark on, as it will last a very long time, and *Oncidium papilio* takes some time to grow into a fine plant, as it sends only one flower spike with each pseudo-bulb.

If a large mass or cluster of pseudo-bulbs be obtained, it is best, when the owner is able to do so, to affix it to a growing tree, for if the flower spike of the older bulbs be healthy, he will soon have a plant which will be in flower all the year round.

Before taking leave of this old friend, the first species of orchid I ever owned, it might be well to mention that it was the sight of this *Oncidium* in bloom, which, it is said, induced the late Duke of Devonshire to become such an enthusiastic orchid fancier, thus stimulating the cultivation of orchids in England, and causing the discovery and introduction of many other beautiful genera and species from all parts of the tropical world.

I find on looking back that this paper has grown to a size to which I never intended it should, therefore I must apologise for its great length, and crave indulgence on the plea that my love of these gems of our native forests, has caused me to trespass extraordinarily on your time, and in concluding, I hope that other members of the Club may be tempted to join the few who are already interested in the study of these curious and beautiful plants, and give us the benefit of their experience also; for orchid growing in this Island is learnt, not so much from books, as from observation, actual experience often taught by sad failure, and a careful study of the native species in their habitat.

At some other time I may be able to give some further notes on the subject of our native species.

Dec., 1894,

THE MEANS OF FREEING CATTLE OF TICKS.

By C. A. Barber, M.A., F.L.S.

(Continued from our last.)

As stated in a bulletin of the Lousiana Experimental Station, 100 cattle can be thoroughly cleansed by a knap-sack sprayer (i. e. one borne upon the back of the operator) in one hour. There surely will be no difficulty in arranging matters so that the animals may be driven one by one through a narrow enclosure where they can be reached by this powerful machine.

But the method obviously requires that the substance to be applied, in order to penetrate to the skin, shall be in a liquid form. Have we washes which will effectually destroy the parasitic animal life upon our cattle without injuring their delicate skins? To this query an unhesitating affirmative can at once be given.

What are the solutions which are thus effective? The difficulty truly is to select from among them, so many are they, and so well authenticated are the results of their application. To the following washes and emulsions it may be at once noted that many of them are equally applicable to the different animals of our estates, and also may be successfully used against almost all the pests attacking them externally. But prominence is throughout given to such as are applicable to destroying ticks upon cattle.

Mr. H. Goodwin, M.R.C.V.S., who has had much experi-experience with Antigua cattle, and has besides carefully studied and written upon the local ticks, has favoured me with the following recipes. I quote Mr. Goodwin's letter in full, as it will be of especial value.

"The following prescriptions have given me every satisfaction : one or two I value more than the others as you will see from the following :

"Raw tar 10 oz., Linseed oil 20 oz., Kerosine oil 80 oz., Sulphur 4 oz. This is *good*. Jeye's fluid 5 oz., water 40 oz. (*better*). Augean fluid 10 oz., water 100 oz. (*best and cheapest*). The Augean fluid is a safe, efficacious and cheap disinfectant and antiseptic preparation. It emulsifies with water in any proportion For the common tick (*Rhipicephalus concinna*) you will find 10 parts in 100 of water act satisfactorily, but for the St. Kitts tick (*Hyalomma venustum*) it will be necessary to make it a little stronger. However sensitive the skin may be there is no danger of its blistering it. As a healing agent I find it second to none."

With the exception of the first, all these could be readily applied by means of a force pump, similar to those used for spraying plants. There should of course be sufficient force used thoroughly to penetrate to the animal's skin, but this would be an easy matter.

London Purple is recommended in the proportions of two table-spoons to a pail of water. This was applied to a pony in Texas literally covered with ticks, and a single application was sufficient to kill every one. But the dangers of using such violent poisons cannot be too strongly urged.

McDougall's Cold water Dip is not infrequently recommended, the strength being 1 in 20. A letter from a farmer in South Africa gives the following details. "One gallon of dip will wash 45 head of cattle. We did this to a span of oxen which returned from East London covered with ticks and a fortnight afterwards not a tick remained on them. The ticks do not appear to die at once, but all fall off in a few days. Last year we lost seven calves; on opening them we found that they had swallowed a quantity of hair by licking themselves. This year we washed them a few weeks after birth with this dip, and I am pleased to say we have not lost any. If the farmers would wash their cattle twice a year to keep off flies, ticks &c., they would soon find that their cattle would feed better, look better and pay better. One penny for dip, with a little labour, will often save a beast's life which otherwise would have died from setwie or ticks."

It is doubtful whether applications at such long intervals would be really effective, especially where there are heavy rains.

Another wash sometimes recommended is a decoction of *Tobacco water mixed with salt.* The strength may be about 1 ℔ of common tobacco with 10 quarts of boiling water. Mr. Goodwin recommends that the tobacco wash to be of any use should be saturated with salt, a formula which finds favour with other writers.

Cooper's Sheep Dip has been recommended for the removal of ticks. It is primarily made up for animals which do not lick themselves; but as cattle do this, it becomes necessary to exercise care because of its poisonous nature. The makers advise that the sediment should be allowed to settle and the clear liquid alone be used.

A good deal may be said of a *Lime and Sulphur dip* made according to the following prescription. Newly slaked lime, 16 ℔s., flowers of sulphur, 25 ℔s., water 50 gallons. Mix the lime and sulphur dry and pass them through a sieve. Boil for 20 to 30 minutes and allow the sediment to fall. Use only the clear liquid. No harm will accrue from the cattle licking themselves, but this will be of benefit to them.

The objection to poisonous liquids being used on estates in large quantities is sufficiently obvious; and some of the above mentioned washes will therefore be useless in these islands. There is another objection which applies to liquids which are not of an oily consistence, namely the frequency of application which is necessary. The effects of a thorough cleansing pass off after a few days in fine weather and more rapidly during the rains. The animal will then be open to the attack of a new set of ticks.

Liquids of an oily nature are as we have already seen, of themselves objectionable to ticks. They will also remain upon the skin a much longer time, and thus extend the period of immunity for the animal.

Several drenchings during the season will probably be sufficient to keep all vermin at a distance.

There are many forms of emulsions—an intimate mixture of oil, soap and water, forming a froth—which can be recommended with confidence. Their preparation is a little troublesome perhaps, but it is well within the reach of any intelligent labourer, and once mastered becomes merely a matter of routine. Much of the success of the preparations depends upon whether there is a proper *froth* and this is simply dependent on sufficient churning or " swizzling."

First and foremost is the *Kerosine Emulsion* which has been found so effective for the destruction of scale insects in plants, and is every day more and more applied for freeing plants and animals of other parasites.

The following is a well recognised formula for ticks :— In 2 quarts of boiling water dissolve ½ lb of soap : remove from fire : immediately add 1 pint of kerosine and agitate. In from three to five minutes the liquid becomes creamy. It may be stored in this form in bottles or barrels. For use add three of water to one of the emulsion, mix thoroughly and apply with a spraying pump.

Another form among many emulsions is made with *Fish-oil*. This is recommended after many trials by the Louisiana Experiment Station. Of six liquids, among which was Jeye's fluid, kerosine and fish-oil emulsion were found to be the only effective ones for the season. The trials were for freeing the cattle from the attack of " horn fly." Fish-oil emulsion was found to keep the cattle longer free from annoyance, by the flies, probably because it evaporated or was washed off less readily. The mixture may be made in the same proportions as the kerosine emulsion with the substitution of fish-oil for the kerosine.

These are a few of the many remedies suggested, after careful study and experiment in countries where stock are a matter of primary importance.

Before leaving this class of remedies for ticks it may be of service to refer to the appropriate *Treatment of Horses*. Greater care will be needed because of the sensitiveness of the animals; but as the horses are more frequently handled than cattle, this will not be a matter of much difficulty.

The following is the method suggested by Mr. Hutcheon, the Cape Colony veterinary surgeon :—" In the case of horses which are grazed and only partly kept in stable, a good plan is to wipe the following parts well with a cloth saturated with a mixture of equal parts of paraffin and oil, viz., from the anus to the sheath between the thighs and around the hind legs, then along the belly and chest and around the fore legs before letting them out to graze. It is difficult to get a cheap preventive which is both harmless to the animal and cleanly to use. To moisten the coat with a weak solution of any of the tar or carbolic acid dips would also answer if frequently applied."

The editor of the Cape *Agricultural Journal* writes " I have tried sulphur with horses, putting a mixed streak of flowers of sulphur and fat all along the back from mane to tail with most capital effect. Sulphur has a peculiar power of diffusing itself all over the animal and the sulphurous acid gas slowly given out is most obnoxious to all insects."

Another authority remarks that when the ticks have effected a lodgement in the horse's ears, nostrils &c., " a little train oil, chloroform, carbolic acid, sweet oil, or dusting with calomel will be effective."

Thus far it has been assumed that the cattle have already been attacked by ticks, and various suggestions have been made to cleanse them from these parasites, and keep others off as long as possible.

If, by other means, it is possible to make the skin of the cattle so revolting to these parasites that they will not in the first instance attach themselves to it, there will then be no need of these spraying operations. This can be effected, as already hinted by introduction of substances into the animal's food.

The addition namely of *small quantities of sulphur* to the solid food appears to have the desired result. Externally applied, sulphur has probably on local action of itself, but part of it is changed by contact with the acid sweat into sulphuretted hydrogen.

The following seems to be the physiological changes which take place during the passage of the sulphur, through the animal's body to its skin. Sulphur taken in with the food passes the stomach unaltered. In the intestines a small portion is changed into sulphides of hydrogen and the alkalies. Part of these sulphides pass into the blood and into the tissues from the blood, and act chiefly upon the central nervous system. The sulphides in the tissues are variously excreted. By the kidneys they are excreted as sulphates : if in excess, part is also excreted in the form of sulphides. By the skin they escape as sulphides, giving the characteristic foul odour to the perspiration and somewhat increasing its amount.

Such is the course of sulphur taken in the food by man, and there is every reason to believe that a similar course is followed in the larger animals. The effect of a constant small exhalation of sulphuretted hydrogen from the animal's skin seems to be distasteful to the ticks, for the practice of mixing small quantities of sulphur with the food, say an ounce or two for each animal per week, seems to have the much desired effect of keeping them free from ticks in regions infested by these parasites.

The addition of the sulphur to the food needs a little management. It may be done by scattering it in the form of a powder over the chopped food. Probably, where the cattle receive only grass, the addition of a small quantity of molasses will cover it sufficiently for them to be eager to take it.

But the best methods appear to be supplying some kind of " lick, " usually with a preponderance of salt. The avidity with which cattle seek out salt left about is well known, and its benefical action upon the nutrition of the animals need not be insisted upon here.

The following lick is quoted from Australia.—214 lbs. of salt, 14lbs. of sulphate of iron, 4 lbs of sulphur and 2 lbs of ginger. But any form would probably be efficient which contained a fair quantity of salt and sulphur.

In conclusion, a summary of the relations between cattle and ticks, and the best means of freeing them from these parasites, as discussed in these articles, may be of use.

(1.)   The ticks in the Leeward Islands are very imperfectly studied, only two species having, been named from Antigua. There are probably other kinds, and efforts should be made to determine these.

(2.)   The influence of ticks upon cattle is disastrous, both from the direct abstraction of blood and the irritation and pain caused to the animal. There are at any rate some cases where ticks are connected with the spread of deadly diseases. This subject is only in its infancy, it is to be feared that continued study will reveal other instances.

(3.)   It is principally the poor cattle that are attacked, partly because of their poverty itself, probably also because a fat or sleek condition of skin is distasteful to the parasites. All means should be adopted to get the animals as well nourished as possible.

(4.) In carrying this out the greatest attention should be paid to the condition of the pastures, so that the animals should be supplied with plenty of grasses, and the natural cover for the ticks should be removed.

(5.) The forcible extraction or maiming of ticks is to be avoided, because this increases the pain of the animal and does not allow the tick time to remove its proboscis. The tick appears to be able to do this voluntarily if so inclined and allowed time.

(6.) It is important to induce the tick to willingly let go. Various smears are referred to. The practice of throwing the animals cannot be too strongly condemned, and this is necessary with smears.

(7.) With liquids, thorough cleansing may be effected by using a good spray-pump such as is employed for freeing plants from scale-insects. 100 cattle may thus be treated in an hour by a knapsack spray-pump.

(8.) A number of types of washes for spraying are selected for description. All poisonous ones should be rejected as there are non-poisonous preparations equally effective.

(9.) Carbolic acid dips and other liquids which evaporate quickly need frequent applications, and should be discarded in favour of the oily liquids or emulsions where the latter are equally effective.

(10.) The best of all these is the Kerosine emulsion regularly used for plants. There are many formulæ for the preparation of this: a useful one is given.

(11.) On the principle that prevention is better than cure, small doses of sulphur with salt are preferable to spraying, in that the animal's skin is thus renderd so offensive to ticks that they will not attach themselves.

(12.) The sulphur taken into the mouth passes through the tissues with various changes, emerging from the skin in the form of Sulphuretted hydrogen, a gas very objectionable to all vermin. The best method of giving the sulphur is in the form of a lick with salt.

(Leeward Islands Agricultural Journal, Jan., 1895.)

NOTES ON BIRDS OBSERVED IN TRINIDAD.

By William Brewster and Frank M. Chapman.

OUR knowledge of tropical birds is so largely derived from the journals of travellers and naturalists, whose arduous explorations in the less accessible parts of the tropics have been attended by hardship and exposure, that most of us are discouraged from even attempting to visit the fascinating regions they describe. The brilliantly colored Trogons, Toucans, Jacamars and Hummingbirds which figure so conspicuously in cases of tropical birds, thus seem to us to be more or less unreal inhabitants of lands forever beyond the bounds of our experience. The truth is, however, that we may be comfortably and safely established in a tropical forest in less time than it frequently takes to reach the nearest European port.

The Island of Trinidad belongs politically to the British West Indies, but faunally it is a small bit of the South American continent which has been detached in recent geological times. Its bird-life therefore is very similar to that of the Venezuelan mainland and is quite unlike the comparatively meagre, insular avifauna of the true West Indian islands to the northward. A visit to Trinidad is thus practically a visit to South America. But it is not alone the richness of the fauna which leads us to recommend Trinidad as an exceptionally favorable field for the naturalist with limited time at his command. Its additional advantages are: accessibility, a healthy, in fact during the dry season, from December to May, perfect climate; the safety and material comforts which one is sure of finding in a British colony; and a Naturalists' Field Club whose members, as we know from pleasant experiences, will cordially receive brother naturalists. It is evident then that a trip to the tropics, far from being an undertaking involving much time and risk of life, may be an excursion from which one may return in two or three months richer both physically and mentally.

From New York to Port-of-Spain, by the direct line of steamers, is a voyage of nine days, or occasionally a steamer of the Windward Island line continues from Barbadoes, the usual terminus, to Trinidad. The latter is by far the more enjoyable sail and, taking only six days longer, gives one an opportunity to land at a dozen or more islands *en route.*

Port-of-Spain possesses fair hotels and stores which will compare favourably with those of our larger cities. Black Vultures swarm in the streets, and many birds, notably the Qu'-est-ce-qu'il-dit (*Pitangus sulphuratus*) and Ani (*Crotophaga ani*), are common in the Botanic Gardens and neighboring

savannas. Indeed the ornithologist will find much to interest
him in the immediate vicinity of the city, but he should lose no
time in hastening to the virgin forests, or 'high woods,' as they
are locally known, where birds may be studied under absolutely
natural conditions. The Government rest-house on the Moruga
Road, kept by Corporal and Mrs. Stoute, was Mr. Chapman's
headquarters during March and April, 1893, and from every
point of view leaves nothing to be desired. In fact, we doubt if
there exists a place elsewhere in the tropics where for a small
compensation a naturalist may find so thoroughly comfortable a
home, with the best of food and attention, at the border of a
primæval forest.

We, however, were even more fortunate, for in accepting the
invitation of Mr. Albert B. Carr to visit him at his cacao estate
in the Caparo district we found not only a delightful home in a
region where birds were abundant, but had also the companion-
ship and assistance of Mr. Carr and his brother, both born
naturalists and skilled woodsmen, with a thorough knowledge of
the country. Every ornithologist knows what this means.
Without the guidance of our hosts we should have seen less in
three months than we did in three weeks. Through their
unceasing efforts every hour of the day, and almost every hour of
the night also, brought some interesting incident. The birds
and mammals of the region were passed in review for our benefit,
and at the conclusion of our stay there were but few species
which had not answered to the roll-call of gun, dog, and trap.

Mr. Carr's home is near the point of a narrow wedge of cacao
estates which penetrates the forests from Chaguanas on the
western side of the island. The limits of the cacao and shading
immortel trees, among which his picturesque, thatched house is
situated, are sharply defined by the dark walls of the virgin
forest, distant only a few hundred yards. In the morning, from
its apparently fathomless depths, came the deep-voiced roaring of
monkeys (*Mycetes*). Toucans, perching on the topmost branches
of the higher trees, croaked defiance at some answering rival half
a mile away. The united voices of cooing Doves (*Engyptila*)
formed a soft monotone to which the ear frequently became
insensible. The sweet, weird trilling of Tinamous arose from the
bordering undergrowth. In the trees about our house were
noisy Qu'-est-ce-qu'il-dits; shrike-like Vireos (*Cyclorhis flavipectus*)
whistled vigorously; active bands of Tanagers (*Ramphocelus* and
Tanagra) flitted restlessly about uttering their weak, squeaky
notes. Five or six species of Hummingbirds were generally
numerous about the blossoming bois immortels, while overhead
were flocks containing four species of Swifts (*Chætura*) whose
twitterings reminded us of other and very different scenes. In
the cool, darkened forest Jacamars were piping, Trogons cooing,

Motmots hooted softly, and the mournful whistle of a Pygmy Owl (*Glaucidium*) told of his partially diurnal habits. The species mentioned were all more or less common. Their voices formed an ever present accompaniment for all other bird-music— a background to the picture of bird-life which we do not intend to attempt describing.

Our stay at Caparo was crowded with events, but the time was too short for us to make many observations sufficiently novel to warrant publication in the pages of a scientific journal, and in this connection we propose to speak of but three species, to the published accounts of whose life-histories, thanks to Mr. Carr's assistance, we think we can make some additions. They are the Bell-bird or Campañero (*Chasmorhynchus variegatus*), a Hummingbird locally called 'Brin-blanc' (*Phaëthornis guyi*), and a large Goatsucker (*Nyctibius jamaicensis*).

To what extent the other three species of the genus deserve the reputation sometimes given them we cannot say, but the voice of *Chasmorhynchus variegatus* would undoubtedly prove a disappointment to those who expect a Bell-bird to be a Bell-bird in more than name. But while its notes bear no resemblance to the "deep tolling of a bell" they proved none the less singular, and we class them among the most remarkable we have ever heard.—*The Auk*, Vol. xii, No. 3., July, 1895.

(To be continued.)

Communications and Exchanges intended for the Club
should be addressed to the Honorary Secretary,
Port-of-Spain, Trinidad, B.W.I.

Price 6d.   Annual Subscription, 3/.

Vol. 2.   DECEMBER, 1895.   No. 11

*J'engage donc tous à éviter dans leurs écrits toute personnalité,
toute allusion dépassant les limites de la discussion la plus
sincère et la plus courtoise.*—LABOULBÈNE

Trinidad·Field·Naturalists'·Club.

NATURA MAXIME MIRANDA IN MINIMIS.

Publication Committee :

HIS HONOUR SIR JOHN GOLDNEY, *President.*
P. CARMODY, F.I.C., F.C.S., *V.P.*: SYL. DEVENISH, M.A., *V.P.*
HENRY CARACCIOLO, F.E.S., R. R. MOLE,
F. W. URICH, F.E.S., *Hon. Secretary.*

CONTENTS :—

In Memoriam—Prof. Riley	261
Report of Club Meetings :	
October	262
New Genus and New Species of	
Proctotrypid—Description	264
Notes on Trinidad Birds (concluded)	266
New Species of Synallaxis	272
Peculiar Feeding Habits—Trin. Ophidia	276
C ... omi Macellari	?

FIRE! FIRE!!
FIRE!

ACCIDENTS WILL HAPPEN

Insure your Houses and Property

of every description against loss by Fire in the

NORWICH UNION
Fire Insurance Society

one of the Oldest English Companies
ESTABLISHED 1797.

Head Offices: London and Norwich.

DIRECTORS.

President: H. S. Patterson, Esq.,

Vice-President: Major-General James Cockburn; M⁻ᵢ⁻
Astley Cubitt; S. Gurney Buxton, Esq; F. Elwin W
William Foster Bart: I. B. Coaks, Esq; R. A, C
George Forrester, Esq; William Birkbeck, Esq
Palgrave Simpson, Esq; G. Hunter, Esq; —— Tuck, Esq.

☞ CLAIMS PROMPTLY AND LIBERALLY SETTLED.

Trinidad Agent,

RANDOLPH RUST,
Marine Square.

Sub-Agent,

L. W. BONYUN,
San Fernando.

☞ All Communications and Exchanges intended for the Club
should be addressed to the Honorary Secretary, Port-of-
Spain, Trinidad, B.W.I.

JOURNAL

OF THE

Field Naturalists' Club.

VOL. II.   DECEMBER, 1895.   No. 11.

IN MEMORIAM.

MANY of the Members of the Trinidad Field Naturalists'
Club will be pained to hear of the sudden death of Professor
Charles Valentine Riley, M.A., Ph. D., through a bicycle
accident in Washington on the 14th September last. Mr. Riley
had just started for the city with his son when his machine ran
over a stone and he was thrown, violently striking his head
against the kerb. When picked up he was insensible and died
a few hours later. Mr. Riley was an Englishman born
at Chelsea on September 18th, 1843. He was educated in
England, France and Germany and at the age of 17 went to the
United States. Mr. Riley was a distinguished entomologist and
frequently identified insects for the members of the Trinidad
Field Naturalists' Club in which he was enrolled as an honorary
member. His advice was sought from all parts of the world on
questions of Economic Entomology and he always promptly
responded to every call made on his valuable time in this way.
Last year he made a tour through the West Indies but did not
come so far south as Trinidad, to the disappointment of many of
our members. He was not known personally to the Club but
several members were in constant correspondence with him. The
Committee feel that they are only doing their duty in placing
on record the Club's regret at the sad accident and in tendering
to Professor Riley's family their sincere sympathy.

REPORT OF CLUB MEETING.

8TH OCTOBER, 1895.

PRESENT : Professor Carmody, V.P., in the chair; Messrs. W. S. Tucker, Syl. Devenish, M.A., J. R. Llaños, T. W. Carr, Dr. F. A. Rodriguez, J. A. Rousseau, M.A., H. Caracciolo, F.E.S., Charles Libert, J. B. Inniss, R. R. Mole, F. W. Urich, Hon. Secretary and Treasurer. After the minutes had been read and confirmed Mr. R. R. Mole proposed and the Vice President seconded Mr. John Joseph, a candidate for town membership. Carried unanimously.—The Secretary read an excuse from Mr. John Barclay for not being present and not being able to forward a paper which he had promised for this meeting. Mr. Mole read some notes " On some peculiar feeding habits in the Trinidad Ophidia" which elicited a lengthy discussion in which Mr. Devenish, Mr. Tucker, Dr. Rodriguez, Mr. Urich and the Vice President took part. Mr. Mole was awarded a vote of thanks for his communication.—Mr. Ewen, having forwarded a spider (found by Miss Mitchell hanging by a filament of its web) which had been killed by some fungoid growth, giving it the appearance of being covered with powdered sugar, Mr. Urich read some passages showing that the fungus was probably a cordyceps. He described some experiments made in the United States with the spores of this fungus which resulted in the death of myriads of a very destructive caterpillar. A discussion followed in which Mr. Caracciolo, Dr. Rodriguez and the Vice President took part and Mr. Urich expressed a belief that sooner or later economic. entomology would receive much more attention from Agriculturists in Trinidad than it did at present.—Mr. Mole showed 25 young mapipires (*Trigonocephalus atrox*) born in his collection, the brood consisted of 26, but 25 were dead when born. He thought this was due to the fact that the mother and a companion of the same species had a family quarrel a few days before the event, during which the snakes furiously bit each other about the mouth, head and neck. As the members knew, the species was a very venomous one, and the consequence was the heads of the combatants swelled to an enormous size—they looked like a couple of pugilists after a hard prize fight where each party had got an equally severe punishing. The swelling went down next day. He thought this was the reason for the great mortality among the young ones. The one survivor, however, was doing well and had gorged a lizard and a frog. The young ones varied in length from 16¾ centimetres to 27½ centimetres. The Vice President said he had recently been making experiments with chloroform and snakes and was surprised how easily they succumbed to its influence.—Mr. Caracciolo exhibited a bat, *Noctilio leporinus*, which differed in colour from the ordinary individuals and he made some interesting remarks upon albinism and melanism in animals and birds. The bat was caught by Supt.-Sergeant Healey in the Police Barracks.—Mr. Caracciolo also drew attention to a Venezuelan jacare (alligator) presented by the American Consul and some young crocodiles (*Crocodilia intermedia*) presented by Mr. Siegert and he pointed out the main differences between the three great groups of these gigantic saurians viz : the Gavials, the Crocodiles and the Alligators. He also described their methods of taking their prey and the structure of their throats which enabled them to hold their victims under water without they themselves being drowned. Mr. Rousseau graphically related how a huge reptile of this class was caught at Cedros by Mr. Kernahan, who harpooned it. It was supposed to have been washed out of the Orinoco in a flood. It was spoilt as a specimen, however, for the sharks had torn away a great portion of its tail. Mr. Caracciolo exhibited a specimen of the elephant sphinx moth which he had bred from a larva presented to him by Dr. Lawrence who had found it in his house in St. Vincent street. The caterpillar was such a peculiar one and assumed such strange positions, which together with its markings made it look like a dangerous reptile, that he had made a series of drawings of it. Its scientific name was *Philampelus lunei*.—Mr. Mole drew attention to a so-called " Two-headed snake " *Amphisbæna alba* which he had been trying to feed on parasol ants but without success. He found, however, it readily ate worms and other soft bodied animals. The " two-headed one " however refused to eat before the meeting.—Mr. Urich exhibited a long thin black worm, *Gordius* species, found by Mr. Ferdinand Maingot. A very nearly allied species deposited its eggs in water and they were swallowed by some soft bodied water larvæ. These were frequently destroyed by water beetles and in the body of the beetle the *Gordius* passed through its transformations before it reached its final stage.—Mr. Devenish exhibited some turned specimens of banyan wood consisting of soda water bottle openers and other articles.—Mr. Guppy had forwarded to the meeting some exquisite drawings of two Trinidad Butterflies *Tithorea flavescens* and *Sais eurymedia* and their food plant.—Mr. Caracciolo showed a collection of insects which he had taken under the electric light, they included a rare *nepa*.—Mr. Caracciolo on behalf of Mr. Borberg exhibited a swallow's (swift's) nest constructed of sticks and a glutinous substance from a vault in the cemetery. Mr. Devenish said he had partaken of the Chinese dish of "bird nest" but there was nothing in it to make him wish to do it again.—Mr. Mole said Mr. Eugene André told him the other day that he had seen a little owl for several nights catching moths under the electric light. The bird darted from a tree near the lamp, seized its fluttering prey and flew with it to a neighbouring tree. Mr. André frightened the owl and secured one of the moths. Mr. Mole added that he had seen bats hunting by the light. Certainly it was very interesting that these animals should have so soon learned the value of the electric light in concentrating their supply of food.—Mr. Caracciolo shewed a number of beetles *Ancistrosoma farinosum*, which he had taken from one cocoa tree. He had counted 350 and then he got tired of his task. He also shewed a very large scorpion from Surinam presented by Mr. Leon Giuseppi, which was like one of our Trinidad species, and also a case of butterflies and an anaconda skin presented by Mr. Siegert.

An interesting meeting adjourned at 10 p.m.

DESCRIPTION OF A NEW GENUS AND NEW SPECIES OF PROCTOTRYPID BRED BY MR. F. W. URICH FROM AN EMBIID.

By William H. Ashmead.

Embidobia Ashm. n.g.

HEAD transverse, very little wider than the thorax, about two and a half times as wide as thick antero-posteriorly, the occiput concave, the face subconvex but with a median sulcus or furrow towards the insertion of the antennæ; ocelli 3, small triangularly arranged, the lateral much closer to the eye margin than to the front ocellus; eyes oblong or subovate, sparsely pubescent.

Antennæ inserted just above the clypeus 12-jointed, in the female terminating in a four-jointed club, the funiclar joints all very minute, except the first, transverse; pedicel obconical stouter and longer than the first three or four funiclar joints united; first three joints of club transverse; in the male the flagellum is filiform with the joints except the last moniliform, the last ovate or cone-shaped.

Maxillary palpi minute, apparently 3-jointed.

Mandibles bidentate,

Thorax short ovoid, the pronotum hardly visible from above; mesonotum smooth without parapsidal furrows; scutellum semicircular or lunate; metathorax short, the hind angles subacute.

Front wings pubescent, the submarginal vein reaching the costa at about two-thirds the length of the wing, the marginal vein short, almost punctiform, only about twice as long as thick, the stigmal vein straight oblique, ending in a minute knob and a little more than twice as long as the marginal vein; basal vein absent.

Abdomen in female fusiform, about one-third (or a little longer) longer than the head and thorax united, the third segment the longest, the basal segment has a more or less distinct hump at base, but it is not greatly developed or produced into a horn as in *Catoleia, Baryconus*, and other genera.

This new genus belongs in the family Proctotrypidæ, subfamily V. Scelioninæ, *Tribe IV Scelionini*, as defined in my Monograph of the North American Proctotrypidæ, p. 208.

It is intermediate between the genera *Cremastobœus* Ashm, and *Hadronotus* Foerst., but with structural characters so distinct as to at once distinguish it from either.

The table of genera of this tribe, as published by me in my monograph, p. 210, line 16, may be modified so as to include it in the following way :

Abdomen without a horn at base.
Abdomen long, fusiform ; mandibles 2-dentate.
Abdominal segments normal ; antennal club 6-jointed......................
 Cacus, Riley.
Abdominal segments strongly constricted; antennal club óval
 5-jointed.........................*Cremastobæus*, Ashm.
Abdominal segments not strongly constricted, the third segment the
 longest; antennal club 4-jointed, the funicle joints very minute,
 transverse, the pedicel as long as the first three or four joints
 united.........................*Embidobia* Ashm. n g.
Abdomen broadly oval, the second segment usually a little the
 the longest ; antennal club 6-jointed............*Hadronotus*, Foerst.

Embidobia urichi, Ashm. sp. n.

Male—Female. Length 0·8 to 0·9 mm. Black, subopaque or slightly shining, feebly microscopically punctate, and sparsely clothed with a fine pubescence, more apparent on the head and abdomen ; eyes sparsely hairy ; antennæ, except three last joints of club, which are brown, and the legs including coxæ brownish-yellow or yellowish. Wings hyaline, pubescent, the venation, except the short marginal vein, yellowish, the latter brown. Abdomen sparsely minutely punctate, the first segment and the second at the suture striate.

The male differs in having the antennæ entirely honey-yellow, the flagellum being filiform, the joints, except the last, all moniliform or rounded, the last ovate or cone-shaped, while the abdomen is shorter, oval, not longer than the head and thorax united, the first and second segments being striated.

The above interesting new genus and species is based upon one male and eight female specimens, bred by Mr. F. W. Urich, of Trinidad, B.W.I., from the eggs of a new Embiid which he intends shortly to describe.

The specimens were transmitted to Mr. L. O. Howard, to whom I am indebted for the privilege and pleasure of working them up.

NOTES ON BIRDS OBSERVED IN TRINIDAD.

By William Brewster and Frank M. Chapman.

(Concluded.)

To hear a Campañero is one thing, to see it quite another. The birds haunt the tree-tops in the virgin forest, where, concealed by the canopy of foliage and intervening parasitic plants and creepers, they can be found even by practiced hunters only under favourable conditions. Mr. Carr had prepared us for the failure which attended our first Campañero hunt. Nevertheless, we actually heard a Bell-bird calling,—sufficient encouragement, if we had needed any, to continue the search. Our persistency, however, was not tested. The following day Mr. Brewster and Mr. Carr discovered a Campañero within a mile of the house and had an exceptional opportunity to study it. After following the sound of the bird's voice for a quarter of a mile, they finally saw it perched on a bare twig at the top of a tree about seventy-five feet from the ground. After watching it there for about fifteen minutes, during which time it uttered its several calls, it was disturbed by two Toucans alighting near it and sought a perch in a strong, clear light about twenty feet from the ground and not over twenty yards from the observers. This, according to Mr. Carr, was an unusual proceeding. It remained in this position for about fifteen minutes, repeating all its notes. The following day we all visited the place and the Bell-bird kept the tryst, appearing on the high perch it had occupied the preceding day. The records of these two occasions were read aloud and endorsed by each member of the party. From them we present the following description of the Campañero's calls. The bird has three distinct notes, the first *bok*, the second *tui*, the third *tang*. The *bok* is by far the loudest and for this reason is the one most frequently heard, and is doubtless the call alluded to by previous writers.[1] It can be heard in the flat forest at a distance of about 600 yards. Waterton, it may be remembered, says the "toll" of *Chasmorhynchus niveus* may be heard at a "distance of three miles." The *bok* is sometimes uttered with much regularity about every ten seconds; at other times longer or shorter intervals may elapse. At a distance of four or five hundred yards it resembles the stroke of an axe on hard, resonant wood. One would now imagine that the bird was within seventy-five yards, so deceptive is the nature of this note. As one approaches, the call does not seem to increase in volume and one is apt to imagine that the bird is retreating slowly from tree to tree. This impression, however, is dispelled when one comes within one hundred yards of the bird, for the sound then becomes much louder until, as one gets directly beneath the caller, its volume is simply tremendous. It now has a slightly rolling quality— *br-r-r-ock*—and is so abrupt and explosive in character that it is nearly as startling as the unexpected report of a gun. At each utterance of this note the bird opens his bill to its widest extent and throws his head forward and downward with a violent, convulsive jerk as if he were in a passion and striking viciously at some rival. This motion is so violent that the bird evidently has some difficulty in maintaining his footing during its delivery as well as in recovering his balance afterward.

The second note, *tui*, is much softer and is delivered from six to eleven times in such rapid succession that the notes form an unbroken series. Despite this, each *tui* is closely followed by a metallic *ting* which sounds exactly like an echo and appears to be of about the same duration and nearly as loud as the note it supplements. The *tui* notes are given so quickly that at first it did not seem possible for the bird to produce another note between them, and it was only after repeated observations we became convinced that the *ting* was an integral part of the *tui* call. While uttering these notes the bird sits rather erect and perfectly motionless save for a slight tremulous movement of the throat and tail which accompanies the delivery of each *tui*.

The third note, *tang*, is also repeated a number of times— eighteen to thirty-three,—in quick succession. It sounds much louder than the *tui* and the intervals between the notes, though short, are well marked. Sometimes the bird began slowly and gradually increased the rapidity of its utterance, at others there were regular intervals between the notes. The *tang* may be likened to the sound produced by striking a piece of bar iron a sharp blow with a hammer. It is accompanied or followed by a distinctly metallic but not clear, ringing vibration. At a distance of one hundred yards the *tang* sounds like a slow strumming on

---

[1] *Cf.* Taylor, Ibis, 1864, p. 88.

the C natural string of a banjo, as Mr. Carr actually demonstrated. It can be heard at a greater distance than the *tui* but not so far as the *bok* and at two hundred yards would attract the attention of only a practiced ear.

While 'tanging' the bird sits rather erect, the head well up, the wings drooping beneath the closed tail. At each utterance the tail vibrates slightly, there is a marked swelling of the black throat, and the mouth is opened to its widest extent, the lower mandible being worked with some apparent effort while the upper mandible and rest of the head are held perfectly motionless.

Although probably an extremely local and not very active species the bird was alert and watchful. Its movements were quick, the head being often turned from side to side, or the wings were twitched nervously, and at more or less regular intervals it would turn squarely on its perch and face in the opposite direction. The fleshy appendages on the Bell-bird's throat resemble bits of leather shoe-string. They hang loosely in the freshly killed specimen and are then so conspicuous that we were surprised to find they could not at any time be distinguished on the living bird.

The greenish plumage of the female Bell-bird renders it so difficult of observation that even Mr. Carr was not familiar with it. It was therefore a rare bit of good fortune for us that a female of this forest-loving species so far departed from its normal habit as to leave the woods and perch on the topmost branch of a bois immortel which shaded the palm-thatch beneath which we prepared specimens—an offered sacrifice we were not slow to accept.

The observations[1] of Mr. Chapman on the song-habit of the 'Rachette' Hummingbird (*Pygmornis longuemareus*) were confirmed by our discovery of a locality to which the birds evidently came to sing, and Mr. Carr directed us to two resorts regularly frequented by *Phaëthornis guyi* for the same purpose. Both were in the forest where the trees were rather small and slender and plentifully undergrown with roseau palms. One locality was not far from the house. We visited or passed it many times always hearing from one to six birds singing within an area one hundred feet square. Each bird seemed to have its own particular perch which we would find occupied day after day. The song of this species is louder and has more character than that of *Pygmornis*. It is an unmusical *yep-yep-yep* uttered very rapidly, and, when the bird is undisturbed, continued for several minutes without break or pause. They sit erect but in an easy attitude with the points of the wings drooping below the tail. With every *yep* the long bill is thrown nearly straight up and the

---

[1] Bull. Am. Mus. Nat. Hist., VI, 1894, p. 55.

mouth slightly opened while the red lower mandible shows conspicuously and the body is twitched convulsively. Each note is accompanied by one or two vertical vibrations of the tail. Rarely, and apparently when under the influence of some excitement, the vibrations are increased in length, force and rapidity until a maximum of motion is attained. Then there is a second's pause, the tail-feathers are spread to the fullest extent and pointed forward over the back until the tips of the long central feathers nearly touch the back of the head. The effect, as may be imagined, is most striking, the birds suggesting diminutive Turkey-cocks.

More or less frequently a rival would approach, buzzing loudly, when the calling bird darted recklessly at the trespasser, and the two birds dashed wildly through the forest, one apparently in pursuit of the other, squeaking loudly and uttering an explosive *tock, tock.* This sound can be closely imitated by pressing the tongue against the roof of the mouth and withdrawing it forcibly. Generally the perching bird returned within a minute and resumed its interrupted song.

It therefore appears that *Pygmornis longuemareus* and *Phaëthornis guyi*—and probably also other species of these genera—have regular resorts which they visit for the purpose of singing and that they evidently sing at no other time. The significance of this habit—unique so far as we know—we cannot satisfactorily explain. All the specimens killed at these singing haunts were males. Whether the females are present we cannot say.

There are few natives of Trinidad who do not know, by name at least, the animal locally termed 'Poor-me one.' This name is given to a small Ant-eater (*Cyclothurus didactylus*) which is popularly supposed to utter the notes serving as the origin of the words. Mr. Carr, however, as quoted by Mr. Chapman[1], definitely proved that Poor-me-one was a species of Goatsucker by shooting the bird in the act of calling, but failing to preserve the specimen, its specific identity could not be determined.

Only a person who has heard Poor-me-one calling from the moonlit forest can understand how ardently one longs to identify the caller. Our curiosity was frequently aroused by the hooting of some, to us, unknown species of Owl, or even the cry of some night-bird whose identity was an entire mystery, but the cry of Poor-me-one is possessed of a human quality which appeals to one as strongly as the voice of a fellow-being. Its tone is so sweet and tender, so expressive of hopeless sorrow, that even the

Bull. Am. Mus. Nat. Hist., VI, 1894, p. 59.

negroes are impressed by it, as its native name, Poor-me-one, meaning " Poor me all alone," clearly shows. To identify Poor-me-one, therefore, became one of our chief objects.

This strange bird calls only on moonlit nights from February to June. The calendar told us the moon would be full March 20, and as the slender crescent grew larger we listened anxiously for the notes of *Nyctibius*. But we neither saw nor heard sign of it until the evening of the 16th when, as we. were strolling homeward from the forest, we saw a large bird, which we at first supposed was an Owl, sitting on the top of a stub about thirty feet in height. We had no difficulty in identifying this bird as a *Nyctibius* and congratulated ourselves on the knowledge that it was probably resident so near our house. For the four succeeding evenings doubtless the same bird appeared about half an hour after sunset and on set wings sailed slowly and majestically from a point of the forest distant some two hundred yards, until directly above the stub upon which we had first seen him. After descending in a broad spiral, which ended a few feet below his perch, he pitched sharply upward, closing his wings as he secured a footing. His position was upright and he seemed a continua-tion of the stub, against which his tail was pressed. He invariably faced the west but kept his head turning from side to side after the manner of Flycatchers. At short, irregular intervals— usually two or three times a minute—he launched out after insects, flying in a perfectly straight, slightly ascending line with firm and vigorous, yet easy wing-beats, his tail wide-spread. At the moment he reached his prey he often turned abruptly to secure it, then wheeled suddenly, and returned to the stub by a long, slow, graceful glide and lit as before described. With few exceptions his sallies were made toward the west, evidently because of the background afforded by the after glow, and he often flew thirty or forty yards before reaching his object.

Interesting as it was to observe a Goatsucker in the rôle of a hawk-like Flycatcher, the certainty of our identification made us earnestly wish to hear the bird call, when the identity of Poor-me-one and *Nyctibius* could be instantly settled. But each night the bird returned to the forest in silence.

March 20 the moon was full and shortly after eight o'clock, to our great delight, we heard Poor-me-one calling from the forest. We at once started in the direction of the sound. Crossing a belt of cacao, leaping some of the drains, stumbling into others, wading knee-deep through the dew-drenched grass, breathless and perspiring, we came at length to the edge of a low, swampy wood whence issued the strange cry. The bird now became silent. We listened anxiously for several minutes and were greeted only by the *cook-er-ree-coo* and startling scream of an Owl (*Megascops brasiliensis*). Finally, after consultation, Mr. Carr whistled an imitation of the cry of Poor-me-one. Almost instantly an answer came from the woods and soon a large Goatsucker, which we at once recognized as the species we had seen on the stub, came sailing directly over us. He circled twice, uttered a low call, and alighted on the topmost twig of a bois immortel distant twenty yards. A moment later, puffing out his throat, he uttered the Poor-me-one call. We suppressed our exultation with difficulty.

After calling a dozen or more times the bird returned to the woods, but several times returned in response to our imitation of its notes. Usually he perched on the topmost, slender twigs of a bois immortel, the last situation one would expect a Goatsucker to select.

The locality was not far from the stub upon which we had originally discovered *Nyctibius,* and we had little doubt that the individual seen there was the one we had heard calling. Indeed, one hour later this bird, which we easily recognized by a peculiarity in its call, came to the vicinity of the stub in response to Mr. Carr's whistle. Here he was joined by his mate, both birds perching in the topmost branches of the forest trees.

The song of Poor-me-one consists of eight notes, which Mr. Carr, in an article[1] on this species, writes :—

At a distance of half a mile only three of these may be heard, and all are not audible until one is quite near the singer. The inexpressibly sad, human quality of Poor-me-one's call affects every one who hears it. Waterton, we have no doubt, refers to this bird when he compares the voice of "the largest Goatsucker in Demerara" to "the last wailing of Niobe for her poor children, before she was turned into stone," and, in describing the call, writes : "Suppose yourself in hopeless sorrow, begin with a high, loud note, and pronounce 'ha, ha, ha, ha, ha, ha, ha,' each note lower and lower, till the last is scarcely heard, pausing a moment or two twixt every note. . . ."

Gosse,[1] on the contrary, in his excellent account of the habits of this species, describes its call as a "loud and hoarse *ho-hoo,*" and adds : "Sometimes the same syllables are heard, in a much lower tone, as if proceeding from the depth of the throat." The

---

1 Journal Trinidad Field Naturalists' Club, II, Dec. 1894, p. 137.

1 Birds of Jamaica, p. 42.

account of so careful an observer is not to be questioned, and it is quite probable that the notes of the Jamaican bird differ markedly from those of the birds which inhabit Trinidad.

It seems little short of murder to kill one of these birds. Certainly to shoot a calling bird was out of the question. Our single specimen was shot as he sailed by one evening near the stub where our first observations were made. He was wing-tipped and before sacrificing him to the cause of science we secured the photograph from which the illustration (Pl. III) accompanying this article was drawn.—*The Auk* xii, No. 3, July, 1895.

NOTES ON TRINIDAD BIRDS, WITH A DESCRIPTION OF A NEW SPECIES OF SYNALLAXIS.

By FRANK M. CHAPMAN.

A SECOND visit to Trinidad during March and April, 1894, while made largely for the purpose of collecting mammals, resulted in the acquisition of notes on birds which supplement those published in the preceding volume of this Bulletin.[1] On this occasion I was accompanied by Mr. William Brewster, and after a brief visit to my former headquarters near Princestown, we became the guests of Mr. Albert B. Carr, on his cacao estate at Caparo, in the west-central part of the island, seven miles east of Chaguanas. The country here is not unlike that about the rest-house where previous collections were made, the primeval forest being broken only by cacao estates. These, however, are younger and smaller, the region having been settled within com-paratively recent years. Probably for this reason certain birds, which are common in the clearings and cacao groves about the rest-house, are as yet comparatively rare or wanting on Mr. Carr's estate ; for example : *Vireo chivi agilis, Ramphocelus jacapa magnirostris, Elainea pagana, Pitangus sulphuratus,* and *Tyrannus mëlancholicus satrapa.*

The month of April was passed in the mountains which form the northern coast of the island. On their northern or seaward side the bases of these mountains are indented by but few bays ; on their southern side, however, they are penetrated by numerous valleys. Our home was near the head of one of the most beautiful of these—the Caura Valley—about seven miles from its opening on the plains. Here we were the guests of Mr. J. E. Lickfold.

---

[1] 'On the Birds of the Island of Trinidad,' Bull. Am. Mus. Nat. Hist., VI, 1894, .pp 1-86.

Mr. Lickfold's house is at an elevation of 500 feet, while the crests of the surrounding hills reach an average altitude of about 2000 feet. The locality has long been devoted to cacao growing, and the primeval forest has largely disappeared. Still there are many large tracts of first-growth timber within a few hours' ride. While I visited them on several occasions, my experience was too limited to render valuable a comparison of the avifauna of the mountains with that of the lowlands; and I leave to future observers the task of explaining the rarity of such common lowland birds as *Glaucis hirsutus*, *Pygmornis longuemareus*, *Phaëthornis guyi*, *Galbula ruficauda*, *Rhamphastos vitellinus*, and *Pionus menstruus;* while the following equally common lowland species were not once observed: *Ostinops decumanus*, *Cassicus persicus*, *Pipra auricapilla*, *Momotus swainsoni*, *Trogon* (three species), *Amazona*, and *Urochroma*. On the other hand, *Euphonia trinitatis* and *Calliste desmaresti* were observed only in the mountains.

In attempting to express my appreciation of the hospitality extended me, I am again impressed by the failure of words to convey a sense of either my indebtedness or gratitude. Mr. Carr and Mr. Lickfold not only placed their homes at our disposal but assisted us in every possible manner. I am also under many obligations to Mr. F. W. Urich.

NOTES ON SPECIES NOT OBSERVED IN 1893.

Merula Phæopyga (Cab.). WHITE-THROATED THRUSH.—One female was taken at Caparo and another at Caura. They agree in color with a Venezuelan specimen.

Euphonia trinitatis (Strickl.). CRAVAT.—Not uncommon in the mountains, but not observed in the lowlands.

Calliste desmaresti (Gray). WORTHLESS.—Observed only on the crests of the ridges in the Caura district where it was not uncommon.

Piranga hæmalea S. & G. RUFOUS TANAGER.—A male in the plumage of the female, but with testes measuring about ·18 in their longer diameter, was taken at Caura, April 21.

Legatus albicollis (Vieill.). BLACK-BANDED PETCHARY.—A male of this species, heard calling from a tree-top, was taken at Caparo.

Chasmorhynchus variegatus (Gm.). BELL-BIRD; CAMPANERO. —This bird was not uncommon in the forests at Caparo, and in the more heavily wooded districts of the Caura Valley I have heard three birds calling at one time. The notes of this species will be found described at length in an article by Mr. Brewster and myself in 'The Auk' for July, 1895.

Synallaxis carri, sp. nov.

Synallaxis cinerascens Léotaud, *nec* Temm.

Char. Sp.—Similar to *Synallaxis terrestris* Ja , but upper parts, wings, and tail darker, throat blacker, rest of underparts darker and more olivaceous.

Description of Type (Coll. Am. Mus., No. 60,614, male, Caparo, Trinidad, March 27, 1894; Frank M. Chapman).—Upper parts mummy-brown;[1] exposed portion of the wing-quills and wing-coverts deep chestnut-rufous, tail decidedly darker; central third of the feathers of the upper throat white, lateral third black; feathers of the lower throat centrally buffy; rest of the underparts bistre with a slight cinnamon tinge, the breast faintly streaked with cinnamon. Wing, 2·88; tail, 2·52; exposed culmen, ·53 inch.

The differences between this bird and a specimen of *S. terrestris* from Tobago are found in its darker coloration throughout, and especially in the restriction of the white of the throat. In this character it resembles the Colombian *S. læmosticta*, from which species it may, however, be distinguished at a glance by its more olivaceous and less rufous color.

The only specimen secured was killed on the ground in the forests at Caparo.

It gives me pleasure to dedicate this species to Mr. Albert B. Carr, of Trinidad, not only as a token of my gratitude for his assistance, but also in recognition of his knowledge of ·the Trinidad fauna.

Chætura cinereicauda (Cass.).—A common species at Caparo, where four species of this genus were more or less abundant—the present, *C. cinereiventris lawrencei*, *C. spinicauda*. and *C. polioura*. Frequently all four would be circling above us at the same time. *C. cinereicauda* has not been before recorded from Trinidad, and this capture extends its known range from Southern Brazil. I have no other specimens of *C. cinereicauda*, for comparison, but my eight specimens differ from six Yucatan examples of *C. gaumeri* as stated by Mr. Hartert (Cat. Birds Brit. Mus., XVI, p. 482).

Lurocalis semitorquatus (Gm.).—Two birds of this species were procured at Caparo. They were observed more or less regularly feeding at dusk near the border of the forest, flying swiftly back and forth over a short circuit and within ten feet of the ground. They thus resembled both a Nighthawk and Whip-poor-will in their feeding habits. A single low, insignificant note uttered in flight, was the only one heard.

Celeus elegans (Müll.). YELLOW-HEADED WOODPECKER.— One of two birds seen at Caparo was secured.´

Falco rufigularis Daud. RED-THROATED FALCON.—A specimen was taken by Mr. Brewster.

---

[1] *Cf.* Ridgway's Nomenclature of Colors.

Cancroma cochlearia Linn. BOAT-BILL.—One immature specimen, in rufous plumage, was killed by Mr. Carr.

ADDITIONAL NOTES ON BIRDS OBSERVED IN 1893.

Thamnophilus major albicrissus (Ridgw.).

Thamnophilus albicrissus Ridgw. Proc. U. S. N. M. XIV, 1891, p. 481.
Thamnophilus major Chapm. Bull, A. M. N. H. VI, 1894, p. 49.

In reviewing my paper on Trinidad Birds,[1] Mr. Ridgway speaks of the "Omission of *Formicarius* [*lege Thamnophilus*] *trinitatis* and *F.* [*lege T.*] *albicrissus* described by me in the Proceedings of the U. S. National Museum, Vol. XIV, No. 871, p. 481." These birds were not omitted, but having overlooked Mr. Ridgway's separation of them, I included them both under the names of the Continental forms. At my request Mr. Ridgway has kindly loaned me the two specimens upon which his descriptions were based. Comparison of the type of *Thamnophilus albicrissus* with seven males from Trinidad and twenty males of true *T. major* from Brazil, apparently proves the Trinidad bird to be a race of the latter distinguished by its larger bill, whiter underparts, narower white edgings on the outer vane of the primaries, and narrower white bars on the rectrices. The character of "remiges entirely without white edgings," given by Mr. Ridgway, appears to be a variable one, dependent probably upon age. Immature specimens with brown wing-coverts, like the type, have no white on the primaries, but fully adult examples have well-developed margins to these feathers.

A male from El Pilar, Venezuela, and also one from British Guiana, agree with Trinidad specimens, and it is probable that all birds from north of the Amazon should stand as *Thamnophilus major albicrissus* (Ridgw.).

Thamnophilus cirrhatus (Gm.).

Thamnophilus trinitatis Ridgw. Proc. U. S. N. M. XIV. 1891, p. 481.

As stated above, Mr. Ridgway has also loaned me his type of *Thamnophilus trinitatis*. The characters assigned to this race prove evidently, in my opinion, to be due largely if not entirely to individual variation. Two of three Trinidad specimens have the back of the same color as Mr. Ridgway's type, while the third agrees in coloration with a Demeraran specimen. The color of the underparts is also variable. Trinidad specimens may *average* darker below, but a specimen from Demerara is fully as dark, if not darker, than one from Trinidad.—BULL. AM. MUS. NAT. HIST. Vol. VII, Art. IX, pp. 321–326, Oct. 7, 1895.

---

[1] Auk, XI, 1894, p. 172.

NOTES ON SOME PECULIAR FEEDING HABITS OF THE TRINIDAD OPHIDIA.

By R. R. Mole.

GENERALLY speaking the Ophidia or snakes can be broadly divided into three great classes by the methods in which they take their food. These are the constrictors or pythonoid snakes which kill the animals on which they subsist by constriction, and which do not absolutely hunt, but wait in a locality where their prey abounds until a favourable opportunity offers. These snakes never follow their victims very far. Then there are the vipers who also wait, but strike their victims down with one fell lightning thrust of their terrific death dealing fangs. The third class is the Colubrinæ which, be they poisonous or not, are for the greater part active snakes, searching for their prey and regularly pursue it either under the ground, on the surface, in the trees or in the water, or all four. Now these three classes invariably devote themselves to some particular order of animal as their particular quarry, although all animals almost would answer the purpose equally well. As a general rule a constrictor will not eat frogs and the smaller colubers will not devour mice. Many snakes will swallow frogs but object to toads, but will take lizards—the reason being presumably that the acrid juices contained in the skin of the toad are too strong a relish. Some snakes will not eat anything but birds while others content themselves with small rodents. Some eat fish. Others eat anything living which comes in their way and which they can swallow. Of course after having kept snakes for some years one now and again begins to imagine he has learned everything to be learned about them. The true constrictors are considered to be the only snakes which constrict; the vipers confine themselves largely to birds and mice or rats &c., &c. But we are constantly being deceived. Now I was, I imagined, irrevocably fixed in my belief that the *Boa constrictors*, except when very young, never ate anything but birds and mammals. In this belief I placed a large lizard (*Tupinambus nigropunctatus*) three feet long, in a cage with an eleven foot boa. That night there was a great row in the box but I did not think it anything unusual, except that the lizard was trying to get out. Next morning to my astonishment the lizard was gone. I could not believe my eyes. There was no means of exit. A few days later the deposits of the boa were plentifully studded with the teeth and talons and the scales too—of the vanished saurian. Yet the boa has never attempted to eat a smaller snake and refused an iguana. Many boas seem to prefer birds to mammals and very often freshly caught snakes will eat a pigeon when they will eat nothing else. The tree boas seem to live chiefly on rats but no doubt they will also take birds. There is a very prevalent belief in Trinidad that the Cribo (*Coluber corais*) is an inveterate snake killer. I have found it to be so to my cost. Sometime ago I bought a couple and a friend brought me a third. I put them in a large box along with some tree boas (*Epicrates cenchris*). One night one of the boas gave birth to a dozen or so little ones. In the morning I found the Cribos busily engaged in gobbling up the newly born boas. The little things, however, made terrific struggles and gave their cannibal cousins no end of trouble. You will remember the Boa which swallowed his mate at the London Zoological Gardens. A similar occurrence took place only last week. But the actors were small serpents of only twenty inches. A small mouse had been place in the box with two of these same little Epicrates. They killed it between them, but, by some accident, one of them caught the other and though they were of equal length, he had succeeded in completely swallowing his brother, only an inch of tail being visible and that was rapidly disappearing. I pulled the swallowed one out but he was quite dead. The cribos I referred to just now are very fond of frogs and the other day when I went into my snake room I saw what looked like an enormous two tailed snake. One had swallowed half of the other. The swallowed one had managed to swell himself out so much that he was apparently as thick as his cannibal chum. I got them apart and then I saw that the half swallowed one had a frog in his mouth on which he had retained his hold even in the stomach of his adversary and which he quickly devoured as he climbed up to his shelf, afterwards flicking his tongue in and out in a fashion which suggested the probability of a serpentine chuckle at how he had scored several points over his friend. I have been accustomed in cases of emergency to make various snakes chum in with each other and through thoughtlessness this has cost me dear. The cribo's cage is a very convenient one and consequently any snake just acquired is usually put into it. My first misfortune was to lose a North American Mocassin (*Tropidonotus fasciatus*) valuable because he was a good show snake and not easily replaced. His rough carinated scales however disagreed with the cribo who disgorged him when half digested. Not so however with a lovely yellow machete (*Herpetodryas carinatus*) which was swallowed while I left the room to fetch the money I had given for him. I felt inclined to kill that worthless cribo after that. On Sunday night he p'ayed me the same trick again swallowing a beautiful little tree boa given me by Mr. Caracciolo two years ago, and which I had carefully reared until it had attained quite a respectable size. One of the most difficult Trinidad snakes to feed is the Tigre and

until very recently I had never seen one do so. Once some time
ago a tigre killed some birds but when attempting to eat them
was accidentally frightened and gave up the task. The other
day, however, I thought I would try again as I had often done
and I threw a young rat to two tigres. One of them suddenly
sprang at it. Now it should be remembered that the Tigre is
not a constrictor, he does not belong to the pythons or the boas.
He is a coluber and hunts his prey. He hunted that rat up and
down the large cage until he caught him and then to my
astonishment killed him by constriction in true boa fashion and
then swallowed him. Mice he would swallow alive without
constricting. Mr. Urich came to see this snake feed but the
lively rat escaped him as the tigre plunged at it and to our
amusement the snake seized a dried piece of excreta of the size
of two walnuts and bolted that. Evidently snakes have little
sense of taste. I was afraid this would kill him but it didn't.
Since then he has fed readily. Snakes occasionally overgorge
themselves, especially young ones, and this has happened several
times lately. For some time I have been desirous of getting the
Trinidad Trigonocephalus (*Atrox*) to feed and though I
have had several specimens, without success. The other day
much against my will—because it seemed so impossible that
vipers should eat batrachians and it is against the evidence
afforded by all the *T. atrox* I have ever seen that they should
eat anything but fur or feathers—I tried frogs. I dropped one
in the box. In an hour or two he disappeared. I tried another,
so did he. A third, he vanished but I afterwards found him
under the snake coils alive and well. A fourth I found in the
snake's mouth. Then he ate half-a-dozen. In a week or two's
time he was eating his proper food, mice and rats. Now we have
dealt with the pythons and vipers and tigres, in captivity, let us
see how snakes behave in their native wilds. A short time
before the last peripatus discussion in the public press, I was
near the Botanic Gardens in a ravine which has yielded numbers
of those strange creatures, I heard a rustle and looking a little
way off I saw a little "Machete couesse" (*Coluber boddaerti*), I
stood still. The little creature was as busy as busy could be pushing
his head into every little cranny. Now and again he would rush off
after some small frogs near but I did not see him catch one. He
startled a large cricket which went away with long leaps of some
times eight or nine feet. The snake chased the cricket at least
thirty feet, every time it dropped after one of its flights, he darted
towards the p'ace with a speed which was almost incredible in a
little thing scarce 10 inches long. While I watched I saw a
Beh-belle-chemin (*Liophis-melanotus*) come out of a hole. He too
was hunting little frogs with an assiduity and intelligence
which I never expected to see a snake exhibit. These

little snakes wound themselves up the face of what were
to them huge precipices and hurled themselves into what were
comparatively speaking yawning chasms, with an intrepidity and
a careless freedom which was quite a revelation, and I never
realised until then how splendidly adapted are the ophidia
to the mode of life which they are intended to pursue. Legs
would seem an incumbrance to such lithe elegant active
animals and the most graceful lizard was clumsiness embodied
besides them. I remained perfectly still and these snakes
darted all round my feet, over my boots, under them, going away
and returning with an activity which was perfectly astonishing.
After watching them for nearly an hour I caught both and
carried them home. On the plain board floor of the cage they
were different creatures and I regretted I had taken them
from their native wilds.

THE CATTLE FLY.

* *Compsomyia macellaria.*

By C. W. Meaden.

THE subject of this short paper is one of considerable interest
to those having the care and management of animals both
working and domestic. The specimen flies which I exhibit,
were produced from larvæ taken from the leg of a cow. These
larvæ were removed from the animal on the 10th February and
immediately placed in damp soil, into which they burrowed with
astonishing rapidity, twelve days after the perfect fly emerged.
From this it would appear that a period of about two weeks
elapses between the time the maggot is deposited by the parent
fly and when the fully developed offspring come forth, capable
and willing to go on with the mischief of an evil life. The
larvæ are of a whitey-brown colour, about ½ an inch long, divided

---

* Mr. L. O Howard who kindly determined this fly writes :—" This
" is the well known screw-worm fly of the south-western United States.
" This species is very annoying to cattle and other live stock while laying
" its eggs in sore places It is also one of the species which is noted as
" occasionally attacking man. It will oviposit in running sores and has
" frequently been known to place its living larvæ in the nostrils of persons
" affected with chronic catarrh."

into six segments, and, as already stated, are capable of rapid movement. The colour of the fly is a shining electric blue, with a bright golden head, these colours fade when life is extinct. The insect has a rapid darting flight, as if always prepared to attack, no doubt this is its object in life, and it unfortunately only attacks too well.

The fly appears to be ovo-viviparous and like the blow fly deposits its maggots on any spot likely to secure their successful progress through life. Any slight abrasion of the skin which would pass unnoticed is seized upon by the fly as a home for her numerous brood. It appears that the larvæ must be deposited at various periods, or that other flies assist because in the particular wound I speak of the maggots were in all stages of growth, from those fully developed, down to the minutest size; and this will show how much suffering is inflicted upon animals by this pest as the individuals come to maturity in succession. What is a mere scratch one day, may on the next be the home of hundreds of these larvæ, feeding on the living flesh of the animal and causing a wound it will take weeks to heal and in the mean time the sufferer loses condition and in the case of a milch cow, there is a considerable loss of milk.

Strange to say though the fly is undoubtedly abundant it is seldom noticed. The men employed on the farm, who almost daily see the maggots in wounds, assured me that they did not know the fly, and had never seen it, and of course had no idea that the maggots were produced by a fly. For the future they swear vengeance against them as the cause of many additional hour's work and trouble.

This particular fly appears to confine itself to living animals for its reproduction, as I have never observed it on dead ones or on meat. This point is interesting and should be followed up as it indicates that so long as the larvæ are destroyed during the dressing of wounds, the mischief may be abated and removed. This fly is always with us, but during the dry season, from December to May its efforts at reproduction are more vigorous with a consequent increased suffering to animals. The most efficacious remedy for the destruction of the maggots is "Capuchin Powder" prepared from the dried seeds of *Schænocaulon officinale*. On the wound being filled with this powder, the maggots which are able, quickly take their departure, those that remain soon succomb to the acrid effect of the drug. A remedy equally effectual to the above, called by the coolies "Manar" and locally "Redeye" is found growing in the pastures, the flowers and leaves of which are roughly pounded thus liberating an acrid juice and the wound when filled with this is soon cleared of its unpleasant occupants. Thus nature though permitting an evil, provides a remedy at its side.

The economic side of natural history is not very interesting and not often nice, but it has its compensating value in the alleviation of suffering and of bringing to light such subjects as I have attempted to deal with.

15th March, 1895.

Note.—At the meeting at which the foregoing paper was read, Mr. E. D. Ewen said that in order to throw more light on the matter, he would mention that the plant Mr. Meaden referred to as "Redeye" was *Asclepias curassavica* L., the milky juice of which is used by the people of Tobago to kill jiggers. The *Schœnocaulon officinale* mentioned by Mr. Meaden is the source of the well known insect killing powder called Capuchin powder and the plant is common about Caracas. Crabwood Oil made from the seed of *Crapa guianensis* Aubl. is a valuable application to wounds in cattle both to prevent the attack of insects and to kill any larvæ which may have established themselves therein.

DIRECTIONS FOR PREPARING SMALL MAMMAL SKINS.

1. With the freshly killed carcase before you, write the label. This should bear a number, locality altitude above sea, sex, date, and the following measurements in millimetres, taken in the flesh ; (1), length, of head and body ; (2), of tail without end hairs ; and (3), hind-foot without claws. In the case of the first two measurements, the body should be straightened out as much as possible, and the tail bent upwards at a sharp angle, and the measurements should then be taken from a point in the angle. The label should also have on its back any notes that may strike you about the place where the specimen was caught,

Example of label :—

FRONT	BACK
24/2/94 (Sex.) (Space for scientific name.) Cheadle 200ft. Staffordshire E. BLAGG.	No. 51. H. & B. 81 mm. Trapped in Wood. Tl......54 mm. Hf ....22 mm. Very common.

The positions of the different items should be as far as possible as in the example, so that skins from different sources may all be similarly labelled.

2. Open the skin by cutting down the belly from the breast-bone to the anus ; first push one and then the other knee through the opening, and cut through the legs at the knee joints ; clear off the chief muscles of the leg-bones and separate away the skin from the body all round the tail ; then holding the skin at the base of the tail firmly between the finger and thumb nails, or in

the fork of a cleft stick, pull out the vertebræ from inside with the forceps ; then gradually turning the skin inside out, skin it up over the body, shoulders, and head, separating the fore-limbs at the elbow joint, and taking great care not to cut it in passing over the eyes ; skin it entirely off over the mouth, cutting carefully round the lips. Throughout the operation plenty of fine sawdust will be found of great assistance in keeping the hands, and consequently the fur, dry and unsoiled.

3. Clean the inside of the skin from blood, fat, &c., and then brush it all over with arsenical soap*, being especially careful that the insides of the limbs, get some put on them.

4. Turn the skin back right side out and fill the cavity of the body with cotton wool, putting it in as far as possible in one piece, (in tropical climates a few drops of carbolic acid or other disinfectant should be put on the wool to keep off insects). Take care just to fill out the skin without overstretching it, and try to get all your skins filled out to about the same degree. Take a piece of straight wire long enough to extend from the front end of the belly opening to the tip to the tail, sharpen if necessary one end of it, and wind round it enough cotton wool to fill out the skin of the tail, then brush it with arsenical soap and push the pointed end down to the extreme tip of the tail-skin, and fit the near end into the belly, packing it round with the wool of the body. Put a small piece of wool into the empty skin of the arms and legs. Then stitch or pin up the opening down the belly.

5. If at all oily or greasy, the fur may be cleaned by being wiped with a rag dipped in benzine, and then having fine sawdust gently rubbed into it, this being afterwards brushed out when dry.

6. Lay the skin on a board or piece of cork, draw out the fore paws forwards and pin them down to the board by a pin passed boldly through the middle of the paw. Take care that they are pinned as close in to the sides of the neck or head as they possibly can be, in order to prevent their claws catching in other skins when all are packed together in boxes. Similarly pin back, soles downwards, the hind feet close by the sides of the tail. It is of considerable importance that neither fore nor hind feet should project laterally outwards nor should curl up in drying.

---

* In damp climates powdered oxide of arsenic should be used, as it helps to dry the skin. but caution should be exercised that it is not inhaled during the operation, or allowed to get under the nails.

7. As the skin dries, try to get the face to assume as natural a shape as possible, and the ears to stand up in their natural position. Tie the label on to the ankle before pinning the skin down.

8. Disarticulate skull from trunk, *roughly* clean it, but do not boil it or separate the lower jaw, and then let it dry. Be very careful not to cut or injure it in *any* way; if there is not time enough to get the brain out through the natural hole at the back, it must be left in to dry up. Label the skull with a corresponding number to that of the skin, and afterwards when both are quite dry, it may be tied on to the leg of the skin, or the skulls may be packed and sent home separately if so labelled as to prevent any possible confusion.

9. Pack the skins up carefully in small boxes when they are dry, with enough paper, or better still, wool, to prevent them shaking about ; but if possible do not roll them up separately in paper *before they are dry*, as drying in paper gives the fur an unnaturally sleek appearance.

10. Bats should have their wings closely folded up on each side of the body, but in such a way as not to hide the fur of the belly. The thumbs should be made to point inwards or downwards, not outwards.

The skinning of larger animals must necessarily be somewhat different to the above, but the make up of skins should be as described, except that when the combined lengths of body and tail exceed 30 inches, the tail should be bent up underneath the belly, while the fore as well as the hind feet should be directed backwards. The total length over all of middle-sized skins, such as foxes, &c., should not if possible exceed 30 inches, any excess over this length being reduced by directing the hind feet forwards, or even by folding the skin up across the belly.

WANTED, generally, all small mammals, *however common*, so long as they are not domesticated or inhabitants of houses, *i.e.* squirrels, rats, mice, shrews, moles, bats, weasels, stoats, &c., &c. Do not be afraid of sending too many of the same sort, if carefully prepared and labelled according to the directions. Series of skins representing the different seasons are always of interest.

Such animals are to be obtained chiefly by trapping, gins, break-back, or other traps being placed about in likely places, runs and mouse holes being specially looked for. Suitable traps may be obtained from Mr. Spong, 106, Fulham Road, London, S.W. Pitfall traps, made out of a glass or metal jar sunk in flush with the ground, are also very often successful.

SALE OF BRITISH LEPIDOPTERA.

THE first portion of the well-known collection of British
butterflies and moths formed by the late William Machin,
who was a compositor in a London printing firm, was sold by
auction on February 26th, at the rooms of Mr. J. C. Stevens,
King Street, Covent Garden. The collection, formed during a
period of fifty-eight years, was chiefly rich in a long series of
rare and now extinct British species. The specimens were
thoroughly authenticated as British, and as there was a large
attendance of buyers, the prices generally were higher than the
usual average, varieties especially fetching high rates. Among
the earlier lots of butterflies, four specimens of *Pieris daplidice*,
taken in Kent, reached 16s to 18s. each. A bred variety of
Argynnis paphia, with confluent spots on the under side, sold,
with seventeen other specimens, for £2 2. A variety of *Vanessa
cardui*, taken on Hackney Marshes, fetched £3 10. A fine
variety of the purple-emperor (*Apatura iris*), with yellow under
wings, £3 5. Six specimens of *Polyomniatus dispar*, "from Mr.
Henry Doubleday," fetched from £2 to £5 5. each, according to
size and condition, the latter price being for females, the finest
males only fetched £4 8. Sixty-three typical "blues," including
a specimen of small copper butterfly (*P. phlœas*), with the blue
spots on the hind wings larger than usual, reached £3 10.
Lycœna acis, £2 for a pair, and £2 10. for three specimens.
Among the moths a specimen of *Sphinx pinastri*, from the late
Mr. F. Bond, went, with nine *Chaerocampa elpenor*, for £1 10.
Two *Sesia scoliæformis*, a yellow-banded variety of *S. culiciformis*
and eleven others, sold for £2 2. Four *S. sphegiformis*, labelled
"from Tilgate Forest," with six *S. chrysidiformis*, fetched £2 5.
Varieties of the common garden tiger, *Arctia caia*, were not
specially fine, the highest price being £2 2. for three. *Laelia
cœnosa* sold in pairs, at £1 7. 6. to £2 2. Seven specimens of
Bombyx trifolii, one being a fine variety bred by the late Mr.
Machin, fetched £3 10. A fine series of eight specimens of
Lasiocampa ilicifolia, taken at Cannock Chase, £2 5. to £4 5. a
pair. A fine female *Noctua subrosea* sold for £4 10., the six
other specimens fetching £4, £4 5. and £5 5. per pair. *Cleora
viduaria*, which has become valuable during the last few years,
£1 15. to £3 5. a pair. The thirty-eight drawer mahogany
cabinet that contained the collection, sold for nineteen guineas.
The total amount realised for the collection of macrolepidoptera,
or butterflies and larger moths, being £363. We understand
that Mr. Machin's still more celebrated collection of British
macro-lepidoptera has been purchased as a whole by a well-
known London amateur naturalist.—*Science Gossip.*

Arnott, Lambie & Co.

FISHING APPLIANCES,

Hooks, Wire, Lines, Twines &c. &c.

always on hand a large supply of

SPORTING GOODS,

Guns, Rifles, Revolvers, Gunbags, Powder, Shot, Caps, Wads, Cleaning Rods.

CARTRIDGES Loaded to Order.

HARNESS & SADDLERY GOODS,

PAINTS, OILS, TURPENTINE & VARNISHES.

ESTATE SUPPLIES & SHIP CHANDLERY.

HOUSEHOLD IRONMONGERY,

ABERCROMBY STRET,

Telephone 133.   (Opposite Ice House.)

PEDRO PRADA,

36, King Street,   Port-of-Spain,

TRINIDAD,

WHOLESALE AND RETAIL.

DRY GOODS MERCHANT.

Ready Made Clothing, Boots and Shoes, Hats, Umbrellas, Guns, and Ammunition, Furniture, &c., &c., &c.,

---

CRONEY & CO.,

GENERAL MERCHANTS,

Port-of-Spain,

TRINIDAD.

Communications and Exchanges intended for the Club should be addressed to the Honorary Secretary, Port-of-Spain, Trinidad, B.W.I.

Price 6d.   Annual Subscription, 3/.

Vol. 2.   FEBRUARY, 1896.   No. 12

J'engage donc tous à éviter dans leurs écrits toute personnalité, toute allusion dépassant les limites de la discussion la plus sincère et la plus courtoise.—LABOULBÈNE.

Trinidad·Field·Naturalists'·Club.

NATURA MAXIME MIRANDA IN MINIMIS.

Publication Committee:

His Honour Sir JOHN GOLDNEY, *President.*
P. CARMODY, F.I.C., F.C.S., *V.P.:* SYL. DEVENISH, M.A., *V.P.*
HENRY CARACCIOLO, F.E.S.,   R. R. MOLE,
F. W. URICH, F.E.S., *Hon. Secretary.*

CONTENTS :—

Report of Club Meeting:
     December   285
Trinidad Scorpions and Pedipalpi—
     Description of   288
Two Embidæ from Trinidad   292
Some Economic Plants and their
     Planting Prospects   294
Coccidæ from the Island of Grenada—
     Notes on a small collection of o6

☞ All Communications and Exchanges intended for the Club should be addressed to the Honorary Secretary, Port-of-Spain, Trinidad, B.W.I.

JOURNAL

OF THE

Field Naturalists' Club.

VOL. II. FEBRUARY, 1896. No. 12.

REPORT OF CLUB MEETING.

10TH DECEMBER, 1895.

PRESENT : His Honour Sir John Goldney, President ; Mr. Syl. Devenish and Professor Carmody, Vice-Presidents ; Messrs. C. W. Meaden, J. R. Llanos, R. R. Mole, J. Hoadley, A. T. Reoch, H. Caracciolo, E. D. Ewen, J. T. Rousseau, T. I. Potter, Dr. Rodriguez, F. W. Urich, Hon. Secretary, and Mr. T. E. Miller (visitor).

Before the minutes were read Mr. Devenish said this was the first occasion the Club had met since Sir John Goldney had returned to the Colony. During his absence the Club had elected Sir John, President, and in the name of the club he heartily welcomed His Honour and congratulated the club upon having such an able and amiable man at their head.

Sir John thanked Mr. Devenish and said that it was a great honour to him to have been elected President of the club. Although he was not at all a Naturalist he felt very proud of being President of an Association which was doing such useful work in Trinidad. He also mentioned that when in England, he had had several enquiries about the club from eminent scientific men and it seemed that the club was well known outside of Trinidad, not only in England, but he might venture to say in the whole world. (Applause).

The minutes of the last meeting were then read and confirmed. After an amendment to the rules, proposed by Mr. Rousseau, had been agreed to, Mr. Potter read a useful paper

"On some Orchid Pests" dealing with some of the insects which are injurious to Orchids and the means of getting rid of them; amongst these he mentioned the Cockroaches as being the chief evil-doers, but there were also scale insects, bugs, beetles and caterpillars that were not to be despised. Among the beetles one species turned out to be new belonging to the *Stethobaris*, a reference to which in a letter from Mr. L. O. Howard was read. The paper was illustrated by specimens of the insects and the havoc they work. A discussion took place in which all present took part.—Mr. Caracciolo then followed with a paper of an economic nature dealing with the insect enemies of some of our most useful fruit trees such as orange, guava, star apples, &c., the sugar cane disease was also referred to. A long discussion took place at the conclusion of which Dr. Rodriguez, Mr. Devenish and the President referred to the presence of maggots and worms in most of our fruit, a fault which was not known in the neighbouring colonies and a hope was expressed that Mr. Caracciolo would endeavour to find out in the interest of the colony a remedy for this evil.—Mr. Caracciolo said although this did not exactly come within the scope of his researches but belonged more to the Botanical Department, yet he would endeavour to find out a method which would tend to lessen this liability of attack from maggots. Perhaps what had been said would induce others to take up the subject.

Mr. Mole said that he had found the following extract in Oliver Goldsmith's History of the Earth and Animal Nature, Vol. IV. "Father Labat took a serpent of the viper kind, that "was nine feet long, and ordered it to be opened in his presence. "In this creature there were six eggs, each of the size of a goose "egg, but longer, more pointed and covered with a membranous "skin by which also they were united to each other. Each of "these eggs contained from thirteen to fifteen young ones about "six inches long and as thick as a goose quill. Though the "female from whence they were taken was spotted, the young "seemed to have a variety of colours very different from the "parent, and this led the traveller to suppose that the colour "was no characteristic mark among serpents. These little mis- "chievous animals were no sooner let loose from the shell than "they crept about and put themselves into a threatening posture, "coiling themselves up, and biting the stick with which he "destroyed them. In this manner he killed 74 young ones; "those that were contained in one of the eggs escaped at the "place where the female was killed by the bursting of the egg "and their getting among the bushes." Father Labat, Mr. Mole continued, was, he believed, for a long time in Grenada, and it was interesting in this connection to note that Mr. Broadway had forwarded to Trinidad a tree boa nearly allied to our Cascabel

Dormillon (*Corallus cookii*) and Mr. Broadway had also sent
with it its young ones numbering between twenty-five and thirty,
which had since been put into spirits. The young ones it would
be found differed so much in their markings that scarcely two of
them were alike. These snakes were not venomous but were
very fierce and struck very readily, and doubtless Father Labat
was misinformed when he was told the snake he saw was a
viper, and very probably it belonged to this species. But the
curious fact about all this was this: Mr. Arthur Carr of Caparo
recently caught a large female Mapepire Balsayn (*Bothrops·
atrox*) one of our most deadly snakes. The captive was in an
interesting condition and shortly afterwards brought forth fifty-
seven young ones, all of which were marked exactly like each
other and presented no differences at all except in size and
shade. These snakes he begged to exhibit in Mr. Arthur Carr's
name and members would see for themselves how all the harmless snakes differed from each other in spots and stripes and colour,
&c., while the venomous ones were all uniformly alike, the exact
reverse of the statement by Father Labat. He further drew
attention to a water snake, probably *Helicops angulatus*, found
near the Toll Gate, which, although a harmless species, bore an
extraordinary resemblance to *Lachesis muta*, a resemblance which
none but a very close observer could detect.

The Secretary produced a beetle (Family: *Lamellicornia*;
Group: *Orycides*) from Montserrat, which was sent by Mr.
John Guilbert. The President said that it looked very much
like the coconut beetles which had done such damage in
Singapore and which had necessitated an Ordinance being
passed to check their ravages. Mr. Urich remarked that, so far
as he knew, Mr. Guilbert's beetle was not very numerous and not
injurious, as the larvæ was supposed to live in decaying wood.—
Mr. Potter laid on the table a piece of a cocoa tree which was
attacked by a Borer which Mr. Blandford had determined as a
probable new species allied to *Xyleborus capucinus*.

Mr. Potter also showed a caterpillar found on a papau tree
which Mr. Caracciolo said would produce a species of Heliconia.
Mr. Urich exhibited two species of *Mutillæ* from Arima, also a
rather dark specimen of the snake *Leptodira annulata*. Mr. Urich
said that through the kindness of Professor Brauer he was able
to give the determination of a very large brown fly, found at St.
Ann's by Mr. Potter, which was supposed to be the imago of a
mosquito worm, but whose larvæ, according to the Professor,
lived in rotten wood, possibly devouring the larvæ of longi-
corn beetles and other wood-boring insects. The fly's name was
given as *Acanthomera championi*, Osten Sacken (*Biol Cent
americana.*) "It showed a broader face than Osten Sacken's
"type. Perhaps it might turn out to be a new species, as Osten

" Sacken's specimen came from the Mainland."—Mr. Urich said that when coming down from St. Ann's one dark night he noticed a patch of phosphorescent light of about six inches diameter on a tree near Coblentz. Thinking it might be a jumbi (he was very anxious to catch one) he struck a match and found the light came from part of a deserted termite's nest, no doubt it was caused by decaying vegetable matter or possibly a fungus. He then read the following extract from an article on ants' nests in Brazil by Frederick Knab in the Entomological News. " The termites on " the contrary, live in large irregular conical mounds, hard as " rock and often ten feet or more high. In the day time there " is no sign of life, but if one enters the forest at night, the sight " is a beautiful and startling one—the darkness is intense. Here " and there in the blackness may be seen clusters of glittering " phosphorescent light. These are the Termite hills. No doubt " the light proceeds from the insects as the particles of the light " mass move and change. The light is greenish and soft, and the " effect is indescribable. In marked contrast is the gleaming red " light of the Elaters as they dash rapidly through the foliage." Mr. Urich said the description of the light agreed with his observation, but it did not move and there were no insects in the nest he saw.—On the motion of Professor Carmody it was decided the Club meetings should take place monthly, as formerly, instead of every two months, as had been the case lately. The meeting came to a close at 10 p.m.

DESCRIPTIONS OF TRINIDAD SCORPIONS AND PEDIPALPI.

Reprinted from " Contributions to our knowledge of the Arthropod Fauna of the West Indies," by R. J. Pocock, of the Natural History Museum. Journal of the Linnæan Society— Vol. xxiv.—Zoology No. 155, pp. 374 to 407. August 26, 1893.

Genus : ISOMETRUS. *(Hemp & Ehrb.)*

Isometrus maculatus, (De Geer.)

This small and slender Scorpion, which is widely distributed throughout the tropical and sub-tropical countries of the Old and New Worlds, is perhaps the best known species of the Order.

The British Museum has examples from the following West Indian Islands :—St. Domingo ; Jamaica ; St. Thomas, St. Croix (A. Newton) ; Barbados (H. W. Feilden) ; Union Island, Grenada (H. H. Smith) ; and Trinidad (W. E. Broadway.)

Tityus androcottoides (Karsh) (Pl. XXIX. figs. 3-3b.)

Isometrus americanus var androcottoides, Karsch, Mithh Münch. ent. Ver. 1879, page 113.

Isometrus androcottoides, Pocock, Ann. Mag. N. Hist. (6) IV. page 57.

The British Museum has received very many examples of this species from Trinidad (Messrs. Hart and Broadway). It is common in British Guiana (Brit. Mus., W. L. Sclater.)

Although Dr. Karsch looked upon this scorpion merely as a variety of *T. americanus*—an opinion in which he has been followed by Professor Kraeplin—I think there can be little doubt of its distinctness. When first I put forward this suggestion, I had only seen a few examples of the form to which the name *androcottoides* would apply ; but during the past three years the British Museum has received many others, all of which justify the belief in the distinctness of *androcottoides* from *americanus*. Apart from sexual characters which are very distinctive, this species may be recognized from *americanus*, as from all the Antillean species of the genus, by the fusion of the inferior keels of the posterior caudal segments.

Tityus melanostictus, sp. n.   (Pl. xxix. figs 4-4b.)

Colour flavous or fulvous, fusco-maculate, the ocular tubercle and anteocular portion of carapace fuscous, mesially flavous, the posterior and lateral portions of this plate fusco-maculate and lineate ; each of the tergites, except the last, furnished in front with a transverse row of fine fuscous spots, the external spot on each side is situated on the very margin of the plates, the median one is divided by a flavous spot marking the median keel ; posteriorly the tergites are adorned with three fuscous spots, the median of which is divided, like the spot in front of it, by a clear flavous spot, and the two lateral ones are united by a fine fuscous line, which marks the transverse granular ridge ; the last abdominal segment, above and below, fusco-maculate ; the rest of the sternites concolorous ; the upper surface of the tail mostly concolorous, sometimes obscurely fusco-maculate, the lower surface of the first three segments mottled with flavous, the fourth and fifth segments and the vesicle generally uniformly infuscate or reddish brown ; palpi subfuscous, mottled with round clear flavous spots above, the digits fuscous at the base, becoming gradually pale distally ; legs externally fusco maculate.

Female : The upper surface of the body subtly granular, the normal keels not strong but visible ; the *sterna*, except the last, smooth, marked with more or fewer large punctures ; the last subtly granular with black keels. The *tail* about 5½ times as long as the carapace, subtly granular, the keels visible but weak and subtly granular ; the first segment with ten keels, the second with eight and a trace of the median lateral ; the first segment a little wider than the fifth. The vesicle armed with an acute tooth beneath the aculeus.

The *palpi* with well-expressed granular keels, the intercarinal spaces coriaceous ; the manus small, internally rounded and produced, normally costate ; the costæ scarcely distinctly granular ; its width considerably less than half the length of the movable digit and less than the width of the brachium. The *digits* long and slender, contiguous, neither lobate nor sinuate ; the movable digit about twice the length of the "hand back" and furnished with 14 rows of teeth.

The pectines shorter than the posterior coxæ, furnished with 15-17 (usually 16) teeth ; the proximal lamella of the intermediate series produced internally into a rounded prominence.

Male : *Tail* parallel sided, the fifth segment being equal to the first in width and a little more deeply excavated above, considerably smoother than in the female ; the keels almost obsolete, about six times as long as the cephalothorax. The *manus* larger than in the female, the width about equal to half the length of the movable digit and much greater than the width of the brachium ; the movable digit sinuate at the base with a distinct lobe, less than twice the length of the "hand back" the immovable also sinuate so that when closed the two are separated at the base.

Pectines longer than in the female, just reaching the end of the 4th coxæ ; the teeth longer than in the female, 17 in number.

Length of female 47 mm. of carapace 5, of tail 28, width of 1st segment 2.8, length of fifth 5.2, width of fifth segment 2.5, of vesicle 1.8, length of "hand back" 2.7, of movable digit 6.

Length of male 43 mm., of carapace 43, of tail 26, width of tail 2.5, of vesicle 1.5, length of vesicle 1.8. Locality : Trinidad.

This pretty little species seems to be tolerably common in Trinidad. The British Museum has received examples from Messrs. W. E. Broadway, J. H. Hart, R. L. Guppy and Lady Broome.

Diplocentrus Gundlachii, Karsch, Zeits Naturwissen, (3) v. pp. 407-408 (1880).

This species has been very briefly described.

The upper surface of the trunk is entirely smooth. The dorsal surface of the hand is evenly arched, and nearly smooth. The first four segments of the tail have 10 complete keels.

Pectinal teeth 6-8.

Length 30-32 millim.

Locality : Trinidad and Cuba.

Broteochactas nitidus, Pocock. (Pl. xxix. figs. 7-7a.)

Female : Colour castaneous ; legs, vesicle, and lower surface a little paler.

Carapace perfectly smooth and polished, marked with a Y shaped groove, the two upper arms of which embrace the ocular tubercle ; the anteocular portion not mesially depressed, the lateral portion sloped away.

Tergites perfectly smooth and polished, marked in front on each side of the middle with a shallow depression ; the last tergite with four tubercles (sometimes six) corresponding to the four superior keels of the tail segments.

Sterna smooth and polished.

Tail robust, narrowed towards its distal end, the upper surface smooth and polished, the first segment scarcely excavated longitudinally, the excavation increasing in the depth to the fourth, the fifth flat posteriorly, excavated in front, the supero-lateral keels well developed on the four anterior segments, shining but obsoletely crenulate ; the superior keels represented by a single posterior tubercle on the first segment, visible on the others and obsoletely tuberculate, or crenulate, the upper angles of the fifth not sharp but squared ; the lateral and inferior surfaces of the first three segments smooth and polished, sparsely haired, not keeled ; of the fourth obsoletely keeled and granular, the lower surface of the fifth somewhat coarsely but irregularly granular, its posterior border denticulate. *Vesicle* piriform coriaceous beneath, smooth and flat above, the aculeus stout at the base, somewhat strongly but evenly curved in its distal half.

Palpi robust ; *humerus* with its keels weak but granular, its anterior surface flat, weakly granular ; brachium smooth, obsoletely costate ; *manus* large convex above, its inner portion produced, smooth, obscurely and very feebly granular towards its inner edge, and very obscurely costate above, its width about equal to the length of the "hand-back" which is furnished with an oblique row of four piliferous pores. Digits short, curved, in contact, the movable a little longer than the length of the "hand back" the immovable furnished with five piliferous pores; four more of these pores in a line on the external surface of the hand, lying between the bases of the two digits.

Legs smooth, exeept the inferior edge of the anterior two femora, which are granular; the distal tarsal segment with two rows of hairs beneath.

Pectines short, furnished with 7 teeth (one specimen with eight.)

Male : Slenderer than female, the tail longer, being about 4½ times the length of the carapace. The upper surface of the hand very finely granular. Pectines larger, the teeth much longer, 8 in number.

Measurements in millims : Female total length 38, of carapace 5, of tail 20; width of 1st candal segment 3, length 2; length of 4th segment 3, width 2.5, length of fifth 5.5, width 2.5; length of palp 15.5, width of *brachium* 1.8, of manus 4, of "hand back" 3.8, of movable digit 4.5.)

Male : Total length 31, of carapace 4, of tail 19.

Locality : Trinidad (W. E. Broadway and J. H. Hart.) This species is very nearly related to *Chactas Gollmeri*, Karsch from Caracas, which probably also belongs to the same genus. Karsch, however, makes no mention of the presence of keels and granules upon the lower surface of the 4th caudal segment. Nor in *Broteochactas nitidus* is the manus marked with many ocelliform punctures arranged in three rows.

I also refer to this genus *Chactas delicatus* Karsch of which the British Museum has a large number of specimens from Demerara (British Guiana) a few ticketed South America, and one from Colombia. The variety named *opacus* by Karsch is nothing but the male of *delicatus*.

PEDIPALPI.

Very few species of this group have been recorded from the West Indies, as may be seen from the subjoined list, and neither of these are restricted in range to this area of the Neotropical Region.

Tarantula reniformis (Linn.).

Phalangium reniforme, Linn.

Colour : Carapace reddish brown or almost black, with some faintly indicated lateral marginal flavous spots and some fine stripes of the same colour radiating from the fovea externally and posteriorly; upper surface of the abdomen in well coloured specimens ornamented with black or deep brown and reddish or flavous spots, the spots alternating like the pattern of a chess-board, each tergite bearing 10 spots, 5 anterior and 5 posterior, the anterior row consisting of 3 black and 2 yellow spots and the posterior of 2 black and 3 yellow spots; palpi the same colour as the carapace; legs ferruginous or fuscous, with a faintly defined flavous spot on the external surface of the femur. Lower surface ferruginous or fulvous.

Carapace coarsely but not closely granular, its anterior border lightly emarginate and conspicuously dentate, the rest of the border denticulate. The upper surface of the abdomen granular like·the·carapace.

Palpi rather short, but varying in length from about three times the length of the carapace to only a little more than twice the length; the brachium a little longer than the humerus, longer than the length of the carapace but shorter than its width. The humerus granular above and below, more weakly granular in front, its upper edge armed in its proximal half with from 5 to 6 (8) larger spines and some smaller ones; the second and third are the largest and the first rises from the base of the second ; its inferior edge armed with about 8 larger and smaller spines, of which the first and second are considerably the largest. Brachium granular like the humerus, its upper edge armed in its distal half with 7 spines, of which the first and seventh are the smallest, the rest are very long, but the fourth and sixth are a little shorter than the rest ; its lower edge armed with 2 long and 3 or 4 shorter spines. *Manus* armed above with 3 spines, of which the second is much the longest, and some spinules ; its lower edge armed with a long spine in the middle and 1 very much shorter one in front and behind it.

Legs thickly granular.

Measurements of largest specimen :

Total length 34 mm., length of carapace along the middle line 12, its greatest length 14, width 19 ; length of abdomen 22, of humerus 15, of brachium 16.5, of manus and dactylus 13.

This species is very widely distributed in the Northern parts of the Neotropical Region. The British Museum has examples from the following West Indian Islands :—Cuba, Jamaica, Hayti, Bahamas, Montserrat, Martinique, Dominica *(Nicholls)* St. Lucia *(Ramage)* St. Vincent *(H. H. Smith)* Barbados *(H. W. Feilden)* and Trinidad.—It has also been recorded from Porto Rico *(Karsch)* St. Bartholomew *(Thorell)* and Antigua *(Brown)*.

TWO EMBIDÆ FROM TRINIDAD.

BY H. DE SAUSSURE, M.D., HON. F.E.S.

THE Embidæ form a small Tribe composed of insects which in their habits remind one somewhat of the Termitinæ, although they are much separated from the insects of this tribe, and more approximate to the Orthoptera, principally the Forficulinæ amd Blattidæ. The whole tribe is formed by a single genus. The sexes are very different from each other in most of the species. Some seem to be winged in both sexes. Others are winged only in the males, the females not going through the last transformation but remaining quite larviform although full-grown. In others still, both sexes remain larviform, neither of them transforming to take wings.

The two species here described belong to the second group. The habits of the Embidæ are not better known than their species ; some live on sandy ground between stones, others seem to prefer rotten wood, others live on Orchidæ, plants which they seem to affect in all the different countries, by cutting galleries into them, to establish their abode and perhaps to eat the roots (?) They live generally in colonies on the same plant and the insects envelope themselves, in a silken case. The organ with which they spin the silk is in the first joint of the anterior tarsi, which is on account of that extraordinarily tumefied.

The two species found by Mr. F. W. Urich in Trinidad can be described as follows :

Embia urichi, n. sp.

Rather large for the genus, of a brown colour, with the anterior part of the head more testaceous, eyes small, slightly prominent; seen from the side, narrowly reniform, more distant on the vertex one from the other than the antennæ.

Antennæ, long, brown, formed of 21 joints, of which the second very short, the third longer than the following joints, the last three or four luteous white, the last with a black dot, all very hairy. Legs rather stout, principally the hind femora. The anterior metatarsus much swelled.

Male—Head more elongate, occiput being somewhat lengthened. Antennæ as long as the thorax. Pronotum with a strong ridge in front, wings transparent blackish with fuscous veins and with five white longitudinal lines between the veins. The medial vein (sector) bifurcate in the middle; its posterior branch itself bifurcate; the 2 ulnar (posterior) veins distinct and of a brown colour. Some scattered transverse venulæ in all parts of the wing, abdomen slender, slightly dilated at its end. Supra-anal plate broadly triangular with its angle obtuse. Appendages (cerci) brown, rather long, compressed ; their second joints somewhat longer than the first, but variable, sometimes cylindrical and slender. The infra-anal plate broadly triangular or slightly asymetric. The titillatores not prominent.—Length 11 mill., wings 10 mill.

Female—Completely apterous, head a little shorter than in the male, the whole body covered with scattered hair ; antennæ shorter than in the males, shorter than the thorax, with shorter joints, sometimes rufous. Pronatum marked with 2 transverse ridges. Mesonotum quadrate, elongate. Metanotum transverse quadrate. First abdominal segment rather broad, with the epimeri produced laterally, yellow; the following epimeri of the same yellow colour. The 5th, 6th segments somewhat narrower, the 7th, 9th somewhat broader again. The supra anal plate triangular, somewhat rounded. The appendages cylindrical, but sometimes compressed, legs somewhat thicker than in the males often of a pale colour.—Length 17 mill.

Embia trinitatis, n. sp.

Rather small, of a brownish black colour, eyes largeal, prominent, seen from the side broadly reniform ; on the vertex more approximate than the antennæ. Antennæ formed of 21 to 22 joints, all brown.

Male—Head stouter than in *E. urichi*. The occiput somewhat shorter. Pronotum brick-red, or orange-red, wings infuscate, but slightly dark, with fuscous veins and between the veins with hyaline lines. The medial vein bifurcate near its base with its posterior branch itself bifurcate in the middle. The two ulnar veins very slight or obliterated. The transverse venulæ wanting in the hind part of the wings. Legs somewhat less inflated than in *E. urichi.* The cerci slender, with the second joint cylindrical, nearly twice as long as the first. The infra-anal plate as in *E. urichi.*—Length 7 mill., wing 6 mill.

Female—Apterous, brown, antennæ paler, with the two last joints pale, annulated with brown, eyes smaller, not prominent, Pronotum rufous parted in three lobes by two transverse ridges. Supra-anal plate rounded triangular, two joints of the cerci not much elongate.—Length 8 mill.

[For the information of readers who may not be able to follow the above technical description, we may state that the insects referred to are those earwig like forms which are found underneath a white web covering the interstices in the bark of the trunks of large trees.—P.C.]

SOME ECONOMIC PLANTS AND THEIR PLANTING PROSPECTS.

By E. D. Ewen.

There are a very great number of economic and valuable plants which here, are chiefly conspicuous by their absence ; and some others which might as well be absent for all the notice they attract and all are exceedingly interesting from the fact that there is perhaps no place in the whole world comparable with this colony in its capacity for growing successfully various economic plants native to the most diverse tropical and sub-tropical countries and climates.

For instance there are not many other places where one can see growing and fruiting freely in the open air within a few yards of one another the Coca from the Andes, the Loquat from Japan, the Oil Palm from West Africa, and the Mangosteen (which won't fruit in India) from Malaya, not to mention others from Australia, Chili and South Africa.

One of these former which from its evident adaptability and its valuable product, highly merits at the present juncture introduction and trial, is the Guaraná, N.O. *Sapindaceae,* of which there are (at least) two species, viz. : *Paullinia Sorbilis (?)* and *P. Cupaná,* (H.B., and K).

The Guarana of commerce, a substance resembling chocolate, and containing Theine as an active principle, is principally made from the fruit of *P. Sorbilis*, which is found wild from Bolivia to the mouth of the Amazon in Brazil, and is cultivated both in Bolivia and the Brazilian Province of Amazonas.

The appearance of the plants will be familiar to most of us when it is stated that they are lianes or vines of the same family as the "Supplejacks," various kinds of which are common in all parts of the colony. They are therefore naturally woody climbers, but under cultivation their appearance is quite changed, and instead of spreading over all sorts of bushes thus, they are trained to stakes like the grape vine. The white flowers are borne in terminial pannels and would hardly be noticeable if they were not massed together. The flowers appear in July and the fruit is ripe in November-December.

The cultivation is as follows :—*Soil*, the plants are said to grow equally well in hilly districts or rather dry sandy-clay soil, and on rich level black soil. *Sowing.* The seeds are sown in beds and transplanted as soon as they are 2″-3″ high, or, another plan is to sow the seeds in small baskets filled with earth, each of which is placed in the hole it is to fill as soon as its seedling is 3″ high. *Planting out, &c.* The seedling are planted at the foot of and trained to stakes placed 18 to 22′ apart. The cultivation demands care and the first crop is not collected till the 3rd year (in Amazonas) or the 4th or 5th year (in Bolivia.) (But as it is probable that the Guarana like most vines can be easily grown from cuttings, a more rapid mode of propagation might be obtained.) From that time forward the plant has to be pruned yearly, and its cultivation closely resembles that of the grape vine. The mature plant is said to fruit for over 40 years, producing six to eight lbs. *per annum.*

Crop and preparation. The fruits begin to ripen in November-December, as indicated by their beginning to burst open and expose the seeds. The fruit is a small pod consisting of an outer shell containing white pulp and seeds (like cocoa) but the latter each in a shell of its own ; each pod contains 4-6 seeds, horse shoe shaped, and about the size of a hazel nut. When the pods begin to burst, no time must be lost in gathering the crop, for the seeds are liable to fall out and get lost. The seeds are then extracted from the pods, and the white pulp is either washed off, or they are sun-dried until it can be rubbed off by the hand. The seeds are then carefully washed and placed in a sack to remove their cartilaginous shell. They are then milled in a hand-mill, or a wooden mortar to a rather fine powder. To each pound of this powder 12-14 tablespoonfuls of water is added, and the whole kneaded into a doughy mass. This mass is then moulded by hand into the balls or rolls like sausages, which are known to European druggists as 'Guarana paste.' These are then baked in the sun or in an oven, and when thoroughly dry and hard, are sent to market wrapped in Banana leaves or mats, a form in which it keeps remarkably well. In the crop season, each person is said to be able to prepare from 150 to 300 ℔s. of Guarana.

In Bolivia and parts of Brazil, Guarana is used in place of tea ; it is grated from the rolls as required, like chocolate, and drunk dissolved in water. The Indians use large quantities as a preventative of fevers. It has been introduced into European *Materia Medica*, and is employed to special advantage in headaches and disorders of the stomach. Its use has also been suggested instead of quinine for intermittent fevers.

A gentleman who has recently made some exploring and prospecting trips in British Guiana, informed me that he found a pleasant but bitterish flavoured drink, like chocolate, much used amongst a certain tribe of Indians there. And his descriptions of the substance, and of the plant which produced it, have left no doubt in my mind that what he met with was Guarana made from the beans of *P. Cupana*. The range of this plant is from Trinidad along the banks to the head waters of the Orinoco (according to Don) I am not aware however of its ever having been found in Trinidad. But there is little or no doubt that both species are found all over the country between the basins of the Amazon and the Orinoco, and most probably both and perhaps yet other species are used in the preparation of Guarana in Brazil.

Codazzi says that on the Rio Negro *P. Cupana* is abundant, and from its beans is made an extract or paste called Guarana from which the inhabitants make a bitter stomatic drink. The husks yield a fine and strong yellow dye.

The writer now begs to suggest that plants of either or both species would be obtainable without difficulty by the Botanic Garden, that they are well worth getting from the value of the product, that if distributed thence they have every chance of being a success here on account of the luxuriance and frequent occurrence of plants of the same order, and that for the extended use of Guarana in the near future no apprehensions need be entertained, for the price for "paste" at Para is said never to have been under 80 cents per lb., while the *minimum* price at Paris (which is the chief European market for it) keeps according to the most recent authorities at about $1.50 per lb. for first quality "Paste." Spon's Encyclopedia, quotes its value at Santarem (in /82) at /8d. to /9d. per lb.,—an obvious mistake for 80-90 cents per lb.—while a few lines lower its occasional sale at the extraordinary price of £1 per ounce is recorded. The latest New York quotation is $1.00 per ℔.

In Europe Guarana is at present principally used in Medicine, as the basis of medicated tonic wines, &c. But a firm of Chocolate Manufactures in Holland is stated to be selling a mixture of Guarana and Chocolate under the name of Guarana-Chocolate with great success. Not only is it held in high esteem as a food as well as a medicine through all South America from Brazil to Mexico; but there are not wanting signs of its beginning to win its way into popular favor both in Europe and North America, as an alimentary beverage worthy to rank with Tea, Coffee and Chocolate.

Another plant, which would be a very promising introduction is the *Star-Aniseed tree (Ilicium Anisatum N.O. magnoliaceæ'* called *Badiane* in Tonquin, and *Qua-hoi* in China. This is one of the most valuable spice plants of the present day and the seeds cost 75/-95/ shillings per cwt. The seed is largely exported from Tonquin, and both the seed and its oil from China and they are largely used in Europe as a spice, and in Perfumery, and Liqueur manufacture, while there is a large and increasing market for this and other spices in the U.S.A.

So close to us as Florida, there is a wild and poisonous species of Star-Aniseed (the "Poison-Bay") *I. Floridanum;* and where a notoriously difficult plant like the Mangosteen from Malaya (where there are several wild specimens of *Ilicium*) fruits here, there can scarcely be any doubt, this plant of similar climatic origin, though of much wider *habitat* (Malaya to Japan) will be comparatively easy of propagation.

This plant grows to about the height and size of a cherry tree. The branches are angular, spreading in an upward direction. The leaves are alternate on the stems and also grow in tufts of 3-4 together at the ends of the small branches, they are evergreen and about 2 inches in length. The flowers grow from the *Oxillæ* of the branches on solitary pederneles. They are yellow in color, and the corolla is composed of 16 petals. Eight or more germs are contained in each flower, and the fruits bear the exact shape of a star with six to eight points about the size of a sixpence, nearly $\frac{1}{4}$ inch thick, pale brown in color, and of a stiff, leathery consistence. The capsules, as well as the seed they contain have a strong flavor of anise. The seeds are smooth, glossy and cinnamon colored, and eliptical in form. The whole plant has a pleasant aromatic odor.

In Tonquin, the seeds are sown in manured soil and the young plants issue from the ground in 20-30 days: in the second or third month they attain the height of 7-8 inches, they are then transplanted and pricked out a distance of 19-20 feet from each other, and always on slopes free from stones,

Trees grown from seed, begin to bear about the sixth year, and are in full bearing about the tenth year, when they are about 20 feet high. In good soil, 20-35 lbs. per tree is considered a good average yield, and it includes the fact that (in Tonquin) these trees bear a "bumper crop" every second year.

In Southern China, according to a Chinese,—Dr. Sang Lo, whom I met within India in 1879,—it is considered a waste of time to grow this plant from seed, and, as plants from layers or cuttings strike freely, and make trees nearly as large as those grown from seed, while they begin to bear well about the third year. He said that in China they are heavily manured, and require a free soil (not clayey), and a well drained site. The Chinese neglect its cultivation in favour of gambir, &c., which thrive with less cultivation.

It is stated that so long ago as Queen Elizabeth's reign, Sir Thomas Cavendish brought her from the Philippine Islands some dried branches bearing capsules and seeds of this spice. And in Japan according to Thurnberg (where one or more species of it grow) the plant has almost a sacred character. The people "wear aromatic garlands of its foliage, which they offer in their "temples, and on the tombs of their departed friends. They also "burn the powder bark as incense. This bark when very finely "powdered, is likewise used in a kind of time-piece, which denotes "to the watchmen of the night the regular intervals as they "elapse. For this purpose the powder is strewn on some ashes, "in small winding grooves, in a box secured from the weather. "This powder being lighted at one end, burns slowly and "regularly, and when it arrives at certain marked stages, the "watchmen strike a bell, being able by means of this indicator "to proclaim with accuracy the stated period."

During my residence in Tobago, the value of this spice and the probability of its easy acclimatization there, induced me to take considerable trouble on two different occasions, to obtain parcels of the seed, it being nearly impossible for me to have obtained living plants, from China. But I am sorry to say, that the seeds were packed in earth, and I did not succeed on either occasion in getting a single one to germinate. No doubt, however, it would be a matter of no very great difficulty for the Botanic Gardens to procure a wardian case or two with an ample supply both of seedlings and plants raised from cuttings: The Botanical Station at Hong-Kong, if not the Singapore Garden, would be likely places to get them from.

Among the things which, during a residence of nearly twenty years in the Eastern and Western tropics, have been forced upom my observation, is the fact that one of the greatest weaknesses of the tropical planter is his custom of "putting all his eggs in one basket." It bore sorely upon the West Indian Indigo planter when his industry was crushed by oppressive taxation; when the Exeter Hall Party through their tool, Sir Ashley Edin, interfered between the labourers and the planters, it was the bane of the East India Indigo planters; practically the only ones surviving, being those who were Zenimdars (i.e. general farmers,) in addition to being Indigo planters. It is now making itself strongly felt both upon the Indian tea planters and the West Indian sugar planters. Once even within the memory of some of our older friends, a time occurred when the value of cacao fell to nothing, estates were valueless, and even merchants threw bags of cacao into the Gulf rather than ship them. Were it not so evident, it would hardly be credible that to-day our cacao planters heedless of the grim axiom that "history repeateth itself" are mostly if not all, committing the same economic error.

No farmer at home in his senses would plant up his whole farm in roots or grain at once. No fruit grower would fill all his orchards with apple trees alone. If the market for roots fell, if the farmer grew no other crop, what would he have to depend on? If the blight spoiled the fruit grower's apple crop, and he had no pears or plums, what would he have to fall back upon? If the history of 1844-6 repeats itself, what has the Trinidad Cacao Planter to look to? Sugar estates are being abandoned right and left, and the remaining planters base their hopes for the future on rather doubtful central factories and cane farming, which at the best promise but a scant measure of success, should Providence—in the shape of the British Government,—not afford them some protection. And that,—to say the least,—is doubtful in the extreme.

The only crop I have ever heard a Cacao Planter talk of going in for in earnest besides his Cacao, is Coffee. This, if "gone in for" by itself with nutmegs or spices among it, and not planted mixed with Cacao, will be profitable in localities where (1) suitable labor can be depended on, and (2) where there are facilities for "rapid" or "stump" planting, which gives a first picking in 18 months.

It is here, and for this purpose that my suggestions for the introduction and trial of Guaraná and Badiane are made. Coffee has no chance of being profitable in many localities, where perhaps either one of these plants might be suitable to endow the Cacao Planter with a mixed cultivation.

In suggesting those plants for trial I do not wish to detract from the merits of other plants which are somewhat better known, and which also tend to fill the same useful purpose. These, at least all that have hitherto seriously been considered suitable, are few in number, and I shall refer to them briefly.

Nutmegs.—These trees are very sparsely cultivated here at present, and are well worth a great deal more attention than is bestowed on them. One great reason of their present scarcity is no doubt the fact that they take so long a time to come into bearing—beginning about the seventh year at the earliest ; and a second perhaps is that any nutmegs hitherto shipped from this island have not brought very satisfactory prices. The cause of this is that nutmegs sell at so much a pound of so many of such a size, and I have never seen any nutmegs grown in Trinidad that came up to the average of middle sized Grenada ones, which is about " 90 to 80 per lb., 2/1 to 2/5." I do not think our soil is to be blamed for this, but the fact that the seed disseminated here has not been carefully selected ; both plants and seed of the very best and largest kinds can easily be obtained in Grenada. I know of an instance in Grenada where nutmegs were successfully planted among young cacao, and the cacao was beginning to be cut away to give more room to the bearing nutmegs when I saw it. For Trinidad, however, I think any cacao planter who is suitably located, and makes up his mind to put part of his land in coffee, would do very good business, if he planted up nutmegs from selected seed about 30 feet to 40 feet apart among his coffee. For the rest, a method of rapid propagation of nutmegs, has, I believe, yet to be discovered. But it would be a discovery fraught with such value and benefit to the planting community that it would be well worth while for the Agricultural Society to offer a substantial *bonus* for the discovery of a satisfactory mode of rapid propagation of Nutmeg trees by any species of grafting or layering. Porter, in his " Tropical Agriculturist " says they are capable of propagation by layer, but it does not appear to have been a satisfactory process, else it would now be extensively used, and about 50 years ago Dr. Otley of Singapore is said to have succeeded in grafting some by approach. But the local wild nutmeg *(M. sebifera)* which is common in the woods, near Cunupia, for instance, is said to be a quick growing plant, seedlings of it could be easily obtained in large numbers, and the grafting of these nutmegs on them by 'budding' or 'approach,' seems to me a very promising line of experiment.

Cloves. The clove trees, though introduced to the West Indies as long ago as 1795, never seems to have taken hold as a staple cultivation here although plants of it have been long in the Botanic Gardens. But now there is a special reason, in addition to our need of new objects of culture, for drawing attention to it, that is the fact that the greatest modern commercial source of clove production, is the Sultanat of Zanzibar on the East coast of Africa which has now become a British protectorate ; and in consequence, the slaves who have hitherto cultivated the cloves in Zanzibar Pemba are now being freed. The result is that the price of cloves which is about 1/0¼ to 1/3¼ per ℔. just now, is certain to rise soon while the demand like that for nutmegs is ever on the increase. Probably cloves like nutmegs would be very suitable for cultivation among coffee, but they require a very well drained situation. "Clove trees may be grown either "from seed or from layers. The seeds may be sown about 6" "apart in shaded seed beds, and may be transplanted 16 "to 20' apart when about 9 months old. Layers will root "in 5 to 6 months if a proper amount of moisture is "maintained about them. The seedling tree begins to .yield "about the 6th year, layered plants about the 4th and the tree "attains the height of about 20.' Cloves are the dried flower calixes "which are borne in bunches at the end of the branches ; when "ripe to gather, they are of a bright red color, and the crop "season is October to December. The nearest clusters may be "taken off with the hand, and the more distant by the aid of "crooked sticks.........the cloves are prepared for shipment by "smoking them on hurdles over a slow fire till they turn brown, "and the drying is afterwards finished in the sun. They are "then cut off from the flower branches and will be found to be "purple colored within, and fit to be baled for shipment." The yield is about 6 lbs. per tree (in Zanzibar). From the way in which cloves have to be gathered it is evident that the same labor conditions as required for coffee are requisite. In the Molucca Islands the trees are generally topped and kept down to 8 to 9 feet for convenience in gathering, but in Zanzibar, the trees are allowed to gain their full height, the cloves being picked from Bamboo ladders. The yield of both nutmegs and cloves can be much improved by judicious manuring.

Black Pepper, and Gambir *(Uncaria Gambir)* plants have been introduced of late into the Colony ; and some time ago, Mr. Meaden, at Chaguanas, shewed that the Black pepper vine did fairly well on very bad land. But I can only mention their introduction and growth here as an interesting fact tending to support the probability of the success of Badiane, for any extended cultivation of them here will have to meet in competition the cheap Chinese labor of the Straits Settlements, and that of the Dutch Indies. Kola nuts and Tonka beans have very uncertain features, and I do not recommend their extensive cultivation.

It must be remembered that the Cacao planter who wishes to extend, or who wishes to put new land into Guarana, Badiane, Coffee, Nutmegs or Cloves, is saddled (in the absence of contractors) with the sum of at least £7 10 per acre before he can put a plant in the ground (cost of Crown Land 30/ per acre, cost of clearing per acre, say £6) and secondly, most of the standing crops suited for his style of cultivation take a long time before they begin to yield (from 1½ years to about 7 years) and therefore of necessity must be plants yielding a high priced product, and of a kind that requires more or less simple preparation for the market. Is any more than these suggestions and considerations needed to show that the cacao planter who plants up Guarana, or coffee, with spices among them, as well as his cacao, is in a better, safer, and more economic position, than the one who plants cacao only,—especially when the price of cacao falls ?

Some enthusiasts, for the last 20 years have been recommending the cultivation of India-Rubber bearing trees of various sorts. In Ceylon and everywhere elsewhere they have been planted, dissatisfaction has been the result. Ceára Rubber (*Manihot* sp.) is useless, Para Rubber (*Hevea* sp.) cannot be safely tapped before the 10th or 12th year ; Columbian Virgin, can be tapped at the 8th year, but is said not to grow under 600 feet altitude. The Ullé (*Castilloa* sp.) begins to give a return, *in its native forests*, in 8 or 10 years. The Assam Rubber (*Ficus Elastica*) is not at maturity till 50 years of age, and it frequently does not survive the first tapping. I do not think there is a single Rubber tree plantation which has even begun to pay its expenses. The future planters of rubber trees, too, had better before they make up their minds consider the fact that a process for making synthetic rubber has been patented, and any day may see it at work on a commercial scale.

Yet, by far the most rapid bearing, and lucrative cultivation that can be suggested for the owners of abandoned, or about to be abandoned sugar estates, is Rubber cultivation of a kind which the Rubber tree enthusiasts have bestowed little notice on, viz. : the cultivation of Rubber vines, such as *Forsteronia, Cryptostegia, Leuconotis* and *Landolphia*. Probably some of these may be capable of such rapid propagation, and such rapid regrowth when cut down for extraction, and withal yield such a high percentage of good rubber as to make them a reliable annual or short crop. *Forsteronia* is a West Indian plant, and is as likely as any to meet those conditions. This is the suggestion to owners of such estates : Given a Rubber bearing vine meeting those conditions, and cleared land in a suitable locality, it would not be a difficult or costly thing to plant up a large area with long cuttings of Cedar, Plum or other quick growing wood,—cedar for preference—and grow the vines (from seed or cuttings) on them. The cultivation expenses would not amount to much, and if the planter could count on ¼ of a ton of rubber per acre at the end of the 2nd year, he would have but small cause to bewail his defunct sugar cane.

The above suggestions, coupled with the others which follow, could be tried under the best auspices by one or more of the large sugar estate proprietors who are about to cease sugar cultivation; one or a company of them might devote some money to putting an abandoned estate under a suitable manager for say 3 to 5 years, to try on a working scale various promising cultivations. No cultivation I can think of to suggest would cost as much as an equal area of sugar, nor can I think of any which fails to promise a better return at less cost.

It has almost become a proverb that 'nothing is so difficult as to re-establish a cultivation in a country where it has once died out!' But perhaps there is some difference in favour of cultivations which have been *killed* out. The first cultivation the British planter in the W.I. found a profitable industry was Indigo. More rapid fortunes were made out it then,—when it was worth a guinea a pound,—than have been made out of any other West Indian cultivation since. It was killed out suddenly in one year by an import duty of about £20 per cwt. being imposed upon it in England, at a time when there were no other markets open to it.

The industry then became the heritage of the East Indian planters, and down till pretty recently a most valuable one, but the invention of coal tar dyes, as well as other events have much decreased its value, which at present stands about 7/6 to 2/- per lb. according to quality. But even at these prices in suitable localities with improved modes of preparation, it is still a profitable industry.

The plant, in several varieties, is wild in many places in Trinidad and Tobago. And any one acquainted with the industry in the East Indies, seeing the wild plant here is at once struck with its vastly greater luxuriance. It partially explained to me why in Edwards 'History' the West Indian crops of old were so much heavier than those of the present time in India. The average yield there is 50-60 lbs. per acre per first cutting, per second cutting about half that weight, say 25 lbs. There is usually no third ratoon to cut. In the last century in the West Indies the average crop is said to have been 200 to 300 lbs. per acre, and that is what the Central American growers still count on. But in the best lands in India and Java the planter thinks himself lucky to get one good ratoon cutting while as the result of actual experiment with a very few plants on very poor land in Tobago I obtained five good cuttings in one year. Now, suppose an abandoned sugar estate were to devote say 50 to 200 acres to Indigo cultivation and manufacture it by the Sayers and Olpherts' processes, it could be sure at the very least to get 60 lbs. of the best Indigo per acre (and probably three times that quantity) which at say $1.00 per lb. would pay handsomely, as the cultivation and manufacturing expenses on ready cleared land would amount to comparatively little. The next season fresh land would be put into Indigo (so as to avoid insects, &c.) and a crop or two of Dál (or Indian lentils, for example "Bari túvaí *Cytisus Cajan*, or Múng Dál' *Phaseolus Mongo, &c.*) might be taken off the first before growing a second crop of Indigo on it. These Dáls, could be sold at a profitable rate on the spot, in place of some part of the large quantities which are annually imported. They are a favourite food grain amongst all classes of natives in India.

If the Indigo seed necessary to begin with could not be gathered here, it no doubt could be got easily from Venezuela, where there is still, I believe, some Indigo cultivation in the Northern provinces. The apparatus for working the crop of say 200 acres would not be expensive and for the most part would be constructed on the Estate. Dál seed would probably have to be imported, or procured from the importers from India. There are plenty of coolies in the Island who have been engaged in Indigo cultivation in India, who would require no teaching, and who would in all probability be very pleased to take a hand in 'Níl-Ki-Kám' as they call it, again, in fact, a few months ago a coolie drew my attention to some Indigo growing wild in a field beside the road, and asked me why, since it grew so strongly in this country, did no one cultivate and manufacture it? But there are two peculiarities of Indigo which deserve mention; one is that with the exception of Tobacco there is perhaps no other crop which is so preyed upon by insects, but this is minimised to a great extent by never taking a crop two years in succession off the same ground. The second is, out of every three years it gives two average crops, and one bumper or extraordinary crop which is usually a matter of great congratulation when it comes off.

One more suggestion for the use of abandoned estates and I am done. You will smile at this one, perhaps, more than at any of the rest. I was asked some time ago by a correspondent in London whether "Piñones" were grown in Trinidad, and if I could give a quotation for the seed. It took me a day or two to discover that "Piñones" is a Spanish or Portuguese name for the seed of the common angular leaved Physic nut (*Jatropha curcas*). It seems that large quantities of the seed of this plant are imported from the Cape de Verde Islands to Lisbon and Marseilles,—according to Spon about 350,000 bushels annually, where it is expressed, and the oil is said to be frequently imported into England under the name of 'Pulza-seed' oil, and as a substitute for Linseed oil. The color is somewhat paler, but it found to answer equally well. Now, this plant is in common use as a fence plant in most places here, and it can be propagated in almost any soil by cuttings, with the greatest facility, while it grows rapidly and bears profusely. All the cultivation it would need would be planting the cuttings and gathering the seed when ripe, so it would not be hard to plant up a large area with it. The expression is performed in the dry, on the seed slightly roasted and crushed; 1,000 lbs. of seed give 640 barrels kernels, which yield 260 gallons of oil, which is probably worth about £12 to £15 per ton, but I have seen no recent quotations. In addition, the Physic nut plant might yield some economic bye products, such as gum or resin. If the oil were expressed locally no doubt the mare resulting would be useful for manure. Some valuable climbing plant, say Sarsaparilla, might be grown along with it, though perhaps this cultivation may not compare in value with any of the others, an estate yielding 100 or more tons of the seed per annum would do so at a very small cost, the plants being free from damage by ants, animals, and would not be the dead loss to its proprietor that the estate abandoned to bush usually becomes. Several other oil bearing plants are well worthy of notice, but none is so easy of propagation or is likely to pay with as little cultivation as this.

It remains for me to say, that while it is a matter of notoriety that the Pioneer British Planters of Indigo, Sugar, Coffee, Tea, Cinchona, &c., made larger and more rapid profits out of those products than any of their successors, it is worthy of note that the two latter would never have become the sources of benefit they now are, but for the enlightened and persistent efforts of the Government of India to introduce and in every way foster their industrial progress :—An example that in the present depressed state of our agricultural staples, our local Government might follow with advantage.

Bibliography.

Genl. A. Codazzi—"Geographia de Venezuela." H. H. Smith—"Brazil, the Amazon and the Coast" 1879. Dr. H. H. Rusby in "Tropical Agriculturist" (Ceylon). H. Semler—"Die Tropische Agricultur." Spon's "Encyclopedia of Manufactures and Raw Materials." G. R. Porter's "Tropical Agriculturist" Fortune's "Residence among the Chinese." Simmond's Coml. Prods. of the Vegetable Kingdom."

ON A SMALL COLLECTION OF COCCIDÆ FROM THE ISLAND OF GRENADA.

By T. D. A. Cockerell, New Mexico (U.S.A.) Agricultural Experiment Station.

THE following notes are based on a small lot of scale-insects collected by Messrs. W. E. Broadway and G. W. Smith, in Grenada, and transmitted to me by Mr. Urich. They are interesting as coming from a locality of which the Coccids are but slightly known, and one of the species is new.

- (1.) *Asterolecanium pustulans*, Ckll. On twigs of *Leucœna glauca*, Benth., a leguminous plant. Botanic Gardens, July 2nd, 1895.
- (2.) *Ceroplastes* sp. Possibly new, but all the specimens are broken. On *Eugenia*, cultivated in the Botanic Gardens, July 21st, 1895.
- (3.) *Lecanium urichi*, Ckll. With an ant, on plant not identified. May 8th, 1895. Hitherto known only from Trinidad and Brazil.
- (4.) *Pulvinaria broadwayi*, n. sp. On twigs of a cultivated plant not identified, Botanic Gardens, August 29th, 1895.

Female scale reddish-brown, about 1½ mm. long in the shrivelled condition, on and surrounded by an oval cushion of white secretion, about 3 mm. long. The white secretion is similar in character to that of the ovisac of an ordinary *Pulvinaria*. Female after boiling in soda colorless or nearly so, the derm with numerous conspicuous round and tubular glands. Margin with a row of stout but rather short pale brown spines, very numerous, 6 or 7 in a distance equal to the length of the tibia. The spines are about as long as the breadth of the tibia. Rostral loop reaching to insertion of middle pair of legs. Anal plates light greenish-brown, rather broad, their outer angle about a right angle. Tip with two hairs, and outer inferior margin with one or two hairs. Surface of plates with small round glands. Anal ring with several hairs.

Antennæ 8-jointed, very pale brownish, 1st joint fully as broad as the length of the 3rd. 3rd longest, but only about twice as long as broad, and about as long as 4 + 5. 2 and 8 subequal in length, 2 perhaps a little the longer; 2 no longer than broad. 4 and 1 next longest, subequal in length; 4 a little longer than broad. 5 and 6 equal and a little shorter than 4, 7 shortest. Formula 3(28)(41)(56) 7　2 with two hairs near its end.

Legs very pale brownish, rather stout and large. Tibiotarsal articulation unusually distinct. Tibia a little shorter than femur. Tarsus one-third shorter than tibia. Claw strong, much curved; digitules of claw stout, extending beyond its tip, with large knobs. Tarsal digitules slender, nearly twice as long as those of claw, with small but very distinct obliquely-placed button-like knobs.

This species will be readily known by the cottony matter entirely surrounding (but not covering) the scale, by the very numerous marginal spines, &c.

(5.) *Chionaspis minor*, Maskell. Very abundant on bark of
    Erythrina sp. Botanic Gardens, June 4th, 1895.

(6.) *Aspidiotus nerii*, Bouché. On underside of leaves of
    Olea europœa. Botanic Gardens, August 26th, 1895.
    Doubtless introduced from Europe; new to the West
    Indies.

(7.) *Icerya montserratensis*, Riley & Howard.
    On leaf of Hyophorbe Palm, Botanic Gardens, 1893.
    Coll. Smith.

(8.) *Vinsonia stellifera* (Westwood).
    On nutmeg. Coll. G. W. Smith, Com. F. W. Urich.

(9.) *Pulvinaria pyriformis*, Ckll., 1894.
    On Cinnamon. Coll. W. E. Broadway, Com. F. W
    Urich. In the Botanic Gardens, 25 Sept., 1894.
    Hitherto only known from Trinidad.

(10.) *Pinnaspis pandani* (Comst.)
    On *Anthurium crystallinum*. Coll. Broadway, Com.
    Urich. Botanic Gardens, 6 Nov., 1894. The second
    lobes are rather longer than usual in the species.

    Mr. Urich informs me that a scale on *Areca lutescesn*
    looked like *P. pandani*, but he did not send me
    specimens.

(11.) *Ischnaspis filiformis*, Douglas.
    On leaf of Hyophorbe Palm, Botanic Gardens, 1893.
    Coll. Smith.

(12.) *Aspidiotus destructor*, Signoret.
    On leaves of Hyophorbe Palm, Botanic Gardens.

WARNING COLOURS AND MIMICRY.

PROFESSOR Felix Plateau, in the most recently issued part
of the "Mémoires de la Société Zoologique de France,"
gives the results of his experiments to ascertain whether the
Magpie moth is really, as has often been stated, an example of
what is termed "warning colour." This daring Professor
actually himself ate a number of caterpillars and found that the
flavour was very pleasant, reminding him of almonds. It would
indeed be well if all the examples of "warning colouration"
were subjected to as careful an examination. Equally cautious
also should naturalists be before accepting examples of
"mimicry" among animals and plants. In some cases the
so-called "advantageous mimicry" falls to the ground, for the
insect which is supposed to imitate one of its fellows appears at
quite a different time of year from it.—*Science Gossip.*

THE COURTING OF ANIMALS.

THIS subject seeems to prove attractive to many naturalists. In vol. x. of the "Transactions of the Wisconsin Academy of Sciences," there is a highly interesting paper by Mr. and Mrs. Peckham on the "Courtship of Certain Spiders." It seems to be the case that the sharpness of vision in spiders is accentuated by love. A male of *Satis pulex* was put into a box in which was a female of the same species twelve inches away, and the male "perceived her at once, lifting his head with an alert and excited expression, and went bounding towards her." By experiments it was proved that this recognition was really due to sight. These results are interesting because some have affirmed that spiders cannot see nearly as far as twelve inches. Further experiments seem to show that spiders can differentiate colour. M. Racovitza, a Roumanian naturalist, has been studying the courting and marriage customs of the octopus, and in a recent number of the "Archives de Zoologie Expérimentale," he gives us some of his observations. It is satisfactory to know that the octopus does not, as some have thought, behave brutally in its love affairs. M. Racovitza assures us that "there is nothing more than a courteous flirtation," and "that the male behaves with a certain delicacy towards his companion."—*Science Gossip.*

CPSIA information can be obtained
at www.ICGtesting.com
Printed in the USA
BVHW071705061118
532319BV00011B/785/P